DATE DUE

WITHDRAWN

Forsyth Library

Proceedings in Life Sciences

Metabolism and Molecular Activities of Cytokinins

Proceedings of the International Colloquium
of the Centre National de la Recherche Scientifique
held at Gif-sur-Yvette (France) 2–6 September 1980

Edited by
J. Guern and C. Péaud-Lenoël

With 170 Figures

Springer-Verlag
Berlin Heidelberg New York 1981

Professor Dr. Jean Guern
Laboratoire de Génétique et Physiologie
du Développement des Plantes, CNRS
F-91190 Gif-sur-Yvette

Professor Dr. Claude Péaud-Lenoël
Faculté des Sciences de Luminy, Case 901
Laboratoire de Biochimie
Fonctionnelle des Plantes, ER 104 CNRS
F-13288 Marseille Cedex 9

ISBN 3-540-10711-8 Springer-Verlag Berlin Heidelberg New York
ISBN 0-387-10711-8 Springer-Verlag New York Heidelberg Berlin

Library of Congress Cataloging in Publication Data. Main entry under title: Metabolism and molecular activities of cytokinins. (Proceedings in life sciences) Bibliography: p. Includes index. 1. Cytokinins--Congresses. 2. Plant Physiology--Congresses. 3. Bacteria--Physiology--Congresses. I. Guern, Jean, 1936-. II. Péaud-Lenoël, Claude, 1918-. III. France. Centre national de la recherche scientifique. IV. Series. QK898.C94M47 581.19'27 81-8771 AACR2

This work is subject to copyright. All rights are reserved, whether the whole or part of the material is concerned, specifically those of translation, reprinting, re-use of illustrations, broadcasting, reproduction by photocopying machine or similar means, and storage in data banks. Under § 54 of the German Copyright Law, where copies are made for other than private use, a fee is payable to "Verwertungsgesellschaft Wort", Munich.

© by Springer-Verlag, Berlin Heidelberg 1981.
Printed in Germany.

The use of registered names, trademarks, etc. in this publication does not imply, even in the absence of a specific statement, that such names are exempt from the relevant protective laws and regulations and therefore free for general use.

Typesetting and printing: Beltz, Offsetdruck, Hemsbach/Bergstr.
Bookbinding: Brühlsche Universitätsdruckerei, Giessen
2131/3130-543210

Preface

In a number of recent biology meetings, plant growth factors have been a favorite topic. One may ask if a new colloquium devoted to the metabolism and the activities of cytokinins, considered at the molecular level, was really necessary and will add anything important to the previous contributions.

As a matter of fact, the primary purpose of the organizers, of the French National Center of Scientific Research, and of the National Institute of Agronomic Research which sponsored the colloquium was less the edition of a proceedings book than the desire to bring together a number of scientists who had made contributions towards solving problems in cytokinin research and thereby to provide an opportunity for fruitful discussions without restrictions in the timetable. Indeed this primary target was attained. Moreover, important new findings and conclusions were presented by the participants that merited publication in the form of this book. It is beyond the scope of this foreword to advertize or even to classify these new findings. We shall only mention a few of them.

The reader will appreciate the contributions of those authors who worked hard to elucidate the biosynthetic pathways of natural cytokinins. These experiments led not only to a knowledge of certain plant biosynthetic mechanisms, but also to an accurate analysis of the pathways through which the plasmid-bearing bacteria, symbiotes or parasites, provide the host plant with a large quantity of cytokinins (and certain other growth factors as well). This latter topic emphasizes the fact that cytokinin activity is of interest not only to plant physiologists but also to microbiologists and to researchers in animal biochemistry. We mention the importance of the contributions relevant to the role of substituted adenines in animal systems which are treated in this book.

Until now, cytokinins had been studied essentially with the aim of explaining their activity as plant cell division factors. From a number of lectures recorded in the present Proceedings, especially from experiments that utilized "anticytokinins", it now appears likely that several distinct molecular pathways of cytokinins should be considered if one wishes to explain the loosely related physiological responses to this class of plant hormones. The reader will find a special chapter in which chloroplast differentiation and maturation of photosynthetic membranes are considered as responses to cytokinins. This system seems quite promising.

Many contributions were devoted to the question: Which is the primary active metabolic derivative of cytokinins, the free bases, the nucleosides, the nucleotides, some specific cytokinin-receptor complex or any other derivative?

Looking at the mechanism of the hormone activity, the reader may appreciate a large number of facts and comments throughout many chapters. At the present stage of knowledge, two lines of evidence led to the alternative hypotheses that cytokinins either stimulate the gross mechanism of protein synthesis, perhaps at the level of transcription, or that cytokinin impact stimulates specific biosynthetic pathways leading to selective changes in the macromolecule equipment of the cells. It will be interesting to see what answers are provided by the rapid progress in these fields.

May, 1981　　　　　　　　　J. GUERN and C. PÉAUD-LENOËL

Contents

Section 1. Cytokinin Biosynthesis

Analytic Procedures for Cytokinins: Application to
Agrobacterium tumefaciens
R.O. Morris, D.A. Regier, and E.M.S. MacDonald
(With 8 Figures) 3

Cytokinin Biosynthesis and Its Relationship to the Presence
of Plasmids in Strains of *Corynebacterium fascians*
N. Murai (With 4 Figures) 17

Cytokinin Biosynthesis in Higher Plants
H. Maaß and D. Klämbt (With 5 Figures) 27

Biosynthesis and Enzymic Regulation of the Interconversion
of Cytokinin
C.-M. Chen (With 3 Figures) 34

Cytokinin Biosynthesis in Cytokinin-Autonomous and
Bacteria-Transformed Tobacco Callus Tissues
W.J. Burrows and K.J. Fuell (With 2 Figures) 44

Cytokinin Biosynthesis and Metabolism in *Vinca rosea* Crown
Gall Tissue
R. Horgan, L.M.S. Palni, I. Scott, and B. McGaw (With 4 Figures). . 56

Section 2. Cytokinin Metabolism and Physiological Responses in Plant or Bacterial Systems

Cytokinin Translocation and Metabolism in Species of the Leguminoseae: Studies in Relation to Shoot and Nodule Development
R.E. Summons, D.S. Letham, B.I. Gollnow, C.W. Parker, B. Entsch,
L.P. Johnson, J.K. MacLeod, and B.G. Rolfe (With 6 Figures) 69

Uptake and Metabolism of Cytokinins in Tobacco Cells: Studies
in Relation to the Expression of Their Biological Activities
M. Laloue, C. Pethe-Terrine, and J. Guern (With 6 Figures)   80

Genetic Regulation of Cytokinin Metabolism in *Phaseolus*
Tissue Cultures
D.J. Armstrong, S.-G. Kim, M.C. Mok, and D.W.S. Mok
(With 1 Figure).   97

Metabolism of the Cytokinins in Mosses
M. Bopp and U. Erichsen (With 7 Figures) 105

Cytokinin Activities of N-Phenyl-N'-(4-Pyridyl) ureas
Y. Isogai (With 6 Figures) 115

Towards an Understanding of the Mechanism of Cytokinin
Activity in *Mercurialis annua* L. Sex Differentiation
A. Champault, S. Chung, B. Guerin, G. Kahlem, A. Lhermitte,
G. Teller, and B. Durand (With 6 Figures) 129

Relative Activities of Cytokinins and Antagonists in Releasing
Lateral Buds of *Pisum* from Apical Dominance Compared with
Their Relative Activities in the Regulation of Growth of Tobacco
Callus
F. Skoog and A.K.B.A. Ghani (With 8 Figures) 140

Section 3. Cytokinin Hormone Receptors: Methods and Results

Photoaffinity Labeling Reagents to Detect Binding of Cytokinins:
Chemical Limitations
R. Mornet, J.B. Theiler, and N.J. Leonard (With 5 Figures) 153

Uptake of [^{14}C]-8-Azido-N^6-Benzyladenine, a Radioactive
Photosensitive Cytokinin, by the Cells of the Moss
Funaria hygrometrica L.
R. Miassod (With 5 Figures). 162

Cytokinins and the Hormone Receptor Concept
D. Klämbt (With 5 Figures). 172

The Current Status of Cytokinin-Binding Moieties
P. Keim, J. Erion, and J.E. Fox (With 6 Figures) 179

Section 4. Cytokinin in Protein Synthesis and RNA Metabolism

Control of Growth by Cytokinin: An Examination of Tubulin
Synthesis During Cytokinin-Induced Growth in Cultured Cells
of Paul's Scarlet Rose
D.E. Fosket, L.C. Morejohn, and K.E. Westerling
(With 12 Figures) 193

The Cytokinin Control of Protein Synthesis in Plants
A. Szweykowska, E. Gwóźdź, and M. Spychała 212

Cytokinin Action on Transcription and Translation in Plants
O.N. Kulaeva (With 7 Figures)......................... 218

Covalent Insertion of N^6-Benzyladenine into rRNA Species of
Tobacco Cells
J.-P. Jouanneau, and B. Teyssendier de la Serve (With 4 Figures) . 228

Section 5. Cytokinins and Chloroplast Development

Plastids and the Origin of Cytokinin Autonomy in Tobacco Tissue
Cultures
M. Kamínek, V. Hadačová, and J. Luštinec (With 6 Figures) 241

Influence of Cytokinin on Plastid Biogenesis in Rye Leaves
J. Feierabend (With 3 Figures)......................... 252

Light and Fusicoccin as Tools for Discriminating Among
Responses of Cotyledons to Cytokinins
C.P. Longo, G.P. Longo, M. Lampugnani, G. Rossi, and
O. Servettaz (With 3 Figures)......................... 261

Studies on the Action of Cytokinin and Light in RNA Synthesis
of Pumpkin Cotyledons by Autoradiography
D. Neumann and W.A. Khokhlova (With 13 Figures) 267

Plastogenesis and Cytokinin Action. Cytokinin and Light
Interactions in Plastid Enzyme Formation of Detached
Cucurbita Cotyledons
B. Parthier, S. Lerbs, and N.L. Klyachko (With 8 Figures)...... 275

Cytokinin Action on RNA Synthesis in Chloroplasts
T.P. Mikulovich, E.R. Romanko, S. Yu-Selivankina, I.M. Kukina,
and R. Wollgiehn (With 3 Figures) 287

Effect of Cytokinin on Plastid Differentiation in Tobacco Cell Suspensions
A.-M. Lescure and P. Seyer (With 5 Figures) 298

Plastid Proteins of Cytoplasmic Origin as Molecular Markers of Cytokinin Activity
C. Péaud-Lenoël and M. Axelos (With 6 Figures) 308

Section 6. Animal Systems Responding to Cytokinin or Cytokinin Analogs

6-Benzylaminopurine as a Growth Factor for *Drosophila melanogaster* Cells Grown in Vitro
J.L. Becker and J. Roussaux (With 5 Figures) 319

Use of N^6-Benzyladenosine as an Analog of Internal Methylation (N^6-Methyladenosine) in Studies of Post-Transcriptional Regulation of RNA in Human Cells Infected by Herpes Simplex Virus
A. Garcia, A. Epstein, and B. Jacquemont (With 4 Figures) 329

Calcium-mediated Transduction of the Hormonal Message in 1-Methyladenine-induced Meiosis Reinitiation of Starfish Oocytes
M. Dorée and T. Kishimoto (With 5 Figures) 338

Subject Index 349

Contributors

You will find the addresses at the beginning of the respective contribution

Armstrong D.J. 97
Axelos M. 308
Becker J. L. 319
Bopp M. 105
Burrows W.J. 44
Champault A. 129
Chen C.-M. 34
Chung S. 129
Dorée M. 338
Durand B. 129
Entsch B. 69
Epstein A. 329
Erichsen U. 105
Erion J. 179
Feierabend J. 252
Fosket D.E. 193
Fox J.E. 179
Fuell K.J. 44
Garcia A. 329
Ghani A.K.B.A. 140
Gollnow B.I. 69
Guerin B. 129
Guern J. 80
Gwóźdź E. 212
Hadačová V. 241
Horgan R. 56
Isogai Y. 115
Jacquemont B. 329
Johnson L.P. 69
Jouanneau J.-P. 228
Kahlem G. 129
Kamínek M. 241
Keim P. 179
Khokhlova W.A. 267
Kim S.-G. 97
Kishimoto T. 338

Klämbt D. 27, 172
Klyachko N.L. 275
Kukina I.M. 287
Kulaeva O.N. 218
Laloue M. 80
Lampugnani M. 261
Leonard N.J. 153
Lerbs S. 275
Lescure A.-M. 298
Letham D.S. 69
Lhermitte A. 129
Longo C.P. 261
Longo G.P. 261
Luštinec J. 241
Maaß H. 27
MacDonald E.M.S. 3
MacLeod J.K. 69
McGaw B. 56
Miassod R. 162
Mikulovich T.P. 287
Mok D.W.S. 97
Mok M.C. 97
Morejohn L.C. 193
Mornet R. 153
Morris R.O. 3
Murai N. 17
Neumann D. 267
Palni L.M.S. 56
Parker C.W. 69
Parthier B. 275
Péaud-Lenoël C. 308
Pethe-Terrine C. 80
Regier D.A. 3
Rolfe B.G. 69
Romanko E.R. 287
Rossi G. 261

Roussaux J. 319
Scott I. 56
Servettaz O. 261
Seyer P. 298
Skoog F. 140
Spychała M. 212
Summons R.E. 69

Szweykowska A. 212
Teller G. 129
Teyssendier de la Serve B. 228
Theiler J.B. 153
Westerling K.E. 193
Wollgiehn R. 287
Yu-Selivankina S. 287

Section 1
Cytokinin Biosynthesis

Analytical Procedures for Cytokinins: Application to *Agrobacterium tumefaciens*

R.O. Morris, D.A. Regier, and E.M.S. MacDonald [1]

1 Introduction

1.1 Crown Gall Disease

Crown gall disease is a neoplastic condition which affects higher plants and which results in the production of discrete tumors. It is restricted to dicotyledonous species but, within that group, has a very broad range and affects many Angiosperms and Gymnosperms. The causative agent, first identified by Smith and Townsend (1907), is a mesophilic, gram-negative, soil bacterium known as *Agrobacterium tumefaciens*.

All crown gall tumors share some common characteristics. First, they must be initiated by infection at a wound site. Second, after the initial transformation, the continued presence of the bacteria is no longer necessary for the maintenance of the transformed state. Thus, tumors freed from the inciting bacteria (White and Braun, 1942) still retain their capacity for uncontrolled growth. Third, all tumors are independent of added phytohormones when grown in culture (Braun, 1958). For example, normal callus from *Nicotiana* requires both exogenous auxin and cytokinin for growth in culture, but tobacco crown gall tissue does not. It grows rapidly in the absence of both hormones. Fourth, the morphology of intact tumors suggests an overproduction of either cytokinin or auxin. The exact morphology of any given tumor depends both upon the host plant and upon the strain of *A. tumefaciens* used for infection. The primary determinants, however, seem to be bacterial (Gordon, 1980). Strains of *A. tumefaciens* bearing octopine plasmids (see below) usually induce unorganized tumors. Strains bearing nopaline plasmids, on the other hand, produce teratomas characterized by excessive production of abortive shoots. Such shoot production has been classically regarded as a consequence of high levels of cytokinins.

Finally, in the past few years, evidence has been provided for the molecular mechanisms underlying tumorigenesis. Virulent strains of *A. tumefaciens* have been shown (Zaenen et al., 1974) to harbor a class of large plasmids, the Ti plasmids, which are not present in avirulent strains. The Ti plasmid is the ultimate oncogenic agent. Upon infection, a portion (the T-DNA) is transferred to the host plant cells and is stably replicated there (Chilton et al., 1977). Once present in transformed tissues, the T-DNA is transcribed to RNA (Drummond et al., 1977) which is presumably then

---

[1] Department of Agricultural Chemistry Oregon State University, Corvallis, OR, 97331, USA

translated, although as yet no tumor-specific proteins have been indentified. The continued expression of the T-DNA is thought to be responsible for maintenance of the transformed state. There are three categories of Ti plasmid (octopine, nopaline, and agropine) which confer upon the host bacteria the ability to utilize these amino acid derivatives. For a recent review see Gordon (1980).

1.2 Cytokinins in Crown Gall Tumors

In light of the morphological changes and the hormone independence of tumor tissues, elevated levels of endogenous auxins, cytokinins or of both hormones might be expected. For the cytokinins, there is evidence that such is the case. Isopentenyladenosine (i^6Ado), trans-zeatin (t-io^{6}Ade), trans-ribosylzeatin (t-io^{6}Ado), 4'-glucosyl-trans-zeatin (t-io^{6}g$^{4'}$Ade) and 4'-glucosyl ribosyl-trans-zeatin (t-io^{6}g$^{4'}$Ado) have been shown (Chen et al., 1976; Miller, 1974; Peterson and Miller, 1976, 1977) to be present in *Vinca rosea* crown gall tissue. Further, tobacco crown gall contains t-io^{6}Ade and t-io^{6}Ado (Scott et al., 1980) while a *Bryophyllum* line contains dihydrozeatin and ribosyldihydrozeatin (Miller, pers. comm.). Ribosylzeatin levels were high (1.1 µg/g) in *Vinca* tumor lines (Miller, 1974) and were elevated when the tissue was grown on a source of reduced nitrogen (Peterson and Miller, 1976). Stuchbury et al. (1979) also found high levels of ribosylzeatin (300 ng/g) in *Vinca* tumors.

No systematic and accurate measurements have yet been made of each cytokinin. This information would be of particular value for unorganized, shoot teratoma and root teratoma lines.

1.3 Cytokinin Production by A. tumefaciens

Several hypotheses can be described whereby the internal T-DNA of a tumor cell can subvert the normal metabolism of that cell to cause the over-production of cytokinins. The effect could be direct or indirect. Perhaps the most attractive, certainly the most easily tested, is that the T-DNA codes directly for cytokinin biosynthesis. This could be via a plasmid-coded, cytokinin-containing tRNA together with an excision-repair mechanism or via a direct tRNA-independent mechanism such as described by Taya et al. (1978). In either case, the bacteria themselves might be expected to produce cytokinins. The first direct evidence that they do was provided by Klämbt (1967). Using paper chromatography and bioassay, he obtained evidence for the presence of i^6Ade in bacterial culture filtrates. Upper et al. (1970) found similar activity in the virulent strain B6 and provided supporting gas chromatographic evidence for the presence of i^6Ade. Hahn et al. (1976) also detected i^6Ade in alcohol extracts of strain B6 cells. Application of HPLC, bioassay and GLC-MS of the permethyl derivatives of active fractions from the culture filtrates of virulent strain C58 allowed Kaiss-Chapman and Morris (1977a) to identify i^6Ade, t-io^{6}Ade and ms^{2}io^{6}Ado at 200 ng/l with smaller amounts of i^6Ado, c-io^{6}Ade and ms^{2}i^{6}Ado.

A correlation between cytokinin production and the ability of the bacteria to incite tumors was first provided by Romanow et al. (1969). They demonstrated the presence of cytokinins in the culture filtrates of two virulent strains (A6 and B6) grown on an adenine-containing medium but found no cytokinin in the culture filtrate of an avirulent strain. Kaiss-Chapman and Morris (1977b) examined virulent strain C58 and its plasmid-cured derivative NT1. Some change were noted in the i^6Ade levels but the major difference was in t-io^{6}Ade which was detected in C58 filtrates but not in those from NT1. Claeys et al. (1978a, b) obtained similar data. Levels of i^6Ade were similar in C58 and C58-C9 (cured) strains but zeatin was detected only in the filtrate from C58. A similar pattern was noted by McCloskey et al. (1980) for the ribosides. Strain C58 culture filtrates contained c-io^{6}Ado, t-io^{6}Ado and t-ms^{2}io^{6}Ado. None was detected in the filtrates from the corresponding plasmidless strain NT1.

2 Analysis of Cytokinins

2.1 Extraction Procedures

The problem of analysis is basically that of isolation and identification of nanogram amounts of cytokinin in the presence of an excess of carbohydrates, terpenoids, polyphenols, etc., which constitute a primary extract from a plant. Conventionally, extraction protocols have relied upon a sequence of steps including maceration and extraction of the tissue in methanol or ethanol for extended periods, removal of excess organic solvent in vacuo, preliminary chromatography on polystyrene-based ion exchange resins, partition into n-butanol or ethyl acetate and final chromatography on paper or Sephadex LH-20 followed by bioassay. While this sequence of steps is moderately effective for extraction, it has been shown to lack quantitation (Dekhuijzen and Gevers, 1975) and to raise problems of artifacts. Degradation of tRNA and release of cytokinin-active nucleosides may be a real hazard (Rathbone and Hall, 1972).

Alternative protocols have been developed and much recent attention has focused upon high-performance liquid chromatography (HPLC) and the use of mass spectrometry. Alternatives to polystyrene cation exchange resins have been introduced (Dauphin et al., 1977; Parker and Letham, 1974) and the use of organic solvents has been avoided by trace-enrichment from aqueous solution onto a hydrophobic silica matrix (Morris et al., 1976). Cytokinins may be recovered quantitatively from the matrix by elution with ethanol.

2.2 High Performance Liquid Chromatography (HPLC)

HPLC of cytokinins was first introduced by Carnes et al. (1975) and has found wide application to cytokinin bases, nucleosides and glucosides (Horgan and Kramers, 1979). Figure 1A illustrates the degree of fractionation which can be achieved by HPLC on microparticulate (5 μ) octadecyl silica columns. The cis- and trans-zeatins and ribosylzeatins are readily resolved from each other in a low pH acetate buffer by elution of the column with linearly increasing concentrations of acetonitrile, ethanol or methanol. Analysis time is short, column capacity is high and memory effects not great. Tailing of the free bases may be eliminated (Fig. 1B) by the addition of suitably competitive substances such as imidazole or triethylamine to mask free residual silanols of the matrix.

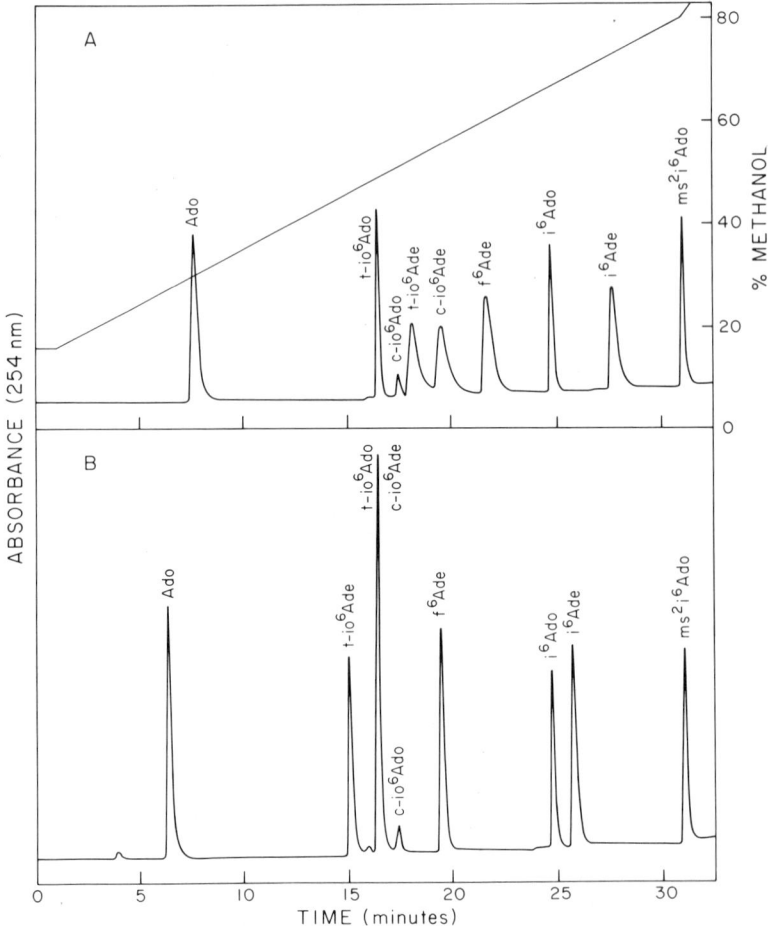

Fig. 1 A, B. High performance liquid chromatography of standard cytokinins. Column, octadecylsilica (Altex Ultrasphere ODS, 5 μ, 25 cm x 4.6 mm). Load, 100 ng each component. Buffer, 0.02 M acetic acid adjusted to pH 3.5 with **A** ammonia, **B** triethylamine. Flow 1.5 ml min^{-1}. Gradient: 15% to 80% methanol over 30 min. Detection, absorbance at 254 nm

2.3 Derivatization and Mass Spectroscopy

The final stage of analysis, conversion to volatile derivatives followed by GLC-MS, has seen the use of the trimethylsilyl (TMS) ethers (Most et al., 1968) and the permethyl derivatives (Chapman et al., 1976). The permethyl derivatives have certain advantages in that they are hydrolytically stable and somewhat lower in molecular weight than the TMS ethers. There is a further advantage in that the perdeuteromethyl derivatives are readily prepared (Morris, 1977) and provide much structural information merely by inspection of the mass spectra. Figure 2 illustrates the mass spectra of permethylated and perdeuteromethylated isopentenyladenine. Comparison of M.$^+$ and fragment ion masses from the two spectra allows immediate calculation of the free base molecular weight and the number of protons accessible to the reagent.

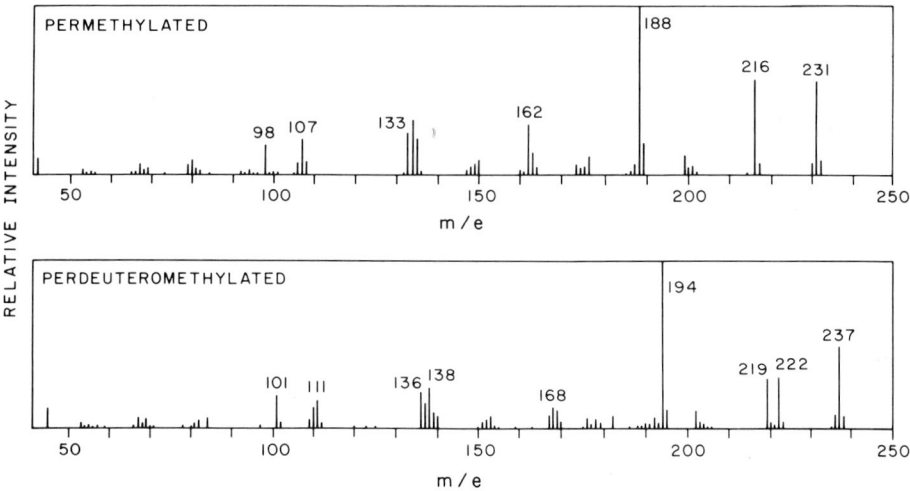

Fig. 2. Mass spectra of permethylated and perdeuteromethylated isopentenyladenine. Permethyl derivatives were prepared by using DMSO carbanion and either methyl or deuteromethyl iodide. Electron-impact spectra, 70 eV

Quantitation is possible via mass spectrometry if internal standards such as kinetin are added early in the protocol. This approach has been carried to its logical conclusion in the elegant use of deuterated internal standards (Hashizume et al., 1979; Summons et al., 1979). Addition of known amounts of deuterated cytokinins at the first extraction step, reisolation and measurement of deuterium/protium ratios in the mass spec-

3 Cytokinins in *A. tumefaciens* tRNA

3.1 Comparison of Virulent and Avirulent Strains

We have previously reported t-io^{6}Ado and ms^{2}io^{6}Ado (of unknown configuration) as components of tRNA from *A. tumefaciens* (Chapman et al., 1976). In order to confirm this work and to compare the cytokinin content of Ti plasmid-containing and plasmidless strains, tRNA was isolated from a number of strains, purified, hydrolysed and the cytokinin-active nucleosides examined by HPLC, bioassay and GLC-MS. Figure 3 illustrates a typical HPLC obtained from strain A596, a virulent strain produced by transformation of the octopine Ti plasmid of virulent strain Ach5 into a cured C58 strain (A136).

Biological activity was noted in fractions 15-17, 23-25, and 30-36 of the chromatogram. Peaks having UV absorbance at 254 nm were present, corresponding in mobility to t-io^{6}Ado, c-io^{6}Ado, ms^{2}io^{6}Ado, i^{6}Ado and ms^{2}i^{6}Ado. Mass spectra were obtained after permethylation of the appropriate fractions. Figure 4 illustrates a typical spectrum from fraction 17 indicating the presence of c-io^{6}Ado. Mass spectral data (not shown) indicated also that other cytokinins present were ms^{2}io^{6}Ado, i^{6}Ado and ms^{2}i^{6}Ado in fractions 23, 25, and 30 respectively. The amount of t-io^{6}Ado was small and a full spectrum was not obtained. However, fragment ions diagnostic for io^{6}Ado (m/e = 390,216) were noted on GLC-MS of fraction 16 having the correct retention times. Because HPLC on C$_{18}$ silica cannot resolve the cis- and trans-isomers of ms^{2}io^{6}Ado, the side chain geometry of the ms^{2}io^{6}Ado was established by TLC in a chloroform, methanol, acetic acid, triethylamine system. The bulk of the ms^{2}io^{6}Ado was present as the cis-isomer (R$_f$ 0.59); no significant amount of trans-isomer (R$_f$ 0.59) was observed.

Comparison of a set of virulent and avirulent strains revealed no striking differences in tRNA cytokinin content. As shown in Fig. 5, cytokinins in the tRNA isolated from virulent strain C58 were not greatly different to those from tRNA from strain NT1 from which the plasmid has been cured. Because the hydrolysates used in these experiments were relatively pure, it was possible to make a quantiative comparison of relative cytokinin levels in the various strains by integration of the HPLC peak areas. Table 1 shows the data obtained. The predominant cytokinins were c-io^{6}Ado, i^{6}Ado, ms^{2}i^{6}Ado and c-ms^{2}io^{6}Ado. The amount of t-io^{6}Ado was low, amounting to only a few percent of the cis-isomer. This is in contrast to our earlier report (Chapman et al., 1976) suggesting that io^{6}Ado was predominantly present as the trans-isomer. There was no obvious correlation of the level of any single component with plasmid presence.

3.2 Distribution of Cytokinins among Isoaccepting tRNA's

In *E. coli*, the cytokinin-active nucleosides are localized within the U-group tRNA's and are distributed evenly across them. Thus, activity has been found within tRNA's

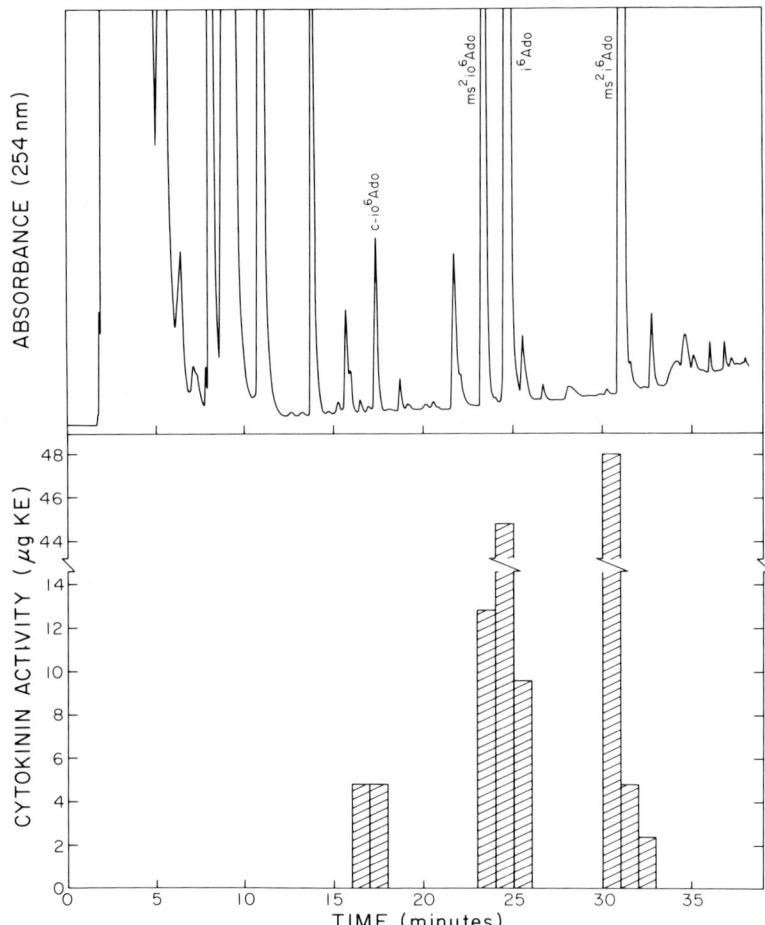

Fig. 3. HPLC of cytokinin-active nucleosides from *A. tumefaciens* strain A596 transfer RNA. Total tRNA from virulent strain A596, (3000 A_{260} units) was hydrolysed and dephosphorylated. The cytokinin nucleosides were partitioned into ethyl acetate t-butanol and applied to a C_{18}/silica column. Elution was as described in Fig. 1. Fractions were collected for tobacco callus bioassay and mass spectroscopy

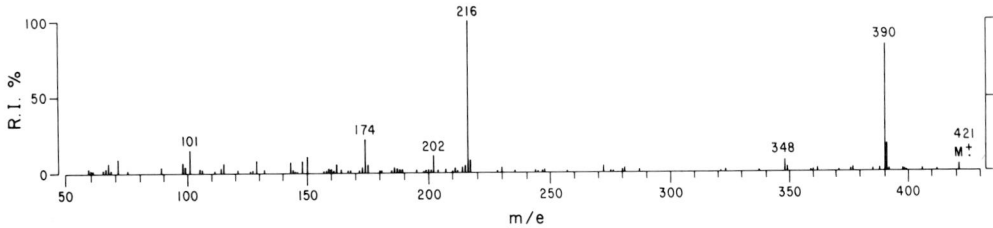

Fig. 4. Mass spectrum of ribosylzeatin isolated from A596 tRNA

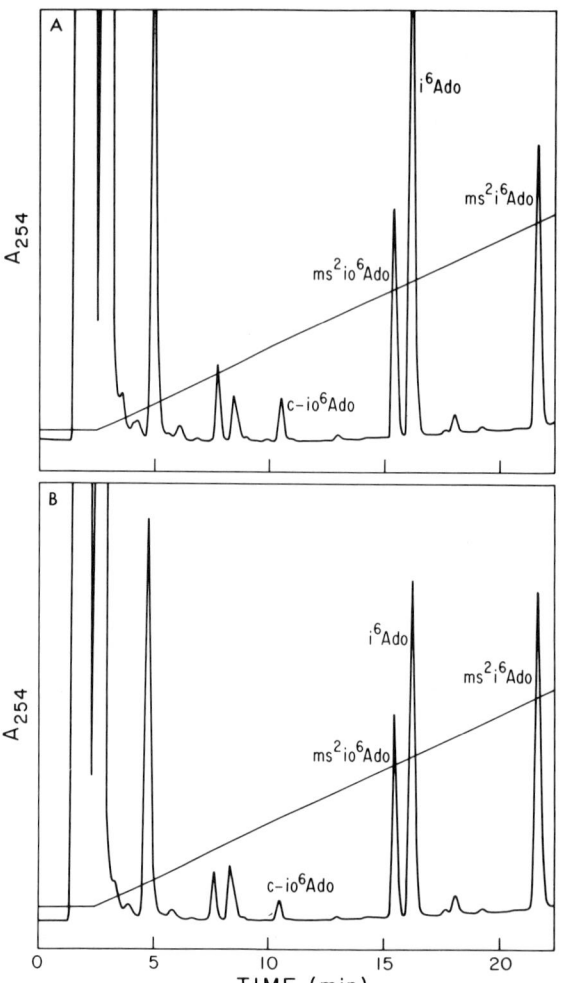

Fig. 5 A, B. HPLC of cytokinin nucleosides in tRNA from strains C58 and NT1. HPLC was as described in Fig. 1. **A.** Strain C58, virulent. **B.** Strain NT1, isogenic with C58 but cured of the pTi C58

Table 1. Cytokinin active nucleosides in *A. tumefaciens* strains

Strain	Plasmid	t-io^6Ado	c-io^6Ado	c-ms^2io^6Ado	i^6Ado	ms^2i^6Ado
Ach5	Ach5	0.23	6.7	29.9	40.3	22.8
Ach5A[a]	–	0.05	2.1	23.2	40.3	34.4
C58	C58	0.18	3.3	24.6	42.3	29.6
A136[b]	–	0.05	1.2	18.6	40.9	39.1
A596[c]	Ach5	0.10	1.2	20.8	38.9	38.9

[a] Cured Ach5; [b] Cured C58; [c] A136 (pTi Ach5)

which can be charged with Cys, Leu, Phe, Ser, Tyr, and Trp (Armstrong et al., 1969). This is not true, however, for the only plant tRNA complex examined so far. The cytokinin-active nucleosides in wheat germ tRNA are found predominantly in two isoaccepting serine species and in one minor leucine acceptor (Struxness et al., 1979). In order to determine whether the distribution of cytokinins within *A. tumefaciens* tRNA resembled that in *E. coli* or in wheat germ, total tRNA from strain C58 was isolated and fractionated by chromatography on BD-cellulose and RPC-5. Individual isoaccepting species, typically separated (Morris and Armstrong, unpublished data) as illustrated in Fig. 6, were hydrolysed. The cytokinin nucleosides were separated by HPLC and identified by GLC-MS. Table 2 shows the distribution obtained. As in *E. coli*, activity was found in all the U-group isoaccepting species. There was a similarity, however, to wheat germ tRNA in that cis-ribosylzeatin was found primarily in one tRNA^{ser} with some in a minor tRNA^{leu} while the distribution of c-ms^{2}io^{6}Ado was restricted to tRNA^{ser}, tRNA^{tyr} and tRNA^{phe}.

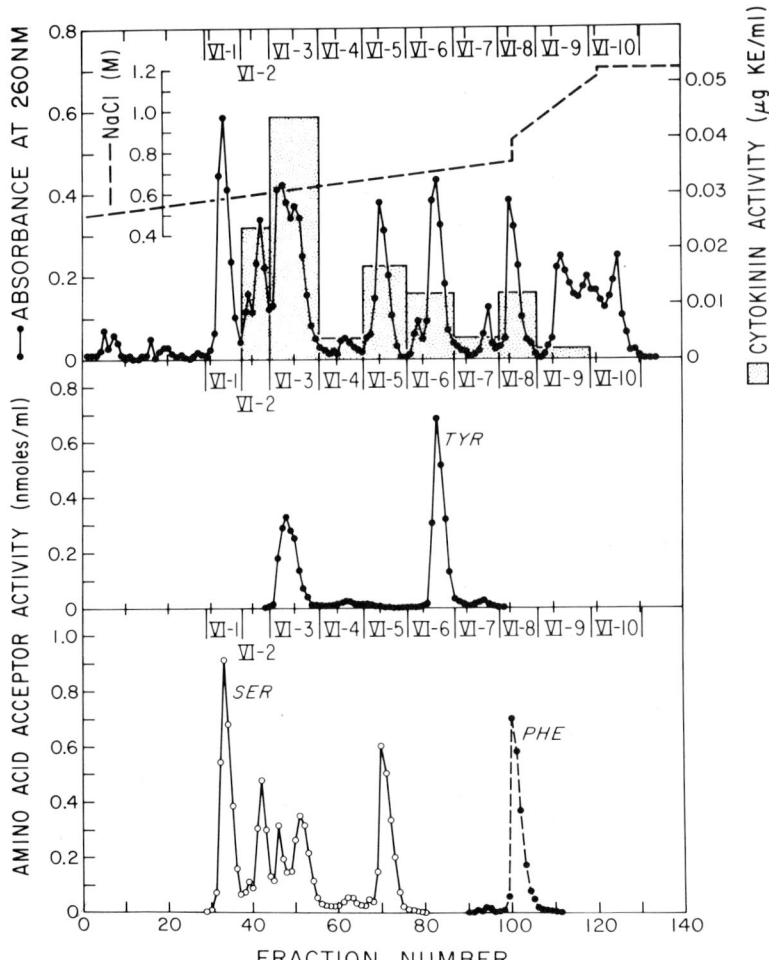

Fig. 6. RPC-5 chromatography of isoaccepting tRNA's from strain C58. Total C58 tRNA was prefractionated on BD-cellulose and the alcohol eluate was applied to RPC-5. UV absorbance and amino acid acceptance profiles determined after fractionation. Cytokinin content determined after hydrolysis of pooled selected fractions

Table 2. Distribution of cytokinins in *A. tumefaciens* isoaccepting tRNA

BD Cellulose fraction	RPC-5 fraction	Amino acid acceptance	Cytokinin present
	IV-1	Leu	i^6 Ado
Salt IV	IV-2	Trp	i^6 Ado
	IV-3	Cys	i^6 Ado
	V-1	Leu + Tyr	i^6 Ado
Salt V	V-2	Tyr + Leu	i^6 Ado
	V-4	Phe	$ms^2 io^6$ Ado
	VI-2	Ser	c-io^6 Ado
	VI-3	Ser + Tyr	$ms^2 io^6$ Ado + i^6 Ado
Alcohol VI	VI-5	Ser	i^6 Ado
	VI-6	Tyr	Unidentified [a]
	VI-8	Phe	$ms^2 i^6$ Ado

[a] HPLC R_f = authentic $ms^2 i^6$ Ado

The BD cellulose column was eluted with a gradient of sodium chloride, giving two fractions (IV and V) having cytokinin activity. This was followed by elution with 15% ethanol in 1 M NaCl giving fraction VI also with cytokinin activity. These were subfractionated on RPC-5 and analyzed for cytokinin content

4 Hybrizidation of Bacterial tRNA to the Ti Plasmid

One method by which one could determine whether the plasmid coded for a specific cytokinin-containing tRNA would be to examine for homology using the Southern blot procedure (Southern, 1975). Intact C58 plasmid was isolated, digested with a number of restriction enzymes and the fragments were separated by agarose gel electrophoresis. Transfer of the fragments to nitrocellulose filters followed by hybrizidation to end-labeled total C58 tRNA (England and Uhlenbeck, 1978) in 50% formamide gave weak hybridization to C58 restriction fragments, Hind III-13, Xba-3, Kpn 1 or 2, Sac 7, Pst 7, Bam 19 or 20 (Morris, Schell and Van Montagu, data not shown). Reference to the total restriction map of the C58 plasmid (Depicker et al., 1980) suggested that these were all consistent with homology localized within fragment Hind III-13. Subsequent experiments using the isolated Hind III-13 fragment cloned into the *E. coli* vector pBR322 indicated, (Fig. 7) that some component of the tRNA preparation of strain C58 was indeed homologous with the Hind III-13 region. Further, when tRNA from the cured strain NT1 was labeled, no hybridization was observed. The position of the homologous region on the total C58 restriction map is indicated in Fig. 8.

Fractionation of the tRNA on BD-cellulose into two major fractions prior to labeling indicated that the homologous component eluted from BD-cellulose prior to the alcohol wash. This fraction includes tRNA's IV-1 through V-4 of Table 2. At this time it is not known whether the homologous species is a tRNA or whether it is a small mRNA coded for by the Hind III-13 region. Two things are clear, however. First, the homologous component cannot be tRNA^{ser} VI-2 or Ser + Tyr VI-3 (Table 2) in which the bulk of the c-io^6 Ado and c-$ms^2 io^6$ Ado resides since these are not present

Fig. 7. Hybridization of *A. tumefaciens* tRNA to pTi C58 Hind III-13. The Hind III-13 fragment of pTi C58 was isolated from the appropriate clone, purified electrophoretically on 0.7% agarose and transferred to nitrocellulose. Total C58 or NT1 tRNA was end-labeled with ^{32}P-pCp and T_4 RNA ligase and hybridized to the blot in 0.3 M Sodium phosphate (pH 7.0), 1 M sodium chloride, 1 mM EDTA, 0.2% w/v SDS containing an equal volume of formamide. Hybridization was at 42°C for 48 h

in the BD-cellulose salt eluate. Second, these data directly exclude the possibility that the T-DNA codes for a tRNA species since the homologous region is at considerable distance from the T-DNA on the C58 plasmid.

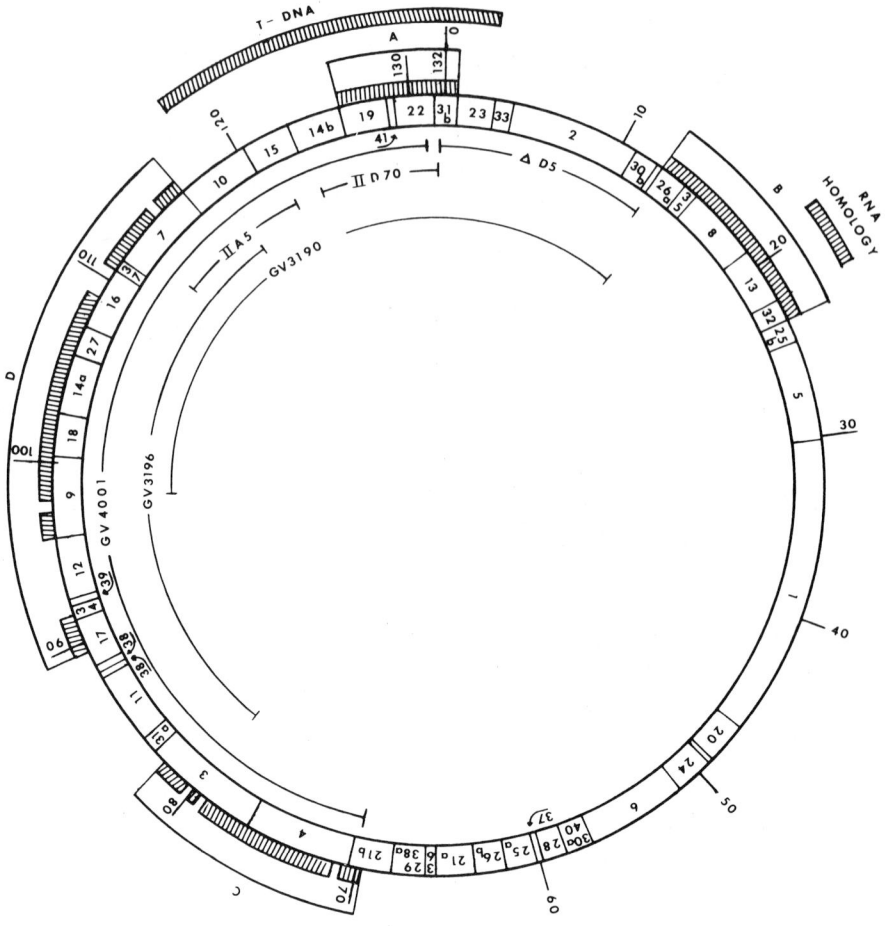

Fig. 8. Restriction map of pTi C58. Hind III restriction map of pTi C58 (redrawn from Depicker et al., 1980) showing location of T-DNA and region of homology with C58 tRNA

5 Conclusion

Agrobacterium tumefaciens contains within its tRNA cytokinins normally found in plant tRNA. The presence of c-io^{6}Ado and c-ms^{2}io^{6}Ado is normally considered to be restricted to plants. In this respect it is similar to *Corynebacterium fascians* (Murai, this vol.). The distribution of cytokinin-active species within the U-group tRNA's also bears some resemblance to that previously observed for wheat germ tRNA, although it is also bacterial in character.

The role, if any, that these cytokinins play in tumorigenesis is still not clear. A T-DNA coded tRNA-mediated pathway does appear to be excluded by the hybridization data, although there is a small RNA homologous to another region of the Ti plasmid present in the tRNA from virulent but not avirulent strains. Current studies with plasmid deletion mutants are underway to identify the route of free cytokinin biosynthesis and to determine whether it is plasmid-coded.

Acknowledgments. This work was supported in part by National Science Foundation Grant PCM 78-22963 and the Science and Education Administration of the U.S. Department of Agriculture Grant 5091-0410-8-0028-0. Receipt of an EMBO Fellowship (to R.O. Morris) is gratefully acknowledged as is the support and hospitality of J. Schell and M. Van Montagu, Rijksuniversiteit, Gent, Belgium.

References

Armstrong DJ, Burrows WJ, Skoog F, Roy KL, Söll D (1969) Cytokinins: distribution in transfer RNA species of *Escherichia coli.* Proc Natl Acad Sci USA 63: 834-841

Braun AC (1958) A physiological basis for autonomous growth of the crown gall tumor cell. Proc Natl Acad Sci USA 44: 344-349

Carnes MG, Brenner ML, Anderson CR (1975) Comparison of reversed phase high pressure liquid chromatography with Sephadex LH-20 for cytokinin analysis of tomato root pressure exudate. J. Chromatogr 108: 95-106

Chapman RW, Morris RO, Zaerr JB (1976) Occurrence of *trans*-ribosylzeatin in *Agrobacterium tumefaciens* tRNA Nature (London) 262: 153-156

Chen C-M, Eckert RL, McChesney JD (1976) N^6-(Δ^2-isopentenyl) adenosine from crown gall tumor tissue of *Vinca rosea.* Phytochemistry 15: 1565-1571

Chilton M-D, Drummond MH, Merlo DJ, Sciaky D, Montoya AL, Gordon MP, Nester EW (1977) Stable incorporation of plasmid DNA into higher plant cells: the molecular basis of crown gall tumorigenesis. Cell 11: 263-271

Claeys M, Messens E, Van Montagu M, Schell J (1978a) GC/MS determination of cytokinins in *Agrobacterium tumefaciens* cultures. Fresenius Z Anal Chem 290: 125-126

Claeys M, Messens E, Van Montagu M, Schell J (1978b) Quantitative determination of cytokinins in culture media of *Agrobacterium tumefaciens.* Quant Mass Spect Life Sci 2: 409-418

Dauphin B, Teller E, Durand B (1977) A new method for the purification, characterization and measurement of endogenous cytokinins extracted from *Mercurialis annua* shoots. Physiol Veg 15: 747-762

Dekhuijzen HM, Gevers ECT (1975) Recovery of cytokinins during extraction and purification of clubroot tissue. Physiol Plant 35: 297-302

Depicker A, DeWilde M, DeVos G, DeVos R, Van Montagu M, Schell J (1980) Molecular cloning of overlapping segments of the nopaline Ti-plasmid pTiC58 as a means to restriction endonuclease mapping. Plasmid 3: 193-211

Drummond MH, Gordon MP, Nester EW, Chilton MD (1977) Foreign DNA of bacterial plasmid origin is transcribed in crown gall tumors. Nature (London) 269: 535-536

England TE, Uhlenbeck OC (1978) Enzymatic oligoribonucleotide synthesis with T4 RNA ligase. Biochemistry 17: 2069-2076

Gordon MH (1980) Tumor formation in plants. In: Stumpf PK, Conn EE (eds) Biochemistry of plants, vol. VI. A comprehensive treatise. Academic Press, London New York, in press

Hahn H, Heitmann I, Blumbach M (1976) Cytokinins: production and biogenesis of N^6-(Δ^2-isopentenyl)adenine in cultures of *Agrobacterium tumefaciens* strain B6. Z Pflanz Physiol 79: 143-153

Hashizume T, Sugiyama T, Imura M, Cary HT, Scott MF, McCloskey JA (1979) Determination of cytokinins by mass spectroscopy based on stable isotope dilution. Anal Biochem 92: 111-122

Horgan R, Kramers MR (1979) High-performance liquid chromatography of cytokinins. J. Chromatogr 173: 263-270

Kaiss-Chapman RW, Morris RO (1977a) *Trans*-zeatin in culture filtrates of *Agrobacterium tumefaciens.* Biochem Biophys Res Commun 76: 453-459

Kaiss-Chapman RW, Morris RO (1977b) Cytokinin production, plasmid status and virulence in *Agrobacterium tumefaciens*. Plant Physiol Suppl Abstr 59: 596

Klämbt D (1967) Nachweis eines cytokinins aus *Agrobacterium tumefaciens* und sein vergleich mit dem cytokinin aus *Cornebacterium fascians*. Wiss Z Univ Rostock Math Naturwiss Reihe 16: 623-625

McCloskey JA, Hashizume T, Basile B, Ohno Y, Sonoki S (1980) Occurrence and levels of *cis*- and *trans*-zeatin ribosides in the culture medium of a virulent strain of *Agrobacterium tumefaciens*. FEBS Lett 111: 181-183

Miller CO (1974) Ribosyl-*trans*-zeatin, a major cytokinin produced by crown gall tumor tissue. Proc Natl Acad Sci USA 71: 334-338

Morris RO (1977) Mass spectroscopic identification of cytokinins: Glucosyl zeatin and glucosyl ribosylzeatin from *Vinca rosea* crown gall. Plant Physiol 59: 1029-1033

Morris RO, Zaerr JB, Chapman RW (1976). Trace enrichment of cytokinins from Douglas-fir xylem extrudate. Planta 131: 271-274

Most BH, Williams JC, Parker KJ (1968) Gas chromatography of cytokinins. J Chromatogr 38: 136-138

Parker CW, Letham DS (1974) Regulators of cell division in plant tissues. Planta 115: 337-344

Peterson JB, Miller CO (1976) Cytokinins in *Vinca rosea* L. crown gall tumor tissue as influenced by compounds containing reduced nitrogen. Plant Physiol 57: 393-399

Peterson JB, Miller CO (1977) Glucosyl zeatin and glucosyl ribosylzeatin from *Vinca rosea* L. crown gall tumor tissue. Plant Physiol 59: 1026-1028

Rathbone MP, Hall RH (1972) Concerning the presence of the cytokinin, N^6-(Δ^2-isopentenyl)-adenine in cultures of *Corynebacterium fascians*. Planta 108: 93-102

Romanow I, Chalvignac MA, Pochon J (1969) Recherches sur la production d'une substance cytokinique par *Agrobacterium tumefaciens*. (Smith and Town) Conn Ann Inst Pasteur Paris 117: 58-63

Scott IM, Browning G, Eagles J (1980) Ribosylzeatin and zeatin in tobacco crown gall tissue. Planta 147: 269-273

Smith EF, Townsend CO (1907) A plant tumor of bacterial origin. Science 25: 671-673

Southern EM (1975) Detection of specific sequences among DNA fragments separated by gel electrophoresis. J Mol Biol 98: 503-517

Struxness LA, Armstrong DJ, Gillam L, Tener GM, Burrows WJ, Skoog F (1979) Distribution of cytokinin-active ribonucleosides in wheat germ tRNA species. Plant Physiol 63: 35-41

Stuchbury T, Palni LM, Horgan R, Wareing PF (1979) The biosynthesis of cytokinins in crown-gall tissue of *Vinca rosea*. Planta 147: 97-102

Summons RE, Entsch B, Parker CW, Letham DS (1979) Mass spectrometic analysis of cytokinins in plant tissues. III Quantitation of the cytokinin glycoside complex of lupin pods by stable isotope dilution. FEBS Lett 107: 21-25

Taya Y, Tanaka Y, Nishimura S (1978) 5'AMP is a direct precursor of cytokinin in *Dictyostelium discoideum*. Nature (London) 27: 545-547

Upper CD, Helgeson JP, Kemp JD, Schmidt CJ (1970) Gas-liquid chromatographic isolation of cytokinins from natural sources. Plant Physiol 45: 543-547

White PR, Braun AC (1942) A cancerous neoplasm of plants. Autonomous bacteria-free crown-gall tissue. Cancer Res 2: 597-617

Zaenen I, Van Larabeke N, Teuchy H, Van Montagu M, Schell J (1974) Supercoiled circular DNA in crown-gall inducing *Agrobacterium* strains. J Mol Biol 86: 109-127

Cytokinin Biosynthesis and Its Relationship to the Presence of Plasmids in Strains of *Corynebacterium fascians*

N. Murai[1]

1 Introduction

Corynebacterium fascians causes fasciation disease or witches' broom in dicotyledonous plants. The disease is characterized by release of apical dominance and outgrowth of lateral buds, thus giving rise to the characteristic witches' broom syndrome (Roussaux, 1965). The symptoms of the disease can be duplicated by treatment of seedlings with cytokinins, indicating that the disease may be caused by cytokinins produced by the microbe (Thimann and Sachs, 1966). This work deals with the kinds and quantities of cytokinins released into culture media and present in tRNA of five strains of *C. fascians,* ranging from highly virulent MW2, moderately virulent Cf2 and Cf1, weakly virulent Cf 15 to avirulent Cf16. It examines the relationships of the presence of plasmids to cytokinin production and to pathogenicity. It also reports the isolation from highly virulent MW2 of activity of cytokinin synthase that catalyzes the formation of N^6-(Δ^2-isopentenyl) adenosine-5'-monophosphate (i^6 AMP) from 5'-AMP and Δ^2-IPP.

All virulent and avirulent strains of *C. fascians* so far examined have been shown to release to the culture medium virtually the same cytokinins, although individual cytokinin contents may vary greatly between strains (Fig. 1). Of the seven cytokinins isolated from Cf1 culture medium, four have been rigorously identified as 6-(3-methyl-2-butenylamino)purine (i^6 Ade), 6-(4-hydroxy-2-methyl-cis-2-butenylamino)purine (c-io^6 Ade), 6-(4-hydroxy-3-methyl-cis-3-butenylamino)-2-methylthiopurine (ms^2-c-io^6 Ade) and 6-methylaminopurine by comparison of their biological activities, chromatographic properties, ultra violet- and mass-spectra with those of authentic, synthetic compounds (Klämbt et al., 1966; Helgeson and Leonard, 1966; Scarbrough et al., 1973; Armstrong et al., 1976). The other three have been tentatively identified as 6-(4-hydroxy-3-methyl-trans-2-butenylamino)purine (t-io^6 Ade), 6-(4-hydroxy-3-methyl-2-butenylamino-9-β-D-ribofuranosylpurine (io^6 A), and 6-(3-methyl-2-butenylamino)-9-β-D-ribofuranosylpurine (i^6 A) on the basis of their biological activities and chromatographic properties (Scarbrough et al., 1973; Armstrong et al., 1976). Two additional cytokinins, 6-(3-methyl-2-butenylamino)-2-methylthiopurine ($ms^2 i^6$ Ade) and 6-(3-methyl-2-butenylamino)-2-methylthio-9-β-D-ribofuranosylpurine ($ms^2 i^6$ A), have been isolated from MW2 culture medium

---

[1] Department of Botany, University of Wisconsin, Madison, WI 53706, USA

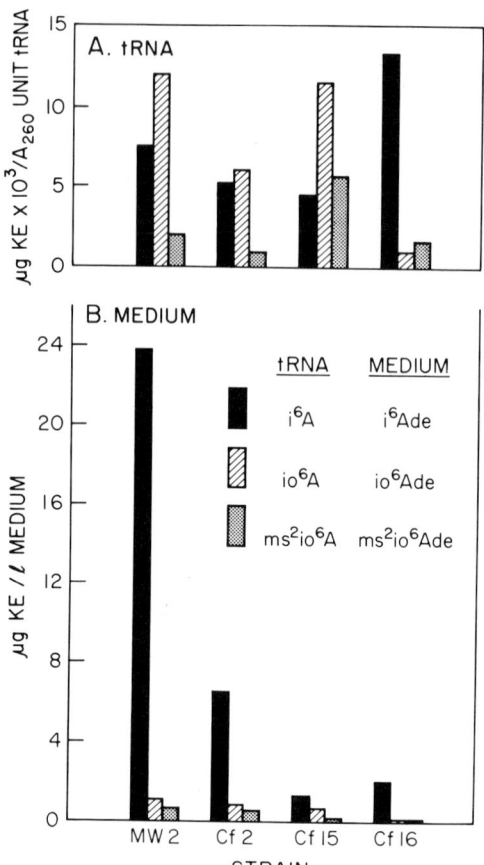

Fig. 1 A, B. Comparison of activities of different cytokinins in A the tRNA hydrolysates and B the culture medium of 4-day old cultures of four strains of *C. fascians*. Purified tRNA was enzymatically hydrolyzed to ribonucleosides, and cytokinin-active ribonucleosides were extracted by ethyl acetate. Cytokinins in the culture medium were purified by Dowex 50-X8 column chromatography followed by ethyl acetate extraction. Cytokinins were separated by Sephadex LH-20 column chromatography in 35%(v/v) ethanol or in H_2O and fractions corresponding to the elution positions of cytokinin standards were pooled and bioassayed for cytokinin activity (Data from Murai et al., 1980).

and tentatively identified (Murai et al., 1979). From each culture medium of Cf2, Cf15, and Cf16, four cytokinins were tentatively identified as i^6Ade, io^6Ade, io^6A, and ms^2io^6Ade (Murai et al., 1980).

The tRNA from each of three virulent strains and an avirulent control yielded four cytokinin-active ribonucleosides which were rigorously characterized as c-io^6A, i^6A, ms^2-c-io^6A, and ms^2i^6A, and a fifth tentatively identified as t-io^6A. These results indicate that all the cytokinin-active bases and ribonucleosides expected from the degradation of tRNA can be recovered from the culture medium of all the tested strains of *C. fascians*. An apparent precursor-product relationship between the cytokinin-active ribonucleosides isolated from tRNA and cytokinin-active bases and ribonucleosides found in the culture medium indicates that tRNA may serve as one source of free cytokinins produced by the bacterium (Table 1). The relatively constant ratios of io^6A and ms^2io^6A in tRNA and their derived free bases in the culture medium of these strains are consistent with this view (Fig. 1). Biosynthesis of the i^6A moiety in tRNA has been shown to involve the transfer of isopentenyl from Δ^2-IPP to specific adenosine residues in preformed tRNA (Kline et al., 1969).

Table 1. List of cytokinins isolated and identified from the culture medium and tRNA of *Corynebacterium fascians*. [Data compiled from Klämbt et al. (1966); Helgeson and Leonard (1966); Scarbrough et al. (1973); Armstrong et al. (1976); Murai et al. (1979); Murai et al. (1980)]

Cytokinin Symbol	Substituent				Source	
	R_1	R_2	R_3	R_4	Medium	tRNA
i^6 Ade	H	H	H	H	#[a]	
i^6 A	H	H	H	$C_5H_9O_4$	+[b]	#
c-io^6 Ade	OH	H	H	H	#	
c-io^6 A	OH	H	H	$C_5H_9O_4$	+	#
t-io^6 Ade	H	OH	H	H	+	
t-io^6 A	H	OH	H	$C_5H_9O_4$	+	+
ms^2-c-io^6 Ade	OH	H	SCH$_3$	H	#	
ms^2-c-io^6 A	OH	H	SCH$_3$	$C_5H_9O_4$	-[c]	#
ms^{2}i^6 Ade	H	H	SCH$_3$	H	+	
ms^{2}i^6 A	H	H	SCH$_3$	$C_5H_9O_4$	+	#
m^6 Ade[e]	–	–	H	H	#	
m^6 A	–	–	H	$C_5H_9O_4$	–	(#)[d]

[a] Identified by comparison of biological activity, chromatographic properties, ultra violet- and mass spectra with those of authentic, synthetic compounds
[b] Identified on the basis of biological activity and chromatographic properties
[c] Not identified
[d] Evidence from other source
[e] 6-methylaminopurine

2 Cytokinin Activity in the Culture Medium

Total cytokinin activity in the culture media of the four pathogenic strains varied from 168 to 0.4 µg kinetin equivalents per liter (µg KE/1) as compared to 0.2 µg KE/1 in an avirulent strain (Table 2). i^6 Ade was the predominant cytokinin in the medium of all five strains. The quantity of i^6 Ade increased with the degree of pathogenicity, from 2 µg KE/1 in the avirulent culture without a plasmid to 24 µg KE/1 in the highly virulent strain MW2, as measured on purified preparations freed from inhibitors (Fig. 1). As shown in Table 2, culture media of Cf15 and Cf16 tested prior to the removal of inhibitors by chromatography on Sephadex LH-20 columns had less total cytokinin activity, and the difference between pathogenic and nonpathogenic strains, therefore, was still more extreme. In media of the virulent strains io^6 Ade was ≈ 10 times more

Table 2. Comparison of virulence, number of plasmids, growth, and cytokinin activity of culture medium in five strains of *Corynebacterium fascians*. [Data from Murai et al., 1980)]

Strain	Virulence[a]	No. of Plasmids[b]	Growth g/liter	Cytokinin activity[c]
MW2	High	2	12.0	168
Cf2	Mod	3	10.0	8.3
Cf1	Mod	3	9.0	2.3
Cf15	Weak	1	11.4	0.4
Cf16	None	0	9.0	0.2

[a] Pathogenicity was tested on *Pisum sativum* seedlings. (Mod. = Moderate)
[b] The number of plasmids was determined from the number of radioactive peaks below the major peak of radioactivity in the alkaline sucrose gradients and confirmed by electrophoresis of purified plasmids on agarose gels
[c] Cytokinins were purified from the supernatant fraction of culture medium by Dowex 50W-X8 column chromatography. Cytokinin activity was determined using the tobacco callus bioassay and expressed in µg KE/1 of culture medium

abundant than in that of the avirulent Cf16 strain. In cultures of the weakly virulent Cf15, io^6 Ade contributed about half as much activity as i^6 Ade, and, therefore, allowing for its ten fold lower specific activity, was the more abundant of the two, on a molar basis, but in culture media of the more virulent strains (Cf2 and MW2) the io^6 Ade accounted for only a minor part of the total cytokinin activity. The other five detected cytokinins io^6 A, i^6 A, ms^2-c-io^6 Ade, ms-$^2 i^6$ Ade and $ms^2 i^6$ A, contributed even less to the total biological activity. It should be noted that ms^2-c-io^6 Ade activity was from 10- to 16-fold higher in the strains with plasmids.

Cytokinin Content in tRNA

The total cytokinin content in mol percent of the three pathogenic strains (MW2, Cf2, Cf15) and the avirulent control (Cf16) did not differ significantly between strains, but the cytokinin composition was very different in the presence and absence of plasmids (Table 3). In each of the plasmid-containing strains the abundance of the different cytokinins decreased in the order ms^2-c-io^6 A, c-io^6 A, i^6 A, $ms^2 i^6$ A and t-io^6 A, but in the plasmidless strain $ms^2 i^6$ A was predominant and was followed in order by i^6 A, ms^2-c-io^6 A, c-io^6 A and t-io^6 A. In each of the strains with a plasmid, the cytokinins with a 4-hydroxylated isopentenyl side chain were more than twice as abundant as cytokinins with an unsubstituted isopentenyl side chain, whereas in the strain without a plasmid this ratio was the reverse. In terms of total cytokinin activity in tRNA hydrolysates as measured in the tobacco bioassay, the observed differences between plasmid-containing and plasmidless strains were less than those between strains which contain plasmids. The evidence indicates that neither total cytokinin

Table 3. Comparison of content and biological activity of cytokinin-active ribonucleosides from tRNA's of four strains of *C. fascians*. (Data from Murai et al., 1980)

Strain	MW2	Cf2	Cf15	Cf16
Cells (g)[a]	71	66	66	65
tRNA (mg)[a]	49	49	49	45
Cytokinins	[b]Cytokinin content [μg (mol %)][a]			
t-io^{6}A	Trace[c]	Trace	Trace	Trace
c-io^{6}A	5.9 (0.8)	4.4 (0.7)	3.9 (0.6)	1.5 (0.2)
i^{6}A	3.2 (0.5)	2.6 (0.4)	2.1 (0.3)	7.8 (1.3)
ms^2-c-io^{6}A	9.8 (1.3)	6.4 (0.8)	16.0 (2.1)	8.1 (1.1)
ms^{2}i^{6}A	3.2 (0.4)	2.6 (0.3)	1.7 (0.9)	9.3 (1.4)
	Cytokinin activity (μg KE)[a]			
Total	21.3	12.4	21.4	16.6

[a] Total per sample
[b] Cytokinin-active ribonucleosides were purified by chromatography on Sephadex LH-20 and/or Porapak Q columns and identified by comparison of their biological activities, chromatographic properties, ultra violet spectra and gas liquid chromatography-mass spectra with those of authentic, synthetic compounds
[c] Trace = detectable in tobacco callus bioassay but <0.1 μg

activity nor i^{6}A content in tRNA hydrolysates of the strains examined are correlated with the degree of pathogenicity. Therefore, it appears that the marked increase in i^{6}Ade content in the culture medium which is associated with pathogenicity is not a reflection of the i^{6}A content in the tRNA of the strains.

3 Activity of Cytokinin Synthase from the Highly Virulent MW2 Strain

The above observations indicate the possible existence of an alternative pathway of i^{6}Ade biosynthesis in virulent *C. fascians* strains other than via tRNA degradation. Activity of cytokinin synthase (Δ^2-IPP: AMP-Δ^2-isopentenyltransferase) has been isolated from the cellular slime mold *Dictyostelium discoideum* (Taya et al., 1978), and from cytokinin-autotrophic tobacco callus (Chen and Melitz, 1979). The in vitro synthesis of i^{6}AMP from Δ^2-IPP and 5'AMP as well as of i^{6}A and i^{6}Ade have been demonstrated in crude enzyme preparations from these sources.

Cytokinin-autotrophic tobacco crown gall callus was used to determine the purification procedure (Chen and Melitz, 1979) and assay method for cytokinin synthase. Radioactive Δ^2-IPP was prepared from 1-^{14}C-isopentenyl pyrophosphate (Δ^3-IPP) by incubation with isopentenyl pyrophosphate Δ^3-Δ^2-isomerase isolated from wheat germ (Banthorpe et al., 1977). Formation of radioactive i^{6}AMP from ^{14}C-Δ^2-IPP and AMP was assayed by Dowex 1-X8 (formate form) column chromatography (Hurlbert et al., 1954; Tchen, 1963). A crude enzyme preparation partially purified by

$(NH_4)_2SO_4$ precipitation from the callus homogenate was fractionated by Sephadex G-100 column chromatography (Fig. 2). The cytokinin synthase activity appeared after elution of the A_{280} peak as a sharp peak at 0.32 to 0.48 column volumes with a shoulder and maximum peak of activity at 0.37 and 0.43 column volumes, respectively. Active fractions were pooled as indicated in Fig. 2 for further purification. By using essentially the same purification procedure, the cytokinin synthase activity was isolated from cell lysates of the highly virulent MW2 strain, and was purified by $(NH_4)_2SO_4$ precipitation and Sephadex G-100 column chromatography (Fig. 3). As in the case of tobacco enzyme, the activity was eluted in column fractions as s sharp peak at 0.32 to 0.45 column volumes with a maximum at 0.41 column volume. About a 30-fold purification of enzyme activity was achieved by Sephadex G-100-chromatography. The enzyme preparation in the pooled fractions was used to optimize assay conditions. After digestion of the isolated reaction products from the assay mixture with alkaline phosphatase, a peak of radioactivity was obtained in fractions corresponding to the elution position of synthetic i^6A from a Sephadex LH-20 column. These results indicate that the highly virulent MW2 strain of *C. fascians* contains a cytokinin synthase that catalyzes the formation of i^6AMP from Δ^2-IPP and AMP. Thus it appears that the marked increase in free i^0Ade in the culture medium which is associated with pathogenicity may result from a high rate of i^6AMP production, catalyzed by cytokinin synthase.

4 Plasmids

All four pathogenic strains (MW2, Cf2, Cf1, Cf15) contained a 63-megadalton (Mdal) plasmid, as estimated by agarose gel electrophoresis (Fig. 4). The first three most virulent strains contained an additional 120-Mdal plasmid. A third plasmid of 5 Mdal has been found in the Cf1 and Cf2 strains. No plasmid was detected in the avirulent Cf16 strain. All *C. fascians* strains so far examined, including Cf16, contained fragments of less than one Mdal that may be DNA or tRNA contaminants. Restriction endonuclease fingerprint analysis of the 63-Mdal plasmids from MW2 and Cf15 revealed that these two plasmids differ to some extent. In the highly virulent MW2, loss of virulence was associated with loss of the 63-Mdal plasmid, but not of the 120-Mdal plasmid. From these results, it appears that pathogenicity is associated with the presence of plasmids. In strains so far tested both the high i^6Ade content in the culture medium and the presence of high amounts of hydroxylated isopentenyl derivatives (cis- and trans-zeatin and their derivatives) in the culture medium and of the corresponding zeatin ribonucleotides in tRNA appear to be associated with the presence of the 63-Mdal plasmids. These results further indicate that cytokinin synthase may be coded for by DNA in the 63-Mdal plasmid.

All four virulent strains, but not the avirulent Cf16 strain, were able to utilize hydrocarbons from octadecane to undecane as a carbon source (Doyle and Hanson, 1980). The latter has been found to be 20 times more sensitive to the presence of cadmium than the virulent strains. These observations are consistent with the view that DNA in the 63-Mdal plasmid codes for the ability to utilize hydrocarbons and for cadmium resistance.

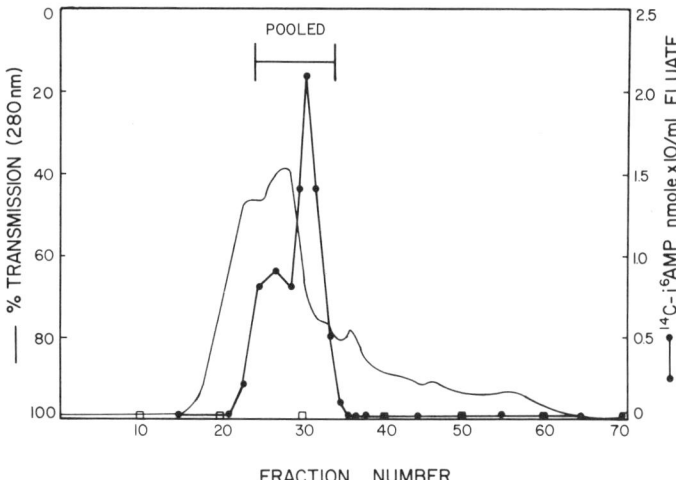

Fig. 2. Purification of tobacco crown gall cytokinin synthase by Sephadex G-100 column chromatography. Cytokinin synthetase was isolated from 500 g of 25-day old tobacco crown gall callus E-1 (octopine-producing) essentially as described by Chen and Melitz (1979). The callus was homogenized in an equal volume of twice concentrated homogenization buffer (10 mM Tris-HC1, pH 7.0, 10 mM Mg-acetate, 6 mM KCl, 1 mM EDTA, 6 mM β-mercaptoethanol, and 25%(v/v) glycerol) with 1%(w/v) of polyvinyl polypyrrolidone. The homogenate was fractionated by $(NH_4)_2SO_4$ precipitation at 40% to 80% saturation. The $(NH_4)_2SO_4$ precipitates were extracted by 5 ml of Sephadex G-100 buffer (10 mM Tris-HC1, pH 7.0, 1 mM Mg-acetate, 0.6 mM KCl, 0.1 mM EDTA, 0.6 mM β-mercaptoethanol, and 25% (v/v) ethylene glycol). The extract was dialyzed against Sephadex G-100 buffer and fractionated on a Sephadex G-100 column (20 g, 364 ml, 2.6 x 66.6 cm) in 5 ml fractions.

Isopentenyl pyrophosphate Δ^3-Δ^2 isomerase was purified from wheat germ as described by Banthorpe et al. (1977) and as modified by Chen and Melitz (1979). The partially purified isomerase preparation (100 μl) was incubated at 37°C for 30 min in a final 200 μl of reaction mixture containing 10 mM Na-maleate, pH 6.3, 2 mM $MnCl_2$, 10 mM KF, and 50 nCi or 45 pM of 1-^{14}C-isopentenyl pyrophosphate (56 mCi/mmol, Amersham/Searl). The isomerase activity was assayed by chromatography on Dowex 1-X8 (chloride form) columns. At the end of incubation, the mixture was adjusted to pH 2.0 and incubated at 37°C for 10 min to cleave the allylic pyrophosphate of Δ^2-IPP. Following a ten fold dilution of the reaction mixture at pH 12.0, liberated Δ^2-isopentanol and the remaining Δ^3-IPP were eluted from Dowex columns with H_2O and 0.3 M KCl, respectively.

The cytokinin synthetase activity was assayed by chromatography on Dowex 1-X8 (formate form) as described (Hurlbert et al., 1954; Tchen, 1963). The enzyme preparation (100 μl) was incubated at 37°C for 30 min in a final 200 μl volume of reaction mixture containing 10 mM Na-maleate, pH 6.3, 2 mM $MnCl_2$, 10 mM KF, 1 mM 5'-AMP, 50 nCi or 45 pM of 1-^{14}C-Δ^3-IPP and 50 μl of wheat germ isomerase. The reaction mixture was further incubated at pH 2.0 at 37°C for 10 min, adjusted to pH 12.0, diluted by ten fold, and applied to Dowex columns. Δ^2-isopentanol, i^6 AMP, and Δ^3-IPP were eluted with H_2O, 2 N HCOOH, and 4 N HCOOH, and 0.8 N NH_4COOH, respectively

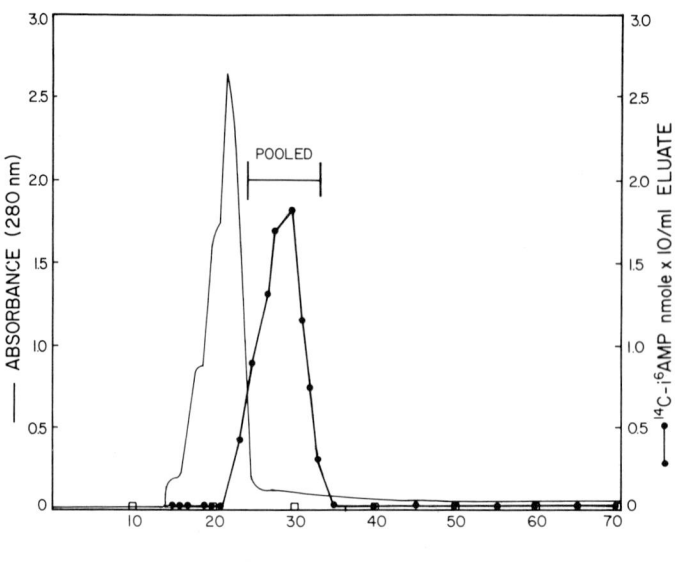

Fig. 3. Purification of *C. fascians* MW2 cytokinin synthase by Sephadex G-100 column chromatography. Cytokinin synthase was isolated from 50 g of 46 h old MW2 cells essentially as described in the legend of Fig. 2. The cells were suspended in 5 volumes of 10 mM Tris-HCl, pH 8.0 and 100 mM EDTA and incubated with 0.2% (w/v) of lysozyme (Sigma Chem. Co.) at 23°C for 30 min. After centrifugation, the cells were suspended in 5 volumes of homogenization buffer, ruptured by two passages through a French press (840-1260 kg/cm^2), and homogenized with 0.5% (w/v) of sodium deoxycholate

Similar correlations have been found between the resence of plasmids and (a) cytokinin activity in the culture medium, (b) presence of zeatin ribonucleosides in the tRNA, and (c) pathogenicity of *Agrobacterium tumefaciens* (Chapman et al., 1976; Kaiss-Chapman and Morris, 1977). High contents of zeatin and/or its derivatives are also present in other pathogenic or symbiotic organisms which cause overgrowths in plants, such as mycorrhizal fungi (Miller, 1967), gall-forming insects (Engelbrecht, 1971; Van Studen, 1975; Green, 1980), even though hydroxylated isopentenyladenines normally are found neither in fungi nor in animals (Skoog and Schmitz, 1979). These results raise the question whether DNA which codes for high cytokinin production and zeatin biosynthesis in prokaryotic plant pathogens, and in the case of *A. tumefaciens* may be transmitted to serve this function in host tissue, is of prokaryotic origin or has been acquired by the pathogens from plants in the course of evolution. They also raise the question whether gall-forming insects and other eukaryotic pathogens which cause overgrowths in plants may possess DNA originally derived from plants.

Acknowledgments. The author expresses his gratitude to Dr. F. Skoog of the Department of Botany and Dr. R.H. Hanson of the Department of Bacteriology, University of Wisconsin, in whose laboratories this research was jointly conducted. He is also grateful to Dr. N.J. Leonard, School of Chemistry, University of Illinois, for synthetic cytokinins.

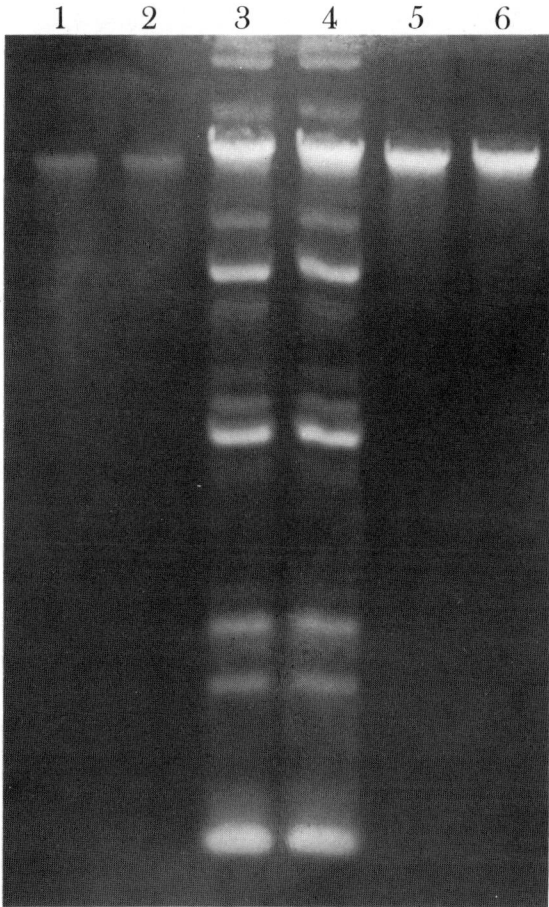

Fig. 4. Electrophoresis of plasmid preparations isolated from strains of *C. fascians* on 0.7%(w/v) agarose. *1-2* 630Mdal plasmid isolated from the highly virulent MW2 strain that have lost 120-Mdal plasmid. The plasmid was purified by CsCl centrifugation. *3-4* 120-, 63-, 20-, and 5-Mdal plasmids isolated from the moderately virulent Cf2 strain. *5-6* Chromosomal DNA contaminants in a crude plasmid preparation

References

Armstrong DJ, Scarbrough E, Skoog F, Cole DL, Leonard NJ (1976) Cytokinins in *Corynebacterium fascians* cultures. Isolation and identification of 6-(4-hydroxy-3-methyl-*cis*-2-butenyl-amino)-2-methylthiopurine. Plant Physiol 58: 749-752

Banthorpe DV, Doonan S, Gwtowski JA (1977) Isopentenyl pyrophosphate isomerase from pig liver. Arch Biochem Biophys 184: 381-390

Chapman NW, Morris RO, Zaerr JB (1976) Occurence of *trans*-ribosylzeatin in *Agrobacterium tumefaciens* tRNA. Nature (London) 262: 153-154

Chen CM, Melitz DK (1979) Cytokinin biosynthesis in a cell-free system from cytokinin-autotrophic tobacco tissue cultures. FEBS Lett 107: 15-20

Doyle ME, Hanson RH (1980) Unpublished observation

Engelbrecht L (1971) Cytokinin activity in larval infected leaves. Biochem Physiol Pflanz 162: 9-27

Greene EM (1980) Cytokinin production by microorganisms. Bot Rev 46: 25-74

Helgeson JP, Leonard NJ (1966) Cytokinins: identification of compounds isolated from *Corynebacterium fascians*. Proc Natl Acad Sci USA 56: 60-63

Hurlbert RB, Schmitz H, Brumm AF, Potter VR (1954) Nucleotide metabolism II. Chromatographic separation of acid-soluble nucleotides. J Biol Chem 209: 23-39

Kaiss-Chapman RW, Morris RO (1977) *Trans*-zeatin in culture filtrates of *Agrobacterium tumefacions*. Biochem Biophys Res Commun 76: 453-459

Klämbt D, Thies G, Skoog F (1966) Isolation of cytokinins from *Corynebacterium fascians*. Proc Natl Acad Sci USA 56: 52-59

Kline LK, Fittler FA, Hall RG (1969) N^6-(Δ^2-Isopentenyl)adenosine. Biosynthesis in transfer ribonucleic acid in vitro. Biochemistry 8: 4361-4371

Miller CO (1967) Zeatin and zeatin riboside from a mycorrhizal fungus. Science 157: 1055-1057

Murai N, Skoog F, Doyle ME, Hanson RS (1979) Correlation between cytokinin production, presence of plasmids, and pathogenicity in strains of *Corynebacterium fascians*. 10th Int Conf Plant Growth Subst, Madison, WI, p 268

Murai N, Skoog F, Doyle ME, Hanson RS (1980) Relationship between cytokinin production, presence of plasmids and fasciation-caused by strains of *Corynebacterium fascians*. Proc Natl Acad Sci USA 77: 619-623

Roussaux MJ (1965) Etude préliminaire des modifications induites chez le pois express Alaska par le *Corynebacterium fascians* (Tilford) Dowson. Rev Gen Bot 72: 21-53

Scarbrough E, Armstrong DJ, Skoog F, Frihart CR, Leonard NJ (1973) Isolation of *cis*-zeatin from *Corynebacterium fascians*. Proc Natl Acad Sci USA 70: 3825-3829

Skoog F, Schmitz RY (1979) Biochemistry and Physiology of Cytokinins. In: Litwack G (ed) Biochemical actions of hormones, vol VI. Academic Press, London, New York, pp 335-413

Taya Y, Tanaka Y, Nishimura S (1978) 5'-AMP is a direct precursor of cytokinin in *Dictyostelium discoideum*. Nature (London) 271: 545-547

Tchen TT (1963) Preparation and determination of intermediates in cholesterol synthesis. In: Colowick SP, Kaplan NO (ed) Method Enzymol, vol VI, Academic Press, London, New York, pp 505-512

Thimann KV, Sachs T (1966) The role of cytokinins in the "fasciation" disease caused by *Corynebacterium fascians*. Am J Bot 53: 731-739

Van Studen J (1975) Cytokinins from larvae in *Erythrina latissima* galls. Plant Sci Lett 5: 227-230

Cytokinin Biosynthesis in Higher Plants

H. Maaß[1,2] and D. Klämbt[2]

1 Introduction

Since cytokinin nucleotides have been detected within different tRNA species there has been continued discussion about the biosynthesis of the free cytokinins (Fig. 1). While a group of authors prefers a de novo synthesis from adenine monomers (Chen et al., 1976; Chen and Petschow, 1978; Chen and Melitz, 1979; Burrows, 1978; Stuchbury et al., 1979; Taya et al., 1978), there are others who sees the pathway via the hydrolysis of polynucleotides (Holtz and Klämbt, 1975, 1978; Klemen and Klämbt, 1974; Helbach and Klämbt, in press; Maass and Klämbt in press. 1980), to be more likely. The first group used slime molds or callus tissues from cytokinin autonomous tobacco or crown gall, which were supplied with radioactive adenine. Their research was restricted to investigation into the possibility for a de novo synthesis by detection of radioactive cytokinins. Chen and Melitz (1979) reported a crude enzyme system being able to synthesize cytokinins directly from 5'AMP and IPP. Taya and coworkers found that 5'AMP is a direct precursor for IPAde in Dictyostelium discoideum. IPAde is an intermediate derivative in the biosynthesis of discadenine, which is involved in spore germination. However, the tRNA and its turnover was not taken into consideration in any of these publications. The second group published results on the turnover rates of tRNA in bacteria and roots of higher plants and compared the content of "bound" cytokinins to the free ones. Moreover, Holtz and Klämbt (1978) isolated the Δ^2 isopentenyldiphosphate: tRNA Δ^2-isopentenyltransferase, which isopentenylated tRNA, oligo N, MS2, poly A, and also oligo A-$(Ap)_{3-7}$A- but not ApApA, ApA or other monomeric adenine derivatives. These data, of course, are not conclusive because the expected isopentenylation of adenine or its derivatives could be catalyzed by another enzyme.

Barnes et al. (1980) reported on the production of cytokinins in potato cells. They showed that up to 40% of the free cytokinins is due to the breakdown of tRNA.

We present some additional experimental data and mathematical models to the two hypotheses of cytokinin production. Comparing these data with the theoretical models we are able to favor the indirect pathway.

Abbreviations: IPA: N^6-(Δ^2-Isopentenyl)adenosine; IPAde: N^6-(Δ^2-isopentenyl)adenine; IPP: Δ^2-isopentenyldiphosphate; oligo N: oligonucleotides; ZR: zeatin riboside

[1] Part of the doctoral thesis, Bonn 1980
[2] Institute of Botany, University of Bonn, Meckenheimer Allee 170, 5300 Bonn 1, FRG

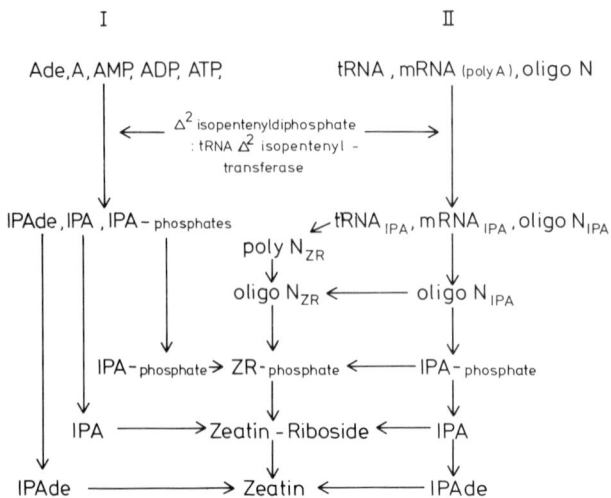

Fig. 1. Two hypotheses for the biosynthesis of cytokinin: *I* de novo from monomers; *II* via breakdown of polymers

2 Material and Methods

Phaseolus vulgaris var. *Saxa* were cultured in liquid medium at 22°C with 12 hrs light-dark changes. Those plants about to flower were used for the tests (the buds had been excised a week beforehand). Five ml of sterile culture medium containing 40 μCi (1.48 MBq) [^{14}C]-adenine (28 μM) were fed to the roots of three plants per test. After pulse labeling for 6 h and a following chase with the application of 28 mM [^{12}C]-adenine for another 6 h the plants were harvested 6, 12, 24, 48, 72, and 96 h after the test started. The preparation procedures for the isolation of free cytokinins from leaves and roots, and the preparation of tRNA and oligo N as well as the detection of their "bound" cytokinins are presented in Fig. 2.

3 Results and Discussion

As already reported (Maass and Klämbt, in press, 1980) the half-life of tRNA was estimated to be about 65 h. That of the oligo N in rapid decay (presumably from mRNA or similar polynucleotides) was determined to be 8 h (Fig. 3). The predominant cytokinin (98%) found within the tRNA was IPA. A ratio of about 1:70 of the total incorporated radioactivity after the pulse labeling with 1.48 MBq [^{14}C]-adenine was found. In contrast ZR could be detected as the predominant cytokinin within the oligo N. A short time after the pulse period ended its ratio to the total radioactivity was about 1:800. A lower ratio was found within the slowly decreasing oligo N. The most frequently occurring free radioactive cytokinin was zeatin. It accounted for 80% of all cytokinin-like radioactive substances.

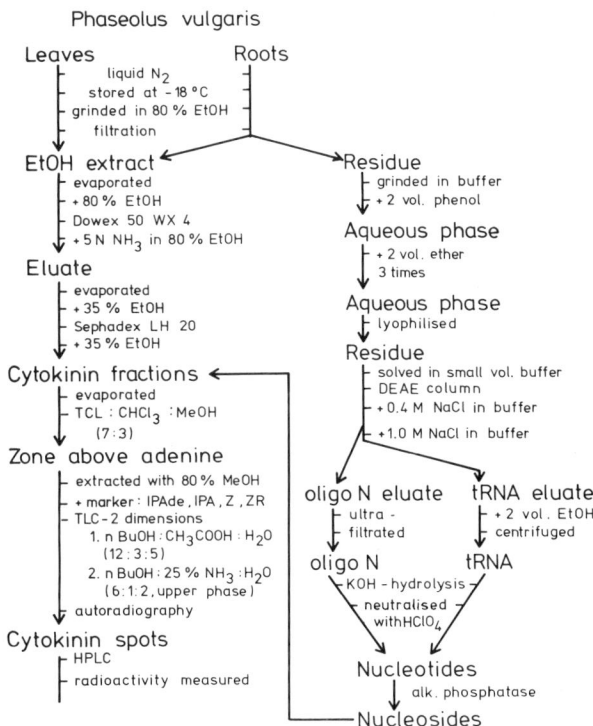

Fig. 2. Preparation procedures

To obtain some information about the turnover rates of cytokinins in leaves, 1 µCi (37 KBq) [^{3}H]-IPAde/ml and 0.435 µCi (16.1 KBq) [^{14}C]-zeatin[3]/ml respectively were fed to a single leaf. After the feeding period of 3 h the radioactivity found in the leaf lamina amounted to 40% of the applied [^{3}H]-IPAde and 60% of the applied [^{14}C]-zeatin on average. The measurements of aliquots revealed that the activities of the given cytokinin bases were reduced to half after about 20 h. The investigations were restricted to the recovery of the applied cytokinins and were not concerned with the resulting products. Therefore this value is only an approximate one, as by the application of rather high concentrations (3.4 and 37 µM resp.) a conversion to ribosides and glucosides are expected. These cytokinin derivatives are not degradation products (which are the basis of the calculation for half-life times), as they are able to change back to the original cytokinin structures. Hence follows that the half-life of cytokinins within a leaf is presumably to be set at much longer than 20 h.

In the light of the given data and with regard to both assumptions of cytokinin biogenesis the following mathematical considerations could be performed.

---

[3] We thank Dr. E. Knegt, Wageningen for the generous gift

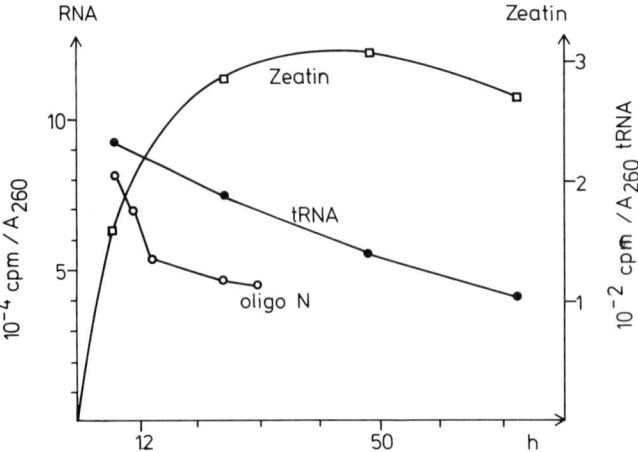

Fig. 3. Kinetics of tRNA and oligo N degradation in roots as well as labeled zeatin arising in leaves. Application of 1.48 MBq [^{14}C]-adenine (28 µM) for 6 h; chase with 1000-fold concentration of [^{12}C]-adenine for another 6 h

3.1 The Cytokinin Production Results Exclusively from the Degradation of tRNA and oligo N

The half-lives of tRNA, oligo N, and cytokinins are to be determined, as the actual amount of cytokinins at the time t yields from the turnover rates of tRNA and oligo N, reduced for that part of already degraded cytokinins. The following formula can be established:

$$N_t = [x \cdot A_0 (1-e^{-at}) + y \cdot B_0(1-e^{-bt})] \cdot e^{-ct},$$

where

N_t = actual amount of cytokinins at the time t,
A_0 = amount of tRNA at time 0,
A_t = amount of radioactive labeled tRNA at time t,
x = cytokinin part of tRNA,
a = degradation constant of tRNA, dated from the half life,
B_0, B_t, y, and b are those figures corresponding to oligo N,
c = degradation constant of the cytokinins.

A_0, B_0, x, y, a, b, and c are experimentally ascertained:
A_0 = 90,000 cpm/A_{260} tRNA,
B_0 = 80,000 cpm/A_{260} Oligo N,
x   = 1:70,
y   = 1:800,
a   = 0.0106638, if half-life is assumed with 65 h,
b   = 0.0866433, if half-life is assumed with 8 h,
c_1 = 0.0144405, if half-life is assumed with 48 h,
c_2 = 0.0288811, if half-life is assumed with 24 h

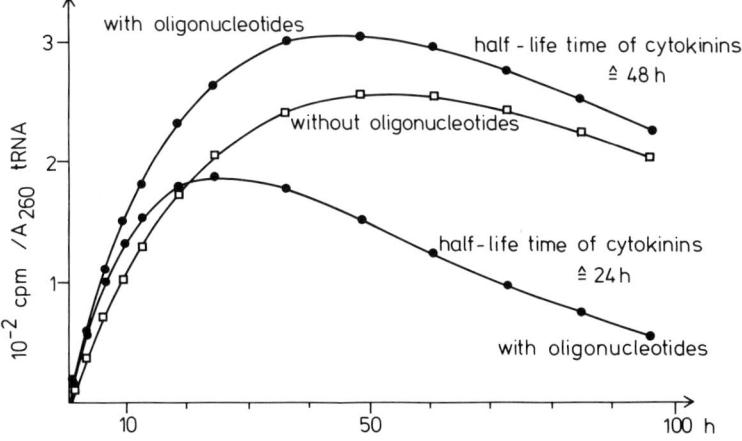

Fig. 4. Theoretical curves referring to the cytokinin biogenesis via RNA breakdown (see text)

Both resulting curves are shown in Fig. 4, whereas Fig. 3 shows the labeled tRNA and oligo N from roots and the labeled zeatin from leaves. The theoretical curve (half-life of cytokinin adopted to 48 h) fits that of Fig. 3 best. As mentioned above, the half-life of cytokinins within leaves is unknown, but is very probably much longer than a day. If the oligo N was not considered the curve runs too low. The early detection of radioactive cytokinins in the leaves is explicable by the more rapid breakdown of oligo N.

3.2 The Cytokinin Production Results Exclusively from a de novo Synthesis

Under this assumption, radioactive labeled cytokinins are synthesized mainly during the pulse incubation with [^{14}C]-adenine. A new synthesis of labeled cytokinins has to be strongly prevented by the following chase with the 1000-fold concentration of [^{12}C]-adenine and the resulting dilution of the radioactive pool. The maximum radioactive amount of cytokinins should be found immediately after pulse ending and then should decrease logarithmically. Corresponding to our results, a maximum cytokinin content of 3000 cpm is expected to be reached a short time after pulse ending. Figure 5 shows the resulting curves which differ completely from the first ones. The maximum detection of labeled cytokinins is to be found early, depending on the velocity, how fast the [^{14}C]-adenine pool is diluted, and how fast the cytokinin is transported. On average, the plants had taken up the [^{14}C]-adenine solution within 3 h. Therefore the maximum of our hypothetical curve was set at 9 h, i.e., 3 h after the end of the labeling phase.

From the mathematical models, the first hypothesis is to be favored, i.e., the biogenesis of cytokinins runs via the turnover of the tRNA plus oligo N.

Very recently Barnes et al. (1980) published results showing that 40% of the biosynthesis of cytokinins in potato tissue might be due to the breakdown of tRNA,

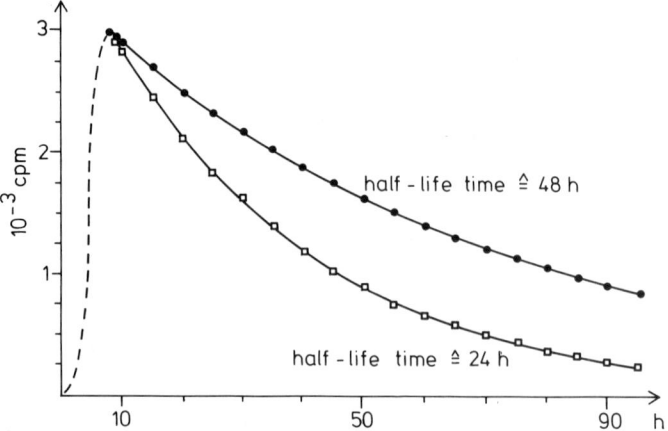

Fig. 5. Theoretical curves referring to the de novo cytokinin biosynthesis (see text)

while the remaining cytokinins should come via another pathway: oligo N as well as de novo.

From our data we doubt that a de novo synthesis is involved in cytokinin production. The oligo N fraction becomes quickly labeled and contains zeatin nucleotides in much higher amount early during the pulse period than later on. Therefore the relative amount of cytokinins and the half-life of the oligo N fraction may be underestimated. On the measured data we calculate the relation of cytokinin production due to oligo N breakdown and tRNA breakdown. The figures we obtained for the relative amount of cytokinins produced from oligo N and tRNA within one hour after the second, the third, and the sixth hour were 51:49, 43:57, and 37:63 respectively.

4 Conclusion

In view of the probable underestimations it seems likely that cytokinin biogenesis may depend on tRNA breakdown at 40% to 50% and for the other 50 to 60% on the breakdown of isopentenylated mRNA and eventually oligo N as their intermediate degradation products, which can also be isopentenylated. In this respect the fairly low $K_m = 3.4 \cdot 10^{-6}$ M measured for the isopentenylation reaction of poly A catalyzed by the highly purified Δ^2-isopentenyldiphosphate: tRNA Δ^2 isopentenyltransferase from corn (Holtz and Klämbt, 1978) has to be considered.

Acknowledgment. This work was supported by the Deutsche Forschungsgemeinschaft. We thank Miss Nicola Vogel for reading the English manuscript.

References

Barnes MF, Tien CL, Gray JS (1980) Biosynthesis of cytokinins by potato cell cultures. Phytochemistry 19: 409-412

Burrows WJ (1978) Incorporation of [³H]-adenine into free cytokinins by cytokinin-autonomous tobacco callus tissue. Biochem Biophys Res Commun 84: 743-748

Chen CM, Eckert RL, McChesney JD (1976) Evidence for the biosynthesis of tRNA-free cytokinin. FEBS Lett 63: 429-434

Chen, CM, Melitz DK (1979) Cytokinin biosynthesis in a cell-free system from cytokinin-autotrophic tobacco tissue cultures. FEBS Lett 107: 15-20

Chen CM, Petschow B (1978) Cytokinin biosynthesis in cultured rootless tobacco plants. Plant Physiol 62: 861-865

Chen CM, Eckert RL, McChesney JD (1976) Evidence for the biosynthesis of tRNA-free cytokinin. FEBS Lett 64: 429-434

Helbach M, Klämbt D (1981) On the biogenesis of cytokinins in *Lactobacillus acidophilus* ATCC 4963. Physiol Plant (in press)

Holtz J, Klämbt D (1975) tRNA-Isopentenyltransferase from *Lactobacillus acidophilus* ATCC 4963. Hoppe Seyler's Z Physiol Chem 356: 1459-1464

Holtz J, Klämbt D (1978) tRNA-Isopentenyltransferase from *Zea mays* L. Characterization of the isopentenylation reaction of tRNA, oligo(A) and other nucleic acids. Hoppe Seyler's Z Physiol Chem 359: 89-101

Klemen F, Klämbt D (1974) Half-life of sRNA from primary roots of *Zea mays*. A contribution to the cytokinin production. Physiol Plant 31: 186-188

Maaß H, Klämbt D (1981) On the biogenesis of cytokinins in roots of *Phaseolus vulgaris*. Planta (in press)

Stuchbury T, Palni LM, Horgan R, Wareing PF (1979) The biosynthesis of cytokinins in crown-gall tissue of Vinca rosea. Planta 147: 97-102

Taya Y, Tanaka Y, Nishimura S (1978) 5'AMP is a direct precursor of cytokinin in *Dictyostelium discoideum*. Nature (London) 271: 545-547

Biosynthesis and Enzymic Regulation of the Interconversion of Cytokinin

Chong-maw Chen[1]

1 Introduction

Plants, as intact organisms, do not have a cytokinin requirement for their growth by virtue of the capacity of their tissues to synthesize cytokinins de novo. Most of the culture cell lines, however, require an exogenous source of cytokinin and other nutrients for cell division and differentiation. Although cytokinins are relatively abundant in shoots and it seems logical that shoots as well as roots are capable of synthesizing cytokinins, little experimental evidence was available for cytokinin formation in shoots. To address this problem, rootless tobacco plants were grown and cytokinin biosynthesis in shoots was examined.

Since plant cells need a balanced supply of cytokinins, mechanisms for maintaining adequate levels of cytokinins in the cells must exist. The most fundamental control mechanisms are presumably operating at the levels of enzymic regulation of metabolism (biosynthesis, interconversion, and degradation). The metabolism of cytokinins has been studied in a variety of plant tissues (see for example Fox, 1969; Hall, 1970; Laloue et al., 1977; Chen et al., 1976) and some of the enzymes which regulate the metabolic pathways of cytokinins have also been isolated (see for example Hall, 1971; Chen and Petschow, 1978b). Data from these studies indicate that there are several major competing enzymic pathways by which cytokinins such as N^6-(Δ^2-isopentenyl)-adenine (i^6Ade) and its derivatives are synthesized, interconverted, and degraded in plant cells. A summary of the enzymic regulation of i^6Ade metabolism is shown in Fig. 1. These metabolic pathways may also be applied to other cytokinins, such as zeatin.

In view of the complexity of the physiological process evoked by cytokinins in plant systems, multilevels of experimental approach are needed to understand the mechanism of cytokinin action. In this paper, we report only two sets of experiments: (a) the biosynthesis of cytokinins in cultured rootless tobacco plants and in cell-free system from cytokinin-autotrophic tobacco tissue cultures, and (b) enzymic regulation of the interconversion of cytokinin base, ribonucleoside, and ribonucleotide.

---

[1] Department of Life Science and Biomedical Research Institute, University of Wisconsin-Parkside, Kenosha, WI 53141, USA

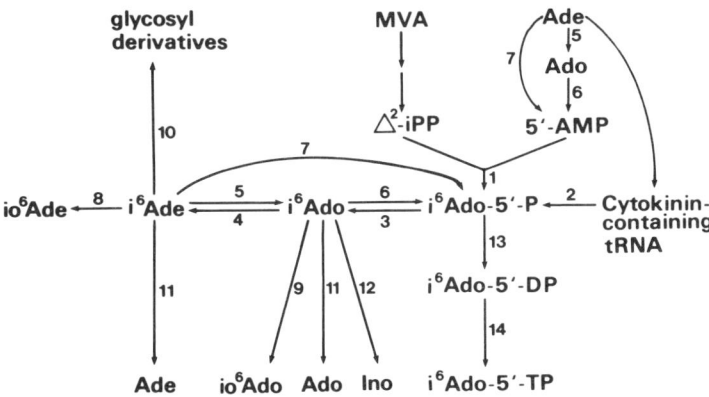

Fig. 1. Cytokinin biosynthesis, interconversion, and degradation in plant tissues. The numbers between compounds refer to the following enzymes or sequences of enzymes: *1* Δ^2-isopentenylpyrophosphate: AMP-Δ^2-isopentenyltransferase; *2* nucleases; *3* 5'-nucleotidases; *4* adenosine nucleosidase; *5* adenosine phosphorylase; *6* adenosine kinase; *7* adenine phosphoribosyltransferase; *8* and *9* enzymes unknown; *10* cytokinin glycosyltransferases; *11* cytokinin oxidase; *12* adenosine deaminase; *13* adenylate kinase?; *14* nucleoside diphosphokinase?

2 Results

2.1 Cytokinin Biosynthesis

2.1.1 Biosynthesis of Cytokinins in Cultured Rootless Tobacco Plants

We are asking a specific question: Is cytokinin biosynthesis restricted to the presumed root site? Although there is evidence that cytokinins from xylem exudate may come from the roots (see review by Torrey, 1976), the question of whether the intact root tips and/or other plant parts serve as a primary site of cytokinin biosynthesis was not answered. The following experiments were designed to address this question.

Rootless tobacco plants were originated from cytokinin-heterotrophic tobacco tissue (*Nicotiana tabaccum* Var. Wis. No. 38). The tobacco tissues were subcultured on a high cytokinin-low auxin medium containing 9.2 μM kinetin and 1.14 μM IAA to induce plantlet formation (Skoog and Miller, 1957). The plantlets formed after 40 to 45 days of incubation were then transplanted to flasks containing "0-1 medium" lacking both cytokinin and auxin. The transplanted plantlets grew into tobacco plants without roots within 20 to 25 days at room temperature (22°-24°C) under fluorescent light. Careful microscopic examination of basal portions from several differentiated shoots which had been cultured for as long as 48 days on 0-1 medium confirmed that no root tissues were formed (Chen and Petschow, 1978a).

To measure cytokinin biosynthesis by rootless plants, plants which had been incubated on 0-1 medium for 20 days were again transplanted into fresh 0-1 medium containing 5 nmol (4.7 x 10^5 cpm) per flask of filtered sterilized [8-^{14}C]Ade. After 24 days of incubation, the average fresh weight for each rootless plant was 0.62 g and the uptake of [8-^{14}C]Ade was 0.19 nmol/g fresh weight. Radioactive materials in the plants were extracted with 50 and 95% ethanol (20 ml ethanol/g fresh tissue) and then with ethyl acetate-water (5:1, v/v; 15 ml/g tissue). The radioactive materials appearing in the ethanol-ethylacetate-water-soluble fraction from the rootless plants accounted for about 4% of the total radioactivity in the medium for a 24-day growth period.

The following methods were employed to separate and identify the biosynthesized radioactive cytokinins (see Chen and Petschow, 1978a): (a) Sephadex LH-20 column eluted with 35% ethanol, (b) paper chromatography in four solvent systems, (c) gas liquid chromatography, and (d) paper electrophoresis. Portions of the ethanol-ethylacetate-water-soluble radioactive components characterized by the chromatographic and electrophoretic methods appear to be i^6Ade, N^6-(Δ^2-isopentenyl)adenosine (i^6Ado), 6-(4-hydroxy-3-methyl-2-butenylamino)purine (io^6Ade) and 6-(4-hydroxy-3-methyl-2-butenylamino)-9-β-D-ribofuranosylpurine (io^6Ado). The quantities of each isolated cytokinin species among the total purified cytokinin samples analyzed by paper chromatography indicated that i^6Ade is about 23% to 32%; i^6Ado, 12% to 18%; io^6Ade, 26% to 37%; and io^6Ado amounted to 9% to 17%.

2.1.2 Biosynthesis of Cytokinins in Cell-Free System

Free cytokinins have been detected in various plant systems, but it was not clear whether the free cytokinins were produced by the turnover of cytokinin-containing tRNA and/or by an alternative biosynthetic pathway. Thus, experiments were designed to test the hypothesis that a tRNA-independent cytokinin biosynthetic pathway exists.

The enzyme, Δ^2-isopentenylpyrophosphate: AMP-Δ^2-isopentenyltransferase (Δ^2-isopentenyltransferase) which catalyzes the formation of i^6Ado-5'-P from Δ^2-isopentenylpyrophosphate (Δ^2-iPP) and 5'-AMP, was partially purified from a cytokinin-autotrophic strain of tobacco tissue initially isolated by Fox (1963). This enzyme was isolated and purified by the following techniques (Chen and Melitz, 1979): (a) differential centrifugations, (b) ammonium sulfate precipitations, (c) DEAE-cellulose column chromatography, and (d) Sephadex G-100 gel filtration. The partially purified enzyme catalyzed the formation of i^6Ado-5'-P at a rate of about 5 nmol/min/mg protein. Δ^2-iPP, not available commercially, was synthesized from Δ^3-iPP using isopentenylpyrophosphate isomerase (isopentenyldiphosphateΔ-isomerase, EC 5.3.3.2) as a catalyst. This latter enzyme was purified from wheat (*Triticum aestivum*) germ cells according to the method of Banthorpe et al. (1977). The purified enzyme catalyzed the formation of Δ^2-iPP at a rate of 3 to 5 nmol/min/mg protein. The isomerase was stable for more than one month at $-70°$C but the Δ^2-isopentenyltransferase lost about 90% of its activity after 3 weeks storage at this temperature. Inclusion of 10 mM 5'-AMP in the cell-free extract during purification increaded the stability of the enzyme by about 40% compared to the extract without 5'-AMP.

To identify the biosynthetic products catalyzed by the Δ^2-isopentenyltransferase, the reaction products were initially chromatographed on Dowex 1-X4 columns by discontinuous stepwise elution techniques with increasing concentrations of KCl as the eluent. Figure 2B shows a separation pattern of the reaction products. The results were compared with the elution profiles of boiled enzyme plus substrates, or authentic i^6Ado-5'-P and isomerase treated Δ^3-iPP control columns (Fig. 2A). The chromatographic comparison with control samples led to the location of radioactive i^6Ado-

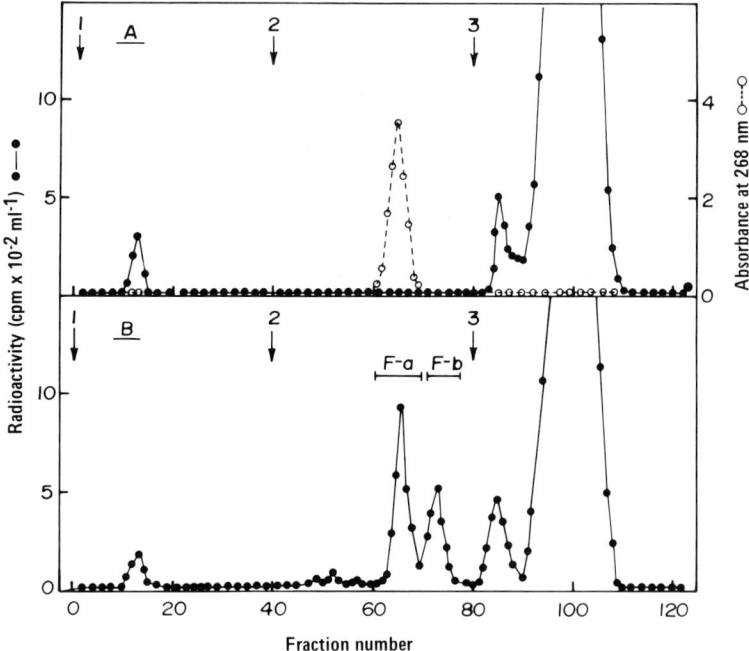

Fig. 2 A, B. Elution profile of biosynthesized and authentic cytokinin nucleotide on Dowex 1-X4 columns (19.2 x 1.5 cm). **A** Isomerase-treated [^{14}C]Δ^3-iPP (●—●) and unlabeled i^6 Ado-5'P (o..o). **B** The reaction products of 5'-AMP and [1-^{14}C]Δ^3-iPP in the presence of isomerase and Δ^2-isopentenyltransferase. The columns were eluted stepwise with alkaline solutions adjusted to pH 11.4 with 1 N NaOH: H_2O (*arrow 1*), 0.1 M KCl (*arrow 2*), and 0.3 M KCl (*arrow 3*). (For details see Chen and Melitz, 1979)

5'-P which was eluted by 0.1 M KCl (Fig. 2B). The two pooled fractions from Dowex column (Fig. 2B, F-a and F-b) were further chromatographed on Sephadex LH-20 columns. The results were compared with relative mobilities of authentic samples. The relative mobility of the F-a sample corresponded to i^6 Ado-5'-P, while the F-b sample remains to be identified.

The suspected radioactive cytokinin ribonucleotide (F-a) from Sephadex LH-20 columns was analyzed further by paper electrophoresis (Fig. 3) and by paper chromatography in five solvent systems. The identity of the radioactive sample characterized by the electrophoresis (Fig. 3) and paper chromatography agreed with the results derived from Sephadex LH-20 column characterization showing that the F-a radioactive sample was i^6 Ade-5'P.

These data provided strong evidence that the radioactive product was a cytokinin nucleotide. However, it remained to be determined whether the product was the 5'-monophosphate or the 3'-monophosphate. The presumed cytokinin nucleotide was digested with 5'-nucleotidase of *Crotalus adamteus* venom and the products were separated by electrophoresis. After treatment, the metabolite migrated with the corresponding unlabeled i^6Ado marker (Fig. 3B). Thus, the experimental evidence indicates that the nucleotide is indeed the 5'-monophosphate.

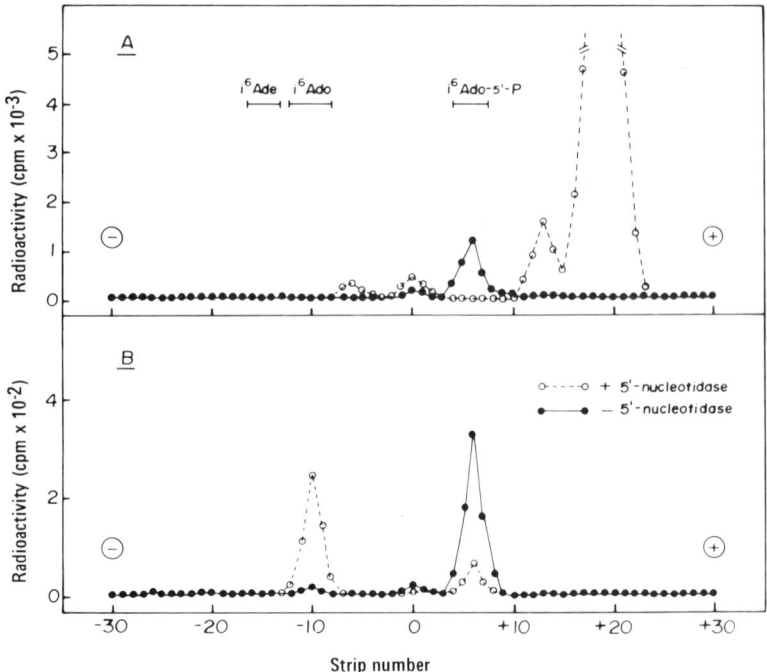

Fig. 3 A, B. Characterization of biosynthesized cytokinin nucleotide by paper electrophoresis. **A** Electrophoretic analysis of radioactive cytokinin nucleotide (●—●) which was initially isolated from Dowex 1-X4 column (Fig. 2B, F-a) and purified further by Sephadex LH-20 column; and isomerase treated [1-^{14}C]Δ^3-iPP (o..o). **B** 5'-nucleotidase treatment of the presumed of the presumed [^{14}C]-i^{6}Ado-5'-P. The ^{14}C-labeled peak corresponding to i^{6}Ado-5'-P on the electropherogram (A) was cut and eluted. A portion of the eluate (1200 cpm) was incubated with (o..o) and without (●—●) 5' nucleotidase

Robins et al. (1967) have shown that i^{6}Ado undergoes a unique reaction in the presence of HCl to produce two major products. Therefore, as a final confirmation in the process of identifying this radioactive nucleotide, the labeled sample was digested with 5'-nucleotidase to form radioactive nucleoside followed by hydrolysis with HCl. The products were analyzed by paper chromatography. Two radioactive peaks that corresponded to the two unlabeled major hydrolytic products were observed. These results show that the biosynthesized radioactive nucleotide is indeed i^{6}Ado-5'-P. If crude enzyme preparation, partially purified by ammonium sulfate precipitations, was used as an enzyme source for the Δ^2-isopentenyltransferase to catalyze the cytokinin biosynthesis, small quantities of i^{6}Ado and i^{6}Ade in addition to the i^{6}Ado-5'-P were also produced. The i^{6}Ado may be derived from i^{6}Ado-5'-P by phosphatases and i^{6}Ade from i^{6}Ado by adenosine nucleosidases.

2.2 Enzymic Regulation of Cytokinin Interconversion

2.2.1 Cytokinin Ribonucleoside from Cytokinin Ribonucleotide

The occurrence of cytokinin ribonucleoside in plant cells is well known (see review by Hall, 1971; Skoog and Armstrong, 1970; Kende, 1971). The ribonucleoside can be formed from the corresponding ribonucleotide (Fig. 1, reaction 3) or from the cytokinin base (Fig. 1, reaction 5). An enzyme system which catalyzes the formation of cytokinin ribonucleoside from its ribonucleotide is 5'-nucleotidase (5'-ribonucleotide phosphohydrolase, EC 3.1.3.5) (Chen and Kristopeit, 1981). These enzymes were partially purified from wheat germ cytosol and appear to be present in at least two forms, the F-I and F-II 5'-nucleotidases naming after the two separated fractions appeared on Sephadex G-100 gel filtration. Following steps were empolyed in purifying these enzymes: (a) differential centrifugations, (b) ammonium sulfate fractionations, (c) DEAE-cellulose chromatography and (d) Sephadex G-100 filtration. The Sephadex G-100 filtration resulted in approximately 79- and 63-fold purification for the F-I and F-II 5'-nucleotidases, respectively, compared to the initial crude extract (20.000 g supernatant). On the basis of Sephadex G-200 filtration these enzyme preparations had a relative molecular mass of 110,000 (F-I) and 57,000 (F-II). Neither sample displayed alkaline or acid phosphatase activity.

The F-I and F-II 5'-nucleotidases both had a pH optimum around 7 (100 mM Tris-HCl buffer) if either AMP or i^6 Ado-5'-P was used as a substrate. Both enzymes functioned optimally at a Mg^{2+} concentration of 12 mM, neither were they inhibited by concanavalin A. These results suggest that the cytosol 5'-nucleotidases are different from membrane-bound 5'-nucleotidases which are inhibited by concanavalin A (Dornand et al., 1978). Studies of the relative activities of the F-I and F-II enzymes toward a number of nucleotides indicate that 5'-AMP was the best substrate; in contrast, the activities of i^6 Ado-5'-P were about 72% (F-I enzyme) and 86% (F-II) of 5'-AMP. These enzymes specifically hydrolyzed purine ribonucleoside-5'-phosphate, but not 3'-AMP or phenylphosphate.

2.2.2 Cytokinin Ribonucleoside from Cytokinin Base

When a cytokinin base was supplied to plant tissues, one of the metabolites formed was the corresponding ribonucleoside (see for example, Sondheimer and Tzou, 1971). An enzyme system must therefore exist in plant cells to catalyze ribonucleoside formation. This enzyme system (Fig. 1, reaction 5), adenosine phosphorylase (EC 2.4.2.1), was purified approximately 23-fold from wheat germ cells (Chen and Petschow, 1978b). The purification procedures included ammonium sulfate fractionations, differential centrifugation and Sephadex G-100 column filtration. The pH optimum for the ribosylation of Ade, i^6 Ade or kinetin ranged from 6.5 to 7.5 (63 mM HEPES buffer). The conversion of cytokinin bases to the corresponding ribosides depends upon the addition of ribose-1-phosphate, while the phosphorolysis of nucleosides requires the presence of inorganic phosphate. At pH 7.2 (63 mM HEPES buffer) and 37°C, the Km was determined to be 32.2, 57.1, and 46.5 µM for Ade, i^6 Ade, and kinetin, respectively; whereas, the Vmax was 193.1, 134.7, and 137.1 pmol/min/mg protein for Ade, i^6 Ade and kinetin. The equilibrium constants of the phosphorolysis of i^6 Ado (1.38×10^{-3}) by this enzyme indicate that the reaction strongly favors nucleoside formation.

This enzyme is distinct from inosine-guanosine phosphorylase based on the differences in the Sephadex G-100 gel filtration behaviors, pH optima, and the product and p-hydroxymercuribenzoate inhibitor studies.

2.2.3 Cytokinin Ribonucleotide from Cytokinin Ribonucleoside

Conversion of cytokinin ribonucleoside to the corresponding ribonucleotide (Fig. 1, reaction 6) was reported to be catalyzed by adenosine kinase (ATP: adenosine-5'-phosphotransferase, EC 2.7.1.20) which was isolated and partially purified from wheat germ cells (Chen and Eckert, 1977). Following steps were employed in purifying the enzyme: (a) centrifugation, (b) pH 5 fractionation, and (c) Sephadex G-100 gel filtration. The degree of purification was approximately 20-fold when compared to the 20,000 g crude extract.

The phosphorylation of i^6Ado depends upon the presence of ATP and Mg^{2+}, and has a broad pH optimum over the range 6.8 to 7.4 (60 mM potassium phosphate buffer). At 4 mM ATP, the optimum Mg^{2+} concentration was 2 to 3 mM and the optimal ratio of ATP to Mg^{2+} was 1.3 to 2. At pH 7 and 37°C, the Km was calculated to be 8.7 μM and 31 μM for Ado and i^6Ado, whereas the Vmax was 46 and 8.3 nmol/mg protein/min for Ado and i^6Ado, respectively. Inosine was not a substrate for wheat germ adenosine kinase. The cytokinin ribonucleotide formed was characterized by paper electrophoresis to be 5'-monophosphate.

2.2.4 Cytokinin Ribonucleotide from Cytokinin Base

Doree and Guern (1973) studied short-term metabolism of some exogenous cytokinins in *Acer pseudoplatanus* cells and observed that one of the major metabolites formed from cytokinin base was the corresponding nucleotide. It is very likely that the formation of cytokinin nucleotide from its base is catalyzed by adenine phosphoribosyltransferase (AMP: pyrophosphate phosphoribosyltransferase, EC 2.4.2.7) (Fig. 1, reaction 7). This enzyme was partially purified from tobacco pith tissue cultures (Chen and Eckert, 1977) and from wheat germ cells. The partially purified wheat germ enzyme had a specific activity of 2.4 nmol AMP formed/min/mg protein. The pH optimum for this enzyme prepared from wheat germ was about 7.5 (50 mM Tris-HCl buffer). Optimum Mg^{2+} concentration required for 5'-AMP or i^6Ado-5'-P formation was 5 mM. The wheat germ adenine phosphoribosyltransferase was competitively inhibited by AMP but not by hypoxanthine, guanine or their nucleotides when adenine was used as a substrate. At 50 μM concentration, i^6Ado-5'-P was only 5% to 10% as effective as 5'-AMP in inhibiting 5'-AMP formation. The Km values for Ade and i^6Ade were 1.4 and 170 μM, respectively. The high Km for i^6Ade suggests that either there is a different form of adenine phosphoribosyltransferase for cytokinin base, or that cytokinin nucleotide is not preferentially formed by this pathway in wheat germ cells. Krenitsky and Papioannou (1969) reported that enzyme binds to the adenine through the 6-amino group and the 3- and 7-nitrogens. Thus, it is not surprising that i^6Ade, an adenine analog which has a modified 6-amino group, would show reduced ability to serve as a substrate.

2.2.5 Cytokinin Base from Cytokinin Ribonucleoside

One of the metabolites formed from cytokinin nucleoside in various plant tissues was reported to be cytokinin base (Chen et al., 1968; Pačes, 1976; Terrine and Laloue, 1980). The enzyme, adenosine nucleosidase (EC 3.2.2.7) (Fig. 1, reaction 4) which catalyzes the irreversible hydrolysis of i^6Ado or Ado to i^6Ade or Ade and ribose, has been partially purified from wheat germ cells. This enzyme was purified by (a) differential centrifugations, (b) heat treatment, (c) ammonium sulfate fractionations, (d) DEAE-cellulose column, and (e) Sephadex G-100 gel filtration. This partially purified enzyme catalyzed 0.48 nmol of i^6Ade formation per mg protein per min. On the basis of Sephadex G-200 gel filtration, a molecular weight of 59,000 was estimated for the native enzyme. The pH optimum is 4.7 (50 mM MES/KOH buffer) for either Ado or i^6Ado. The substrates or the reaction products apparently protected the enzyme against heat inactivation. The Km values for i^6Ado and Ado are 14 and 6.8 μM, respectively. The formation of cytokinin base from the corresponding nucleoside may depend upon the source of plant tissue used. Pačes (1976) reported that one of the major metabolites of i^6Ado in barley leaves was i^6Ade, whereas Terrine and Laloue (1980) observed that i^6Ade was a minor metabolite of i^6Ado in tobacco cells.

3 Discussion and Conclusion

These studies indicate that cytokinin can be synthesized in shoots and in cell-free systems; and that enzyme systems exist in plant cells regulating balanced supply of cytokinin for their growth.

The rootless tobacco plants at some stage in their development synthesize and respond to cytokinin. Although our experimental results support our view that cytokinin biosynthesis is not restricted to roots, it remains to be determined whether every part of shoot tissue or only certain parts of the shoot serves as the site for cytokinin biosynthesis. The free cytokinins extracted from the rootless plants may be derived from the tRNA-independent biosynthetic pathway (Chen et al., 1976; Burrows, 1978; Taya et al., 1978) and/or from the turnover of cytokinin containing tRNA (see review of Hall, 1970). The fact that cytokinin-autotrophic tobacco callus contains an enzyme system catalyzing the formation of cytokinin nucleotide from 5′-AMP and Δ^2-iPP (Chen and Melitz, 1979) indicates the presence of tRNA-independent cytokinin biosynthetic pathway in plant tissues. Using crude extract of the tobacco tissue as an enzyme source for the cell-free cytokinin biosynthesis, cytokinin nucleoside and cytokinin base were also detected in the reaction products. Collectively, these results suggest that the crude enzyme preparation contained both 5′-nucleotidase and nucleosidase which catalyzed the formation of cytokinin nucleoside and cytokinin base.

Cytokinin regulation of physiological processes in a plant system is accomplished by a variety of mechanisms. For example, the plant tissue response to cytokinin is proportional to the concentration of the hormone present; higher concentration is inhibitory whereas removal of this hormone diminishes or stops the response. Thus, by regulating the rate of cytokinin biosynthesis, transport, interconversion and degradation, the tissue responses to cytokinin may be controlled. Consequently, the cytokinin metabolic enzymes shown in Fig. 1 play a crucial role in balancing the supply of "active cytokinin".

When a cytokinin is synthesized de novo, or enters plant cells from an exogenous source, many factors may determine whether it is to be converted to an "active form" (cytokinin base?); a "storage form" (cytokinin glucosides?) (Fox et al., 1973; Letham et al., 1978); a "transport form" (cytokinin nucleoside and/or glucoside?); or a "degraded form" (Ade, Ado, or inosine). One of the important factors in determining the fate of the cytokinin is the relative activities of cytokinin metabolic enzymes, which in turn are affected by the relative concentrations and distributions of the hormone and its precursors, in the plant cells.

Studies of cytokinin metabolic enzymes have already revealed some important information on the physiological function of this hormone; there is, however, much more to learn.

Acknowledgment. This work was supported by the National Science Foundation Grant PCM 76-82158 and PCM 79-03832 and partially be the National Institute of Health Grant GM-22543.

References

Banthorpe DV, Doonan S, Gutowski JA (1977) Isopentenyl pyrophosphate isomerase from pig liver. Arch Biochem Biophys 184: 381-390
Burrow WJ (1978) Incorporation of ^{3}H-adenine into free cytokinins by cytokinin-autonomous tobacco callus tissue. Biochem Biophys Res Commun 84: 743-748
Chen C-M, Eckert RL (1977) Phosphorylation of cytokinin by adenosine kinase from wheat germ. Plant physiol 59: 443-447
Chen C-M, Kristopeit SM (1981) Metabolism of cytokinin: Dephosphorylation of cytokinin ribonucleotide by 5'-nucleotidases from wheat germ cytosol. Plant Physiol (in press)
Chen C-M, Melitz DK (1979) Cytokinin biosynthesis in a cell-free system from cytokinin-autotrophic tobacco tissue cultures. FEBS Lett 107: 15-20
Chen C-M, Petschow B (1978a) Cytokinin biosynthesis in cultured rootless tobacco plants. Plant Physiol 62: 861-865
Chen C-M, Petschow B (1978b) Metabolism of cytokinin; Ribosylation of cytokinin base by adenosine phosphorylase from wheat germ. Plant Physiol 62: 871-874
Chen C-M, Logan DM, McLennan BD, Hall RH (1968) Studies of the metabolism of a cytokinin N^6-(Δ^2-isopentenyl)adenosine. Plant Physiol 43: 5-18
Chen C-M, Eckert RL, McChesney JD (1976) Evidence for the biosynthesis of transfer RNA-free cytokinin. FEBS Lett 64: 429-434
Doree M, Guern J (1973) Short-term metabolism of some exogenous cytokinins in *Acer pseudoplatanus* cells. Biochim Biophys Acta 304: 611-622
Dornand J, Bonnafoud J, Mani J (1978) Effects of con A and other lectins on pure 5'-nucleotidase isolated from lymphocyte plasma membranes. Biochem Biophys Res Commun 82: 685-692
Fox JE (1963) Growth factor requirements and chromosome number in tobacco tissue cultures. Physiol Plant 16: 793-803
Fox JE (1969) Cytokinins. In: Wilkins MB (ed) Physiology of plant growth and development, Chapter C. McGraw-Hill, New York, pp. 85-123
Fox JE, Cornette J, Deleuze G, Dyson W, Giersak G, Niu P, Zapata J, McChesney JD (1973) Formation, isolation, and biological activity of cytokinin 7-glucoside. Plant Physiol 52: 627-632
Hall RH (1970) N^6-(Δ^2-isopentenyl)adenosine: chemical reactions, biosynthesis, metabolism and significance to the structure and function of tRNA. In; Davidson JN, Cohn WE (eds) Progress

in nucleic acid research and molecular biology, vol. 10. Academic Press, London New York, pp 57-86

Hall RH (1971) The modified nucleosides in nucleic acids. Columbia University Press, New York, pp 295-311

Kende H (1971) The cytokinins. Int Rev Cytol 31: 301-338

Krenitsky TA, Papioannou R (1969) Human hypoxanthine phosphoribosyltransferase. II. Kinetinics and chemical modification. J Biol Chem 244: 1271-1277

Laloue M, Terrine C, Guern J (1977) Cytokinins: metabolism and biological activity of N^6-(Δ^2-isopentenyl) adenosine and N^6-(Δ^2-isopentenyl)adenine in tobacco cells and callus. Plant Physiol 59: 478-483

Letham DS, Summons RE, Entsch B, Gollnow BI, Parker CW, MacLeod JK (1978) Regulators of cell division in plant tissues. XXVI. Glucosylation of cytokinin analogues, Phytochemistry 17: 2053-2057

Pačes V (1976) Metabolism of cytokinins in barley leaves. Biochem Biophys Res Commun 72: 830-839

Robins MT, Hall RH, Thedford R (1967) N^6-(Δ^2-isopentenyl)adenosine. A component of the transfer ribonucleic acid of yeast and of mammalian tissue, methods of isolation, and characterization. Biochemistry 6: 1837-1884

Skoog F, Armstrong DJ (1970) Cytokinins. Annu Rev Plant Physiol 21: 359-384

Skoog F, Miller CO (1957) Chemical regulation of growth and organ formation in plant tissues cultured in vitro. Symp Exp Biol 11: 118-131

Sondheimer E, Tzou DS (1971) The metabolism of hormones during seed germination and dormancy. II. Metabolism of 8-^{14}C-zeatin in bean axes. Plant Physiol 47: 516-520

Terrine C, Laloue M (1980) Kinetics of N^6-(Δ^2-isopentenyl)adenosine degradation in tobacco cells. Evidence of a regulatory mechanism under the control of cytokinins. Plant Physiol. 65: 1090-1095

Taya Y, Tanaka Y, Nishimura S (1978) 5'-AMP is a direct precursor of cytokinin in *Dictyostelium discoideum*. Nature (London) 271: 545-547

Torrey IG (1976) Root hormones and plant growth. Ann Rev Plant Physiol 27: 435-459

Cytokinin Biosynthesis in Cytokinin-Autonomous and Bacteria-Transformed Tobacco Callus Tissues

W.J. Burrows and K.J. Fuell[1]

1 Introduction

Cytokinins are one of the potent classes of naturally occurring plant growth "hormones". They occur in plants both as bases adjacent to the anticodon triplet in specific tRNA species and as free bases, ribonucleosides and ribonucleotides.

Earlier work (Burrows, 1976) had identified the "non-tRNA" form of the cytokinin as that active in eliciting the many phenotypic responses observed by plant physiologists in plants and detached plant parts in response to exogenous application of kinetin.

The endogenous level of free cytokinins is the product of balancing their rates of biosynthesis and their rates of conjugation and degradation. There is no information as to the absolute rates of any of these processes. The science is at the stage of defining the products and processes and outlining the various possible biochemical routes.

The two obvious possible pathways that could be employed by plants to synthesise "free" cytokinins are (1) biosynthesis de novo and (2) release of the cytokinin intact during turnover of the cytokinin-containing RNA species. Indirect evidence supporting both pathways has been published (Leineweber and Klämbt, 1974; Klemen and Klämbt, 1974; Chen and Eckert, 1976; Burrows, 1978a). Direct evidence supporting biosynthesis de novo also has been published (Burrows, 1978b; Stuchbury et al., 1979). It was not until Chen and Melitz (1979) successfully demonstrated the in vitro alkylation of 5'-AMP by an enzyme preparation from cytokinin-autonomous tobacco callus that we obtained conclusive evidence that biosynthesis de novo was present in higher plants. Earlier studies (Taya et al., 1978) had demonstrated cytokinin biosynthesis de novo in the slime mould *Dictyostelium*. One of the questions now outstanding relates to the possibility that the endogenous pool of free cytokinin is supplemented by cytokinins released intact from RNA during turnover of the macromolecule.

In this paper we report a quantitative and qualitative comparison of the free cytokinins with those in RNA isolated from both cytokinin-autonomous and bacteria-transformed tobacco callus incubated in medium containing ^{3}H-adenine and ^{14}C-mevalonic acid.

Abbreviations used: Ribosyl-trans-zeatin, t-io^{6}ado; ribosyl-cis-zeatin, c-io^{6}ado; trans-zeatin. t-io^6-ade; cis-zeatin, c-io^{6}ade; N^6-(Δ^2-isopentenyl(adenosine, i^{6}ado; N^6-(Δ^2isopentenyl)adenine, i^{6}ade

[1] Shell Biosciences Laboratories, Sittingbourne Research Center, Sittingbourne, Kent ME9 8AG, England

2 Methods and Materials

The cytokinin-autonomous tobacco callus was a gift from Dr. J. Einset, Department of Plant Sciences, University of California, Riverside, USA, and originated in the laboratory of Professor Folke Skoog. The bacteria-transformed callus (line A66A) was a gift from Drs N. Brewin and A. Johnstone, John Innes Institute, Norwich, and was the product following infection of tobacco cells with the tumour-inducing bacterium *Agrobacterium tumefaciens*.

Tobacco callus, ca. 400 g, grown in mineral medium (Linsmaier and Skoog, 1965) minus kinetin was harvested after ca. 20 days growth and incubated in 150 ml similar medium omitting the agar and with a reduced sucrose concentration (0.1%) containing either 30 μCi ^{14}C-mevalonic acid (DBED) salt, specific activity (18.5 mCi/mCi/mmol) or 30 μCi 8-^{3}H-adenine (specific activity 25 Ci/mmol) for either 1 h or 8 h. In the case of the longest incubation time (672 h) small pieces of tobacco callus (ca. 50 mg) were planted on mineral medium (Linsmaier and Skoog, 1965), minus kinetin, containing either 10 μCi/L ^{14}C-MVA or 10 μCi/L ^{3}H-adenine. After the appropriate incubation time the tissue was harvested and the cytokinins present isolated and purified as detailed below.

2.1 Free Cytokinin Bases and Ribonucleosides

The tissue was harvested and homogenised in an Ultra-Turrax homogeniser (full speed, 2 min) in 1.5 volumes of cold N perchloric acid and left for 1 h at 2° to inactivate completely any phosphatase present (Bieleski, 1964). The resultant slurry was centrifuged at 10,000 g for 15 min and the supernatant was decanted. The pellet was re-extracted twice further with equal volumes of cold N perchloric acid and the combined supernatants were spiked with 100 μg each of io^{6}ado, t-io^{6}ade, c-io^{6}ade, i^{6}ado and i^{6}ade and adjusted to pH 5.0 with KOH. The mixture was centrifuged at 10,000 g for 20 min and the supernatant was lyophilised, dissolved in 10 ml distilled water, adjusted to pH 9.0 and extracted three times with 0.5 volumes of water-saturated butan-1-ol. The butanol extracts were combined and lyophilised, and the solid material was extracted twice with 2.0 ml dry methanol. The methanol extracts were combined and lyophilised, and the solid material was dissolved in 2.0 ml of 35% aqueous ethanol and chromatographed on a Sephadex LH-20 column (44 x 2.0 cm) equilibrated and eluted with the same solvent (Armstrong et al., 1969). The regions of the column eluates known to contain the cytokinins were combined separately and lyophilised, and each was further analysed by reverse phase HPLC (Holland et al., 1978) on Lichrosorb RP8 semi-preparative columns (20 x 10 mm id) equilibrated and eluted with H_2O/MeOH (70:30 for the io^{6}ado and io^{6}ade fractions and 40:60 for the i^{6}ado and i^{6}ade fractions). The column eluates were monitored at 260 nm. The fractions were collected directly into plastic scintillation vials as indicated in the figures. The vials were placed in an oven overnight at 35° to remove most of the eluant. The scintillation fluor (Packard M197) was then added and the samples were counted in a Beckman LS7500 liquid scintillation spectrometer.

2.2 Free Cytokinin Ribonucleotides

The aqueous fraction remaining after butan-1-ol extraction, was stored at $-15°$ prior to use. On thawing the extracts were lyophilised separately at $30°$ and the solids redissolved in 20 ml of distilled water. After centrifugation at $10,000\,g$ for 15 min the supernatant was divided into three and each incubated overnight at $37°$ with:
a) 5 units of 5'-phosphatase (E.C. No. 3.1.3.5) from *Crotalus atrox* venom, 0.01 M $MgCl_2$ in a final glycine- NaOH buffer concentration of 0.1 M at pH 9.0.
b) 5 units of 3'-phosphatase (E.C. No. 3-1-3-6) from ryegrass in a final Tris-HCl concentration of 0.1 M at pH 7.5.
c) Control incubation was as (a) minus the enzyme.

All incubations were saturated with toluene. After the incubation period 50 μg of io^6ado and i^6ado were added as internal standards. All incubations were adjusted to pH 9.0 and extracted three times with 5 ml of water-saturated butan-1-ol. The cytokinins in the combined extracts were further purified by both LH-20 and HPLC as described above.

2.3 Total RNA Extraction

The harvested tissue (ca. 500 g) was stored at $-15°$ prior to freeze drying. After freeze drying the callus was homogenised in a Virtis blender (full speed, 30 s) in a solution of 0.2 M Tris-HCl, pH 8.6, containing 0.1 M NaCl and 1% SDS. After a further 30 s homogenisation (Ultra Turrax, full speed) 250 ml of phenol (saturated with 0.2 M tris-HCl, pH 8.6) containing 0.1% 8-OH quinoline was added and the slurry shaken on a reciprocal shaker for 3 h. After centrifugation at $1000\,g$ for 30 min 2.5 vol of cold ethanol was added slowly to the supernatant and the DNA spooled on a glass rod and discarded. The nucleic acids were precipitated overnight at $4°C$ and recovered by centrifugation at $2000\,g$ for 20 min. The precipitate was washed first with cold ethanol and then with ether and allowed to air-dry for 2 h. The solid material was dissolved in 300 ml of 0.1 M tris-HCl, pH 8.6, containing 0.001 M $MgCl_2$. After centrifugation at $1000\,g$ for 20 min the supernatant was saturated with toluene and incubated overnight in the presence of 30 mg of phosphodiesterase (E.C. No. 3-1-4-1) and 100 mg of alkaline phosphatase (E.C. No. 3-1-3-1). Each incubation mixture was spiked with 50 μg of io^6ado and i^6ado as internal standards and the cytokinins were extracted and purified by the standard solvent extraction and chromatographic procedures outlined above. Prior to chromatography on Sephadex LH-20 a 0.1 ml aliquot was taken from each extract and used to determine the total A_{260} units chromatographed.

3 Results

The results (Table 1a and b) indicate that both ^{3}H-adenine and ^{14}C-MVA are incorporated into free cytokinins by both tissues. In both tissue extracts there was a decrease in the amount of radioactive precursors incorporated into free cytokinins with increase in incubation time (Table 1a and b).

Table 1 a, b. The incorporation of ³H-adenine and ¹⁴C-MVA into free cytokinins by both (a) bacteria-transformed and (b) cytokinin-autonomous tobacco callus tissue expressed as dpm incorporated/100 g tissue

Radiolabelled precursor	cytokinin / Incubation time (h)	t-io⁶ ado	c-io⁶ ado	t-io⁶ ade	c-io⁶ ade	i⁶ ado	i⁶ ade
³H-adenine	1	1021	–	2402	–	2392	126
	8	802	138	1713	–	2959	288
	672	616	171	–	–	206	–
¹⁴C-MVA	1	–	–	–	–	–	–
	8	161	66	–	–	–	–
	672	656	168	–	–	–	–

a

Radiolabelled precursor	cytokinin / Incubation time (h)	t-io⁶ ado	c-io⁶ ado	t-io⁶ ade	c-io⁶ ade	i⁶ ado	i⁶ ade
³H-adenine	1	16896	*	1667	*	396	485
	672	4501	–	–	–	123	–
¹⁴C-MVA	1	10934	*	2064	*	4029	18711
	672	2100	–	–	–	542	1985

b

In the bacteria-transformed tissue only c- and t-io^{6}ado contained radioactivity when incubated for 8 h and 672 h in the presence of ^{14}C-MVA (Table 1a). The ratio of cis: trans isomer decreased from 1:2.5 to 1:4 during this period. The ratio in tissue incubated with ^{3}-H-adenine increased from 1:6 to 1:4.5 (Table 1a). The most abundantly labelled cytokinins in extracts of bacteria-transformed tissue incubated for either 1 h or 8 h in the presence of ^{3}H-adenine were i^{6}ado and t-io^{6}ade. Significant incorporation was also found in t-io^{6}ado and to a lesser extent in i^{6}ade.

There was more ^{3}H-adenine and ^{14}C-MVA incorporated into free cytokinin by the cytokinin-autonomous tobacco callus (Table 1b) than by the bacteria-transformed tissue (Table 1a). After 1 h incubation the predominantly radiolabelled cytokinin was t-io^{6}ado. No radioactivity could be specifically attributed to occurring in either c-io^6-ado or c-io^{6}ade. There was radioactivity present in the eluate denoted with an asterisk but the specific activity in the fraction did not remain constant. We were unable to attribute radioactivity specifically to the cytokinins.

No radioactivity could be detected supporting the presence of a cytokinin 3'-ribonucleotide in the extracts from either tissues (Table 2). There was no evidence to support the occurrence of c-io^{6}ado-5'-phosphate. In the cytokinin autonomous tobacco callus tissue, t-io^{6}ado-5'-phosphate was detected only after 1 h incubation in ^{3}H-adenine. In the bacteria-transformed tobacco callus, radiolabel from both precursors was detected in both t-io^{6}ado-5'-phosphate and i^{6}ado-5'-phosphate. It was thought that this low level of detectable incorporation into cytokinin nucleotides could be due either to the presence of an inhibitor of the specific enzyme or that these cytokinin nucleotides were more resistant to specific enzyme cleavage. To test these possibilities the aqueous fractions remaining after the above butanol extract were recombined, lyophilised and redissolved in 50 ml 0.1 M tris-HCl pH 8.6 and incubated overnight at 37°C in the presence of 50 mg alkaline phosphatase (E.C. No. 3-1-3-1). After incubation the samples were processed as detailed above for the specific phosphatase incubations. Some additional radioactivity was detected in both t-io^{6}ado and i^{6}ado, indicating that incomplete hydrolysis of the cytokinin nucleotides had been achieved with the specific phosphatases.

3.1 Cytokinins in Total RNA

Radioactivity from both ^{3}H-adenine and ^{14}C-MVA was incorporated by both tissues into c-io^{6}ado with incubation times of 8 h or more (Table 3). In both the cytokinin-autonomous and bacteria-transformed tissue radioactivity from ^{14}C-MVA was detected in i^{6}ado after 1 h incubation.

The distribution of radioactivity in the elution profile from the HPLC of the io^{6}ado fraction obtained following Sephadex LH-20 chromatography of an extract of bacteria-transformed tobacco callus tissue incubated for 672 h in the presence of ^{3}H-adenine is shown in Fig. 1A. Not surprisingly, most radioactivity in the RNA preparation was found in the eluate co-chromatographing with adenosine. The second most abundant area of the profile was that eluting after ca. 9-10 min. The eluate in the region of t- and c-io^{6}ado was subdivided into the three fractions, as indicated, and rechromatographed separately under identical conditions. The distribution of the

Table 2. The incorporation of ^{3}H-adenine and ^{14}C-MVA into cytokinin nucleotides by both bacteria-transformed and cytokinin-autonomous tobacco callus expressed as dpm incorporated/100 g tissue

Incubation time (h)	Tissue	Radiolabelled precursor	5'-phosphatase			3'-phosphatase			Alkaline phosphatase		
			t-io^{6}ado	c-io^{6}ado	i^{6}ado	t-io^{6}ado	c-io^{6}ado	i^{6}ado	t-io^{6}ado	c-io^{6}ado	i^{6}ado
1	Autonomous	^{3}H-adenine	–	–	–	–	–	–	36	–	–
		^{14}C-MVA	–	–	–	–	–	–	–	–	–
	Bacteria-transformed	^{3}H-adenine	132	–	100	–	–	–	239	–	123
		^{14}C-MVA	48	–	100	–	–	–	132	–	105
8	Autonomous	^{3}H-adenine	–	–	–	–	–	–	–	–	–
		^{14}C-MVA	–	–	–	–	–	–	–	–	–
	Bacteria-transformed	^{3}H-adenine	–	–	–	–	–	–	–	–	–
		^{14}C-MVA	–	–	–	–	–	–	–	–	60
672	Autonomous	^{3}H-adenine	–	–	–	–	–	–	–	–	–
		^{14}C-MVA	–	–	–	–	–	–	99	–	–

Table 3. The incorporation of ^{3}H-adenine and ^{14}C-MVA into the cytokinins in RNA isolated from cytokinin-autonomous and bacteria-transformed tobacco callus expressed as dpm incorporated/100 g tissue

Incubation time (h)	Tissue	Radiolabelled precursor	A_{260} applied to LH_{20} column	t-io^6 ado	c-io^6 ado	i^6 ado	'shoulder'
1	Autonomous	^{3}H-adenine	10325	—	—	—	Absent
		^{14}C-MVA	24200	—	—	20.4	Absent
	Bacteria-transformed	^{3}-H-adenine	13625	—	—	—	Not determined
		^{14}-C-MVA	16225	—	—	30.3	Not determined
8	Bacteria-transformed	^{3}-H-adenine	11575	—	42.0	169.2	645.6
		^{14}C-MVA	21600	—	168.2	38.8	Not determined
	Autonomous	^{3}H-adenine	9650	—	91.6	7.0	Absent
		^{14}C-MVA	5575	—	47.8	10.4	Absent
672		^{3}H-adenine	20925	—	100.1	28.4	768
	Bacteria-transformed	^{14}C-MVA	18650	—	67.8	5.8	Not determined

radioactivity in the eluate in the t-io^{6}ado region is shown in Fig. 1B. There was radioactivity in the eluate co-eluting with t-io^{6}ado but it is not certain that there is radioactivity which specifically co-eluted with the cytokinin. No radioactivity was present in any of the other t-io^{6}ado fractions. Radioactivity was associated with the "shoulder" of the t-io^{6}ado peak (Fig. 1A) when rechromatographed under identical conditions (Fig. 1C). The activity in the t-io^{6}ado fraction rechromatographing with an elution time of ca. 16 mins (Fig. 1B) co-chromatographs with this shoulder (Fig. 1C). Radioactivity can be specifically attributed to the eluate co-eluting with c-io^{6}ado.

In both autonomous-and bacteria-transformed tobacco callus tissue radioactivity from ^{14}C-MVA was detected in i^{6}ado after 1 h incubation (Table 3). Most radioactivity was detected after an 8 h incubation period. The distribution of radioactivity in the elution profile from the HPLC of the i^{6}ado fraction obtained following Sephadex LH-20 chromatography of an extract of bacteria-transformed tobacco callus tissue, incubated for 672 h in the presence of ^{14}C-MVA is shown in Fig. 2. The "peak of activity", although not abundant, is clearly defined and co-eluted with i^{6}ado. This elution profile was deliberately chosen since it represents the lowest radioacitivty incorporated into a cytokinin (Table 3) and illustrates the resolving power of the analytical techniques.

4 Discussion

The rapid incorporation of both ^{3}H-adenine and ^{14}C-MVA into the free cytokinins t-io^{6}ado, t-io^{6}ade, i^{6}ado and i^{6}ade in both tissue systems indicates biosynthesis de novo of free cytokinins. In the first conclusive evidence for cytokinin biosynthesis de novo using plant material, Chen and Melitz (1979) found that the first product was i^{6}ado-5-phosphate, a result previously reported for *Dictyostelium* (Taya et al., 1978). The large incorporation of radiolabelled precursors into t-io^{6}ado reported above for the cytokinin-autonomous callus tissue indicates that the hydroxylating and phosphatase enzymes are very active. In the bacteria-transformed tissue it would appear that this hydroxylase enzyme is less active.

A comparison of the results obtained above using the cytokinin-autonomous tissue with those published earlier indicates that the mechanism of cytokinin autonomy is not a stable controlled phenomenon. In our earlier studies (Burrows, 1978b) the predominant cytokinin was i^{6}ade. As seen above, in the present results obtained with tissue from the same clone, the predominant cytokinin was t-io^{6}ado.

In earlier studies on the biosynthesis in vivo of cytokinins in plant tRNA it was reported (Chen and Hall, 1969) that the ^{14}C-MVA was incorporated exclusively into i^{6}ado by cytokinin-dependent tobacco callus. In a later study (Murai et al., 1975) it was reported that incorporation was predominantly into c-io^{6}ado by similar tissue. Murai et al. (1975) attributed this difference in results to both age of tissue and incubation conditions. The work reported above shows incorporation into both c-io^{6}ade and i^{6}ado in RNA by cytokinin-autonomous and bacteria-transformed tobacco callus, incorporation into the former cytokinin being the most abundant. In all but three treatments the radioactivity associated with c-io^{6}ado accounted for ca. 75%-85% of that present in cytokinins. These differences in the relative amounts of io^{6}ado and

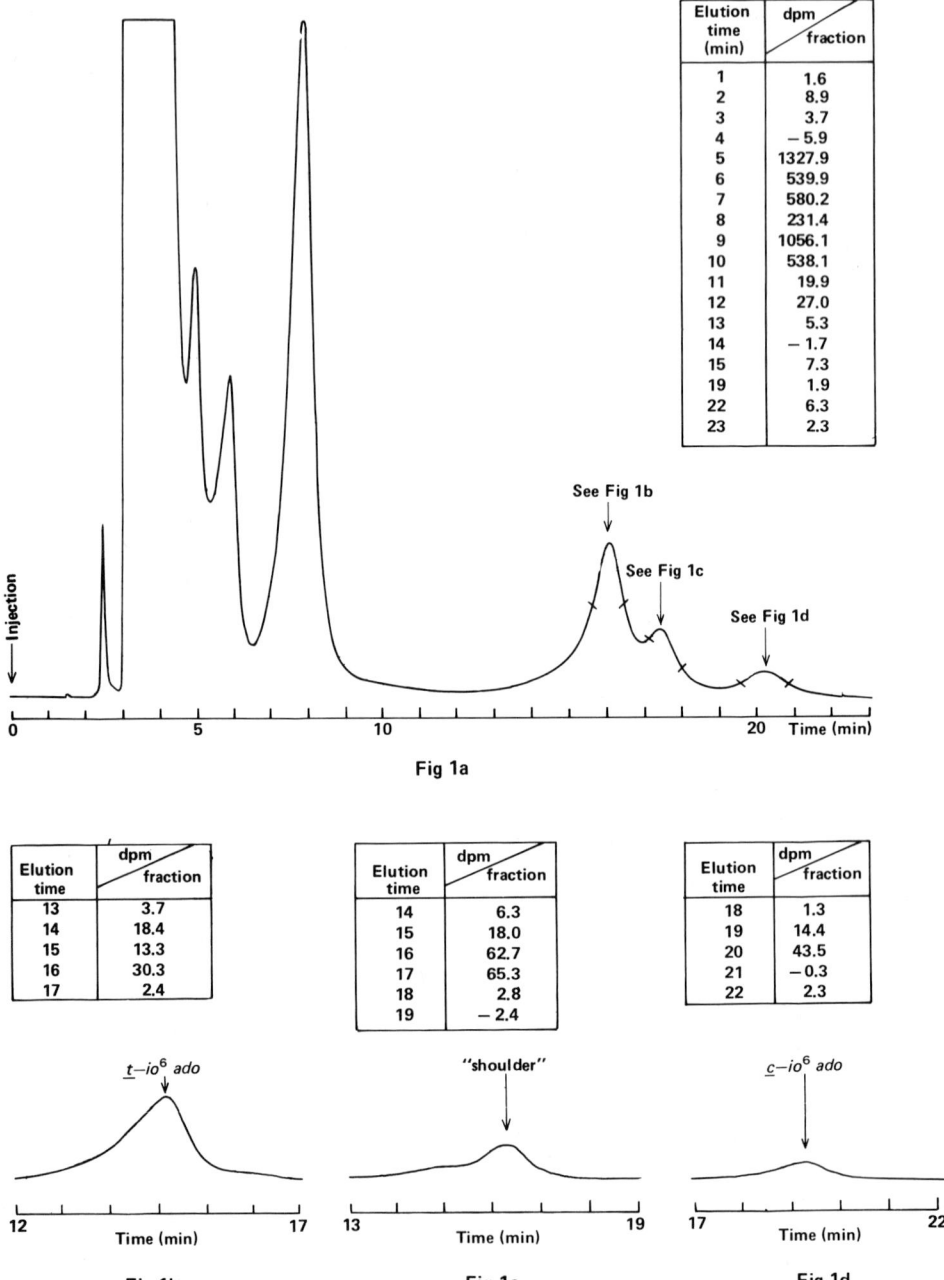

Fig. 1a–d. The distribution of radioactivity in the elution profile from an HPLC of **a** the io⁶ado fraction obtained from a Sephadex LH-20 chromatography of an extract of bacteria-transformed tobacco callus tissue incubated for 28 days in the presence of ^{14}C-MVA. **b** rechromatography of the t-io⁶ado fraction in **a**, **c** rechromatography of the "shoulder" of the t-io⁶ado fraction in **a**. **d** rechromatography of the c-io⁶ado fraction in **a**

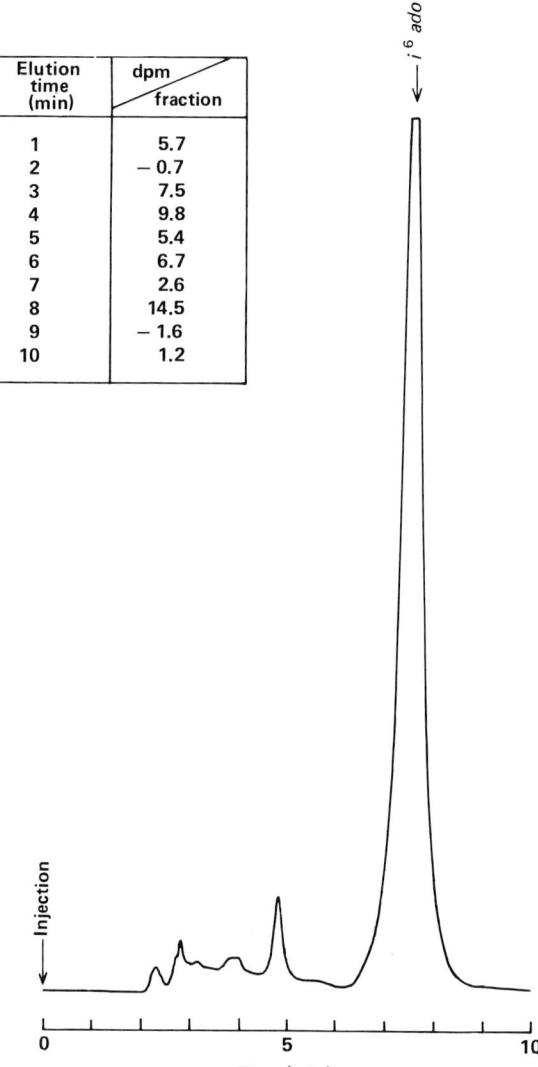

Fig. 2. The distribution of the radioactivity in the elution profile from the HPLC of the i^6 ado fraction obtained following Sephadex LH-20 chromatography of an extract of bacteria-transformed tobacco callus tissue, incubated for 672 h in the presence of ^{4}C-MVA

Elution time (min)	dpm / fraction
1	5.7
2	− 0.7
3	7.5
4	9.8
5	5.4
6	6.7
7	2.6
8	14.5
9	− 1.6
10	1.2

i^6 ado could be accounted for by differences in the levels of hydroxylating enzymes. The incorporation of ^{14}C-MVA into i^6 ado, in the absence of any detectable radioactivity into c-io^6 ado, by both tissues, with a 1 h incubation time, is evidence that i^6 ado in the tRNA is the substrate for the cis-specific hydroxylating enzyme. The absence of any detectable ^{3}H-adenine incorporation into the i^6 ado indicates alkylation of RNA synthesised prior to the incubation period.

The cytokinins t-io^6 ade and t-io^6 ado were identified in the culture filtrate and the tRNA respectively of the causal bacterium *Agrobacterium tumefaciens* (Chapman et al., 1976; Kaiss-Chapman and Morris, 1977). The hydroxylated cytokinins in another pathogenic bacterium *Corynebacterium fascians* have been identified as the cis-isomers (Scarborough et al., 1973; Armstrong et al., 1976). The suggestion is, therefore, that the prokaryote has "picked up" the genes for the specific hydroxylations during evolution. The failure to detect t-io^{6}ado in the RNA from transformed callus following re-infection suggests that the gene is not transcribed very actively, if at all, in the plant cells. The presence of the plant cytokinin form in both non-virulent and virulent bacteria excludes the possibility that the phytohormones are responsible for tumor inception. They are more likely to be involved in the second phase, that of tumour proliferation.

It is tempting to speculate that the unknown cytokinin observed by Chapman et al. (1976) may be the same compound as that radiolabelled with ^{3}H-adenine and found only the hydrolysate of the RNA from bacteria-transformed tissue.

Radiolabel from both precursors was detected in both t-io^{6}ado-5'-phosphate and i^{6}ado-5'-phosphate. In the *Vinca* crown gall tissue (Stuchbury et al., 1979) radioactivity could be detected in io^{6}ado-phosphate but not in i^{6}ado-phosphate. This led Stuchbury et al. (1979) to suggest that the iso-prenoid side chain was hydroxylated prior to attachment to the 5'-AMP. This is unlikely to be the case in the study reported here. The low level of incorporation into the cytokinin nucleotides was surprising in view of the report by Stuchbury et al. (1979) that the level of incorporation in the *Vinca* system into io^6-ado-phosphate exceeded that into io^{6}ado by some sixfold. In our studies rehydrolysis of the samples with alkaline phosphatase after the hydrolysis with the specific 3'- and 5'-phosphatases did not dramatically increase the amount of radiolabel estimated to have been incorporated into the cytokinin nucleotides.

The presence of radiolabel only in the cytokinin nucleosides io^{6}ado and i^{6}ado with the 672 h incubation time could reflect a release intact from cytokinin-containing RNA species during turnover. A recent publication (Barnes et al., 1980) suggests that RNA could act as a source of free cytokinin in excised potato tuber cells. The predominance of the trans-isomer may be a product of a cis-trans isomerase. The presence of these ribosides could reflect equally the release from a bound or conjugated form. To date only the 4'-0-glucoside of io^{6}ado has been identified (Morris, 1977; Peterson and Miller, 1977); perhaps the presence of the ribose moiety at N^9 also prevents the formation of the N^3- and N^7-conjugates.

In general the total radiolabel incorporated into the free cytokinins far exceeded that present in RNA. In the absence of precise data on the turnover of cytokinin-containing RNA species, it is not possible to estimate accurately the contribution to the free pool of cytokinins resulting from RNA turnover, Analysis of the data in this report suggests that this contribution is unlikely to exceed 2% of the total cytokinin content.

References

Armstrong DJ, Burrows WJ, Evans PK, Skoog F (1969) Isolation of cytokinins from tRNA. Biochem Biophys Res Commun 37: 451-456

Armstrong DJ, Scarbrough E, Skoog F, Cole DL, Leonard NJ (1976) Cytokinins in *Corynebacterium fascians* cultures. Isolation and identification of 6- (4-hydroxy-3-methyl-cis-2-butenylamino)-2-methylthiopurine. Plant Physiol 58: 749-752

Barnes MF, Tien CL, Gray JS (1980) Biosynthesis of cytokinins by potato cell cultures. Phytochemistry 19: 409-412

Bieleski RL (1964) The problems of halting enzyme action when extracting plant tissues. Anal Biochem 9: 431-442

Burrows WJ (1976) Mode of action of N, N'-diphenylurea: The isolation and identification of the cytokinins in the transfer RNA from tobacco callus grown in the presence of N,N'-diphenylurea. Planta 130: 313-316

Burrows WJ (1978a) Evidence in support of biosynthesis de novo of free cytokinins. Planta 138: 53-57

Burrows WJ 1978b) Incorporation of ^{3}H-adenine into free cytokinins by cytokinin-autonomous tobacco callus tissue. Biochem Biophys Res Commun 84: 743-748

Chapman RW, Morris RO, Zaerr JB (1976) Occurrence of *trans*-ribosylzeatin in *Agrobacterium tumefaciens* tRNA. Nature (London) 262: 153-154

Chen C-M, Eckert RL (1976) Evidence for the biosynthesis of transfer RNA-free cytokinin. FEBS Lett 64: 429-434

Chen C-M, Hall RH (1969) Biosynthesis of N^6-(Δ^2-isopentenyl) adenosine in the transfer ribonucleic acid of cultured tobacco pith cells. Phytochemistry 8: 1687-1695

Chen C-M, Melitz DK (1979) Cytokinin biosynthesis in a cell-free system from cytokinin-autotrophic tobacco tissue cultures. FEBS Lett 107: 15-20

Holland JA, McKerrell EH, Fuell KJ, Burrows WJ (1978) Separation of cytokinins by reversed-phase high-performance liquid chromatography. J Chromatogr 166: 545-553

Kaiss-Chapman RW, Morris RO (1977) *Trans*-zeatin in culture filtrates of *Agrobacterium tumefaciens*. Biochem Biophys Res Commun 76: 453-459

Klemen F, Klämbt D (1974) Half-life of sRNA from primary roots of *Zea mays*. A contribution to the cytokinin production. Physiol Plant 31: 186-188

Leineweber M, Klämbt D (1974) Half-life of sRNA and its consequence for the formation of cytokinins in *Lactobacillus acidophilus* – ATCC 4963 grown in the logarithmic and stationary phases. Physiol Plant 30: 327-330

Linsmaier EM, Skoog F (1965) Organic growth factor requirements of tobacco callus cultures. Physiol Plant 18: 100-127

Morris RO (1977) Mass spectroscopic identification of cytokinins. Glycosyl zeatin and glycosyl ribosylzeatin from *Vinca rosea* crown gall. Plant Physiol 59: 1029-1033

Murai N, Armstrong DJ, Skoog F (1975) Incorporation of mevalonic acid into ribosylzeatin in tobacco callus ribonucleic acid preparations. Plant Physiol 55: 853-858

Peterson J, Miller CO (1977) Glucosylzeatin and glycosyl-ribosylzeatin from *Vinca rosea L*. crown gall tumour tissue. Plant Physiol 59: 1026-1028

Scarborough E, Armstrong DJ, Skoog F, Frihart CR, Leonard NJ (1973) Isolation of *cis*-zeatin from *Corynebacterium fascians* cultures. Proc Natl Acad Sci USA 70: 3825-3829

Stuchbury T, Palni LM, Horgan R, Wareing PF (1979) The biosynthesis of cytokinins in crown-gall tissue of *Vinca rosea*. Planta 147: 97-102

Taya Y, Tanaka Y, Nishimura S (1978) 5'-AMP is a direct precursor of cytokinin in *Dictyostelium discoideum*. Nature (London) 271: 545-547

Cytokinin Biosynthesis and Metabolism in *Vinca rosea* Crown Gall Tissue

R. Horgan[1], L.M.S. Palni[2], I. Scott[1], and B. McGaw[1]

1 Introduction

Progress in the understanding of the mechanisms and controls of plant hormone biosynthesis and metabolism has to a large extent been limited by the availability of suitable systems for study. Whilst the ultimate aims of studies of this type must be an understanding of the ways in which plants regulate their endogenous hormone complement and the physiological significance of this regulation, we are still very far from achieving these in practice. In general, biosynthetic and metabolic studies of plant hormones are severely hampered by the extremely low levels of these compounds in most plant tissues. Indeed, the recent spectacular progress in the area of gibberellin biosynthesis and metabolism has been very dependent upon the use of fungal and immature seed systems in which gibberellin levels are sufficiently high to permit the use of unambiguous physical techniques for these studies.

Although the sensitivity of physical techniques is continually increasing, as, unfortunately, is their cost, progress in the field of cytokinin biochemistry is sorely hampered by the lack of suitable systems for biosynthetic and metabolic studies, in which these processes can be studies in a physiologically relevant and chemically rigorous manner. Ideally, the following requirements should be met by any system used for studies into plant hormone biosynthesis and metabolism:

1. The endogenous hormones should be completely and unambiguously characterised, both qualitatively and quantitatively;
2. The levels of these compounds should be sufficiently high to allow the feeding of precursors and intermediates, in amounts high enough to permit identification of products by reliable physical techniques, without greatly exceeding the endogenous levels.

The above two points represent a minimum requirement for valid studies. In addition, the lack of information on compartmentation of endogenous compounds and the effects of this on their biosynthesis and metabolism presents considerable further difficulties to the interpretation of results.

---

[1] Department of Botany and Microbiology, University College of Wales, Penglais, Aberystwyth, SY 23 3DA, United Kingdom
[2] Department of Developmental Biology RSBS, Australian National University, P.O.Box 475, Canberra City Act 2601, Australia

In recent years at Aberystwyth we have been investigating the suitability of several systems for detailed studies of cytokinin biosynthesis and metabolism, bearing in mind the above comments. To date, two systems which, in part at least, fulfil the above requirements have been studied. These are rooted cutting and decapitated plants of *Phaseolus vulgaris,* and tissue cultures of *Vinca rosea* crown gall tumours.

This paper describes preliminary studies on cytokinin biosynthesis and metabolism in *Vinca rosea* crown gall and normal tissue. Although the studies described here fall short of the criteria outlined previously, they are directed towards the eventual fulfilment of these criteria. The direction of future work based on these preliminary studies will be discussed briefly.

The origin and nature of crown gall tissues have been extensively covered in the literature and will not be discussed further here. However, from the point of view of studies of cytokinin biosynthesis and metabolism, the relatively high levels of cytokinins in *V. rosea* crown gall tissue and the fact that the compounds have been unambiguously identified make it a suitable system for study. In addition, the fact that cytokinin production in this tissue is in some way under the control of the Ti plasmid adds an additional valuable genetic dimension to these studies.

2 Endogenous Cytokinins of *V. rosea* Crown Gall Tissue and Its Normal Counterpart

The initial reports on the nature of the endogenous cytokinins of *V. rosea* crown gall tissue are very confused and controversial. To the best of our knowledge, the original work on cell division factors in this tissue (Wood et al., 1969; Wood, 1970) has never been substantiated. However, Miller (1974) unambiguously established that the major endogenous cytokinin of *V. rosea* crown gall tissue was zeatin riboside. In addition, Peterson and Miller (1977) and Morris (1977) identified zeatin-O-glucoside and ribosyl zeatin-O-glucoside as major cytokinins when the tissue was grown on ammonium chloride as the nitrogen source.

Although in general we have confirmed these initial findings, our results do differ from those of Miller in some important aspects. Firstly, when care is taken to minimise the effects of phosphatase activity during extraction, by use of the extraction medium of Bieleski, a major portion of the extractable cytokinin activity is found in the nucleotide fraction. Further fractionation of this activity into mono-, di-, and trinucleotides by DEAE cellulose chromatography reveals that the activity is confined exclusively to the mono-phosphate components of the extract. Chemical and enzymatic degradations, and subsequent identification of the biologically active material as zeatin riboside, suggest that a major portion of the cytokinin activity of *V. rosea* crown gall tissue is due to zeatin riboside-5-monophosphate.

Secondly, HPLC fractionation of an extract of *Vinca* tissue, using a HPLC system specifically designed to separate cytokinin glucosides (see Fig. 1) revealed a large UV-absorbing peak at the elution position of zeatin-9-glucoside. UV and MS analysis of this component as the methylated, trimethylsylylated (TMS) and free compound confirmed its structure (Scott et al., 1980). Thus zeatin-9-glucoside must also be considered as a major component of the *Vinca* cytokinin complex. It was probably not observ-

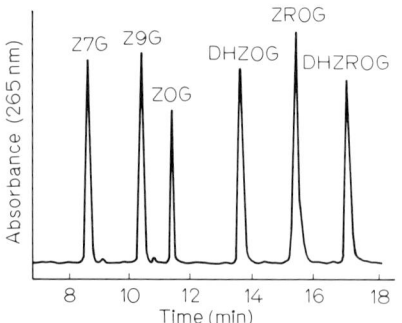

Fig. 1. Separation of a series of cytokinin glucosides by reversed-phase HPLC. Column, ODS Hypersil (150 ; 4.5 mm) flow rate, 2 ml · mm^{-1}; mobile phase, linear gradient of 5% acetonitrile in water (pH 7 TEAB) to 20% acetonitrile over 30 min. Abbreviations: *Z7G* zeatin-7-glucoside; *Z9G* zeatin-9-glucoside; *ZOG* zeatin-O-glucoside; *DHZOG* dihydrozeatin-O-glucoside; *ZROG* ribosyl zeatin-O-glucoside; *DHZROG* ribosyl dihydrozeatin-)-glucoside

ed in previous studies because of its very low biological activity in the soybean test and the difficulty of separating it from the O-glucoside.

Thirdly, we have been unable to reproduce the reduced nitrogen effect of Peterson and Miller. This is probably due to changes in the nature of the tissue culture with continued subculture. In general, we have found significant levels of zeatin-O-glucoside and ribosyl zeatin-O-glucoside in tissue grown on a nitrate medium. Table 1 gives a rough indication of the levels of cytokinin bases, ribosides and glucosides in *V. rosea* crown gall tissue. These values are almost certainly underestimates as measurements were made by UV absorbance following HPLC fractionation.

Table 1. Cytokinin levels in mature. *V. rosea* crown gall tissue. Values expressed as µg/kg fresh weight of tissue

Zeatin	9
Zeatin riboside	80
Zeatin-O-glucoside	25
Ribosyl zeatin-O-glucoside	32
Zeatin-9-glucoside	71

It is clear that any study of cytokinin biosynthesis and metabolism must not only take into account the nature of the endogenous compounds but also their levels and any changes in these levels with the growth and development of the tissue. Thus prerequisites for any meaningful metabolic and biosynthetic study are means of isolating the compounds in question and an accurate method for their quantitative determination. Our early work on *V. rosea* crown gall tissue using bioassay has convinced us of the extreme inaccurary of this technique even when carefully applied, and so we have expended considerable effort on the development of accurate and precise analytical techniques for the endogenous cytokinins of this tissue. The ultimate aim of this work is to completely describe the endogenous cytokinin status of *V. rosea* crown gall tissue in relation to its growth and development. This will provide the essential background information for undertaking meaningful studies of biosynthesis and metabolism.

The development of two techniques provided the key components of our analytical system and will be mentioned briefly here. We have developed HPLC technique methods for the separation of all the cytokinins present in *Vinca* tissue. Details of this technique have been published (Horgan and Kramers, 1979). Figure 1 illustrates the very high resolving power of the method.

Although we have used HPLC to obtain rough quantitative estimates of *Vinca* cytokinins, the need for several purification steps prior to measurement renders the use of internal standards to estimate recovery essential. The most satisfactory method for estimating endogenous cytokinin levels would seem to be mass spectrometry using isotopically labelled internal standards. This technique, using ^{2}H labelled cytokinins, has been described by Summons et al. (1979) and McCloskey et al. (1980).

We have developed a similar technique using [^{15}N]-labelled compounds as internal standards. The particular advantage of these compounds is that there is no risk of isotope exchange in the mass spectrometer as sometimes occurs with ^{2}H internal standards, and the presence of 4 [^{15}N] atoms in the purine ring virtually eliminates cross

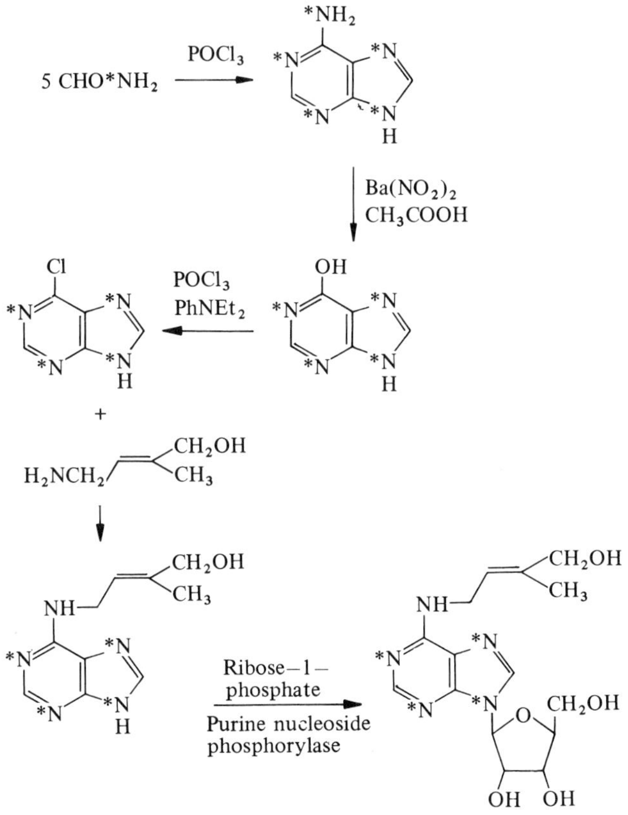

Fig. 2. Synthesis of 1,3,7,9-^{15}N -zeatin and zeatin riboside. * indicates ^{15}N atom

talk between ions and provides a large choice of ions suitable for monitoring. The chemical synthesis of [^{15}N]-labelled cytokinins is shown in Fig. 2. Preliminary studies using this technique have been most satisfactory. Using GCMS and computer controlled Multiple Ion Monitoring (MPM) it is possible to determine cytokinins in extracts down to the nanogram level. Using this technique we have measured the level of zeatin riboside in mature crown gall tissue as 402 ng per g fresh weight, and in normal stem callus tissue of the same age as 2.5 ng per g fresh weight. The technique has also revealed considerable variation in the level of zeatin riboside in crown gall tissue with the age of the tissue. The average level in young tissue is 50 ng per g.f.wt while in mature tissue it is 400 ng per g.f.wt.

In general, we have found methylated cytokinins to be preferable to TMS cytokinins for quantitative determination by this method. Zeatin, zeatin riboside and zeatin-O-glucoside have been measured directly by this technique and ribosyl zeatin-O-glucoside after enzymatic hydrolysis to the riboside.

In addition to their relatively high levels of endogenous cytokinins *V. rosea* crown gall tissues excrete large quantities of cytokinins into the growth medium. This occurs with tissues grown on liquid and solid media. It is possible that the high rate of cytokinin production by this tissue may be due to its general leakiness to cytokinins. A similar situation seems to exist in cytokinin overproducing mutants of the moss *Physcomitrella patens* (T L Wang, pers. comm.). We have unambiguously identified all the cytokinins endogenous to the tissue as being present in the medium and in addition have found small and variable quantities of dihydrozeation-O-glucoside and ribosyl dihydrozeatin-O-glucoside in the medium of agar-grown cultures.

3 Cytokinin Biosynthesis in *Vinca rosea* Crown Gall and Normal Tissue

Peterson and Miller (1976) first reported the incorporation of radioactivity from labelled adenine into cytokinins by *Vinca* crown gall tissue. Although the labelled products were not unambiguously identified good circumstantial evidence was presented that adenine was being incorporated into zeatin and zeatin riboside.

We have extended this work to include a time course study of the incorporation of adenine into putative cytokinins and a more detailed study of the nature of the cytokinins involved. The initial study utilised column chromatography, paper chromatography and 2D TLC as the analytical techniques. These results have in general been substantiated by subsequent experiments utilising HPLC as the final analytical method, with the important difference that in the latter case the levels of incorporation were 30% to 50% lower. This is interpreted as indicating that the putative cytokinins in the preliminary investigation were not radiochemically pure and highlights one of the technical problems commonly found in work of this type. Since co-chromatography even on HPLC does not prove coidentity between the labelled compounds and the endogenous cytokinins, the work is currently being repeated using [^{15}N]-labelled precursors so that mass spectroscopy can be used to establish the identity of the metabolites and determine their specific activities. Accurate determination of the specific activity of metabolites is obviously of key importance to assessing turnover rates. This is one of the main future objectives of this part of the study.

The findings of our preliminary investigations into the incorporation of ^{14}C-adenine into cytokinins (Stuchbury et al., 1979) and the extension of these studies using HPLC as the major analytical method are summarised below:

1. Adenine was rapidly taken up by the tissue and after 3 h of incubation less than 5% of the radioactivity remained in the medium.
2. Most of the radioactivity in the tissue was in the form of nucleotides.
3. DEAE cellulose fractionation of the labelled nucleotides into mono-, di- and tri-nucleotides, followed by chromatographic, chemical and enzymatic analysis of the individual peaks revealed zeatin riboside-5'-monophosphate as the only labelled cytokinin in the nucleotide fraction.
4. Small quantities of zeatin and zeatin riboside were present in the tissue. At all times the amount of zeatin riboside exceeded that of zeatin.
5. At incubation times in excess of 10 h compounds giving zeatin and zeatin riboside on hydrolysis with β-glucosidase appeared in the tissue. These have been tentatively identified as zeatin-O-glucoside and ribosyl zeatin-O-glucoside by HPLC, and enzymatic degradation followed by HPLC of the aglycones.
6. At no time was any isopentenyl adenine (2iP), its riboside (2iPA) or its nucleotide detected. Attempts to "cold trap" radioactivity in these compounds failed and resulted in only a very marked increase in the degradative metabolism of the supplied [^{14}C]-adenine.
7. Labelled zeatin and zeatin riboside were excreted into the medium during the course of the incubation. Small quantities of putative glucosides were also detected but more detailed confirmation of the identity of these is required.

The results of time course studies into the incorporation of ^{14}C-adenine into cytokinins in *V. rosea* crown gall tissue are presented in Fig. 3.

In duplicate experiments utilising normal stem callus tissue of *V. rosea*, no incorporation of [^{14}C]-adenine into cytokinins was observed.

4 Cytokinin Metabolism in *Vinca rosea* Crown Gall Tissue

Since neither endogenous isopentenyl adenine compounds or labelled compounds of this type derived from adenine could be detected in *Vinca rosea* crown gall tissue attempts were made to see if 2iP or one of its derivatives could be a precursor of the zeatin type cytokinins. Thus, [^{14}C]-labelled 2iP was fed to *V. rosea* crown gall tissue to ascertain whether or not it could be stereospecifically hydroxylated to trans-zeatin or one of its derivatives.

[^{14}C]-2iP was rapidly taken up by *Vinca* tissue and after 2 h radioactivity could be detected in trans-zeatin, its nucleoside and nucleotide. Products were identified by HPLC using solvent systems which clearly separate the cis and trans isomers of zeatin and zeatin riboside. However, of the 2iP fed the major portion was broken down to adenine derivatives.

The metabolism of [^{14}C]-zeatin and [^{14}C]-zeatin-O-glucoside have also been studied in *V. rosea* crown gall tissue. [^{14}C]-zeatin was rapidly taken up by the tissue and converted to its 5'- nucleotide and nucleoside and at later times to its O-glucoside and

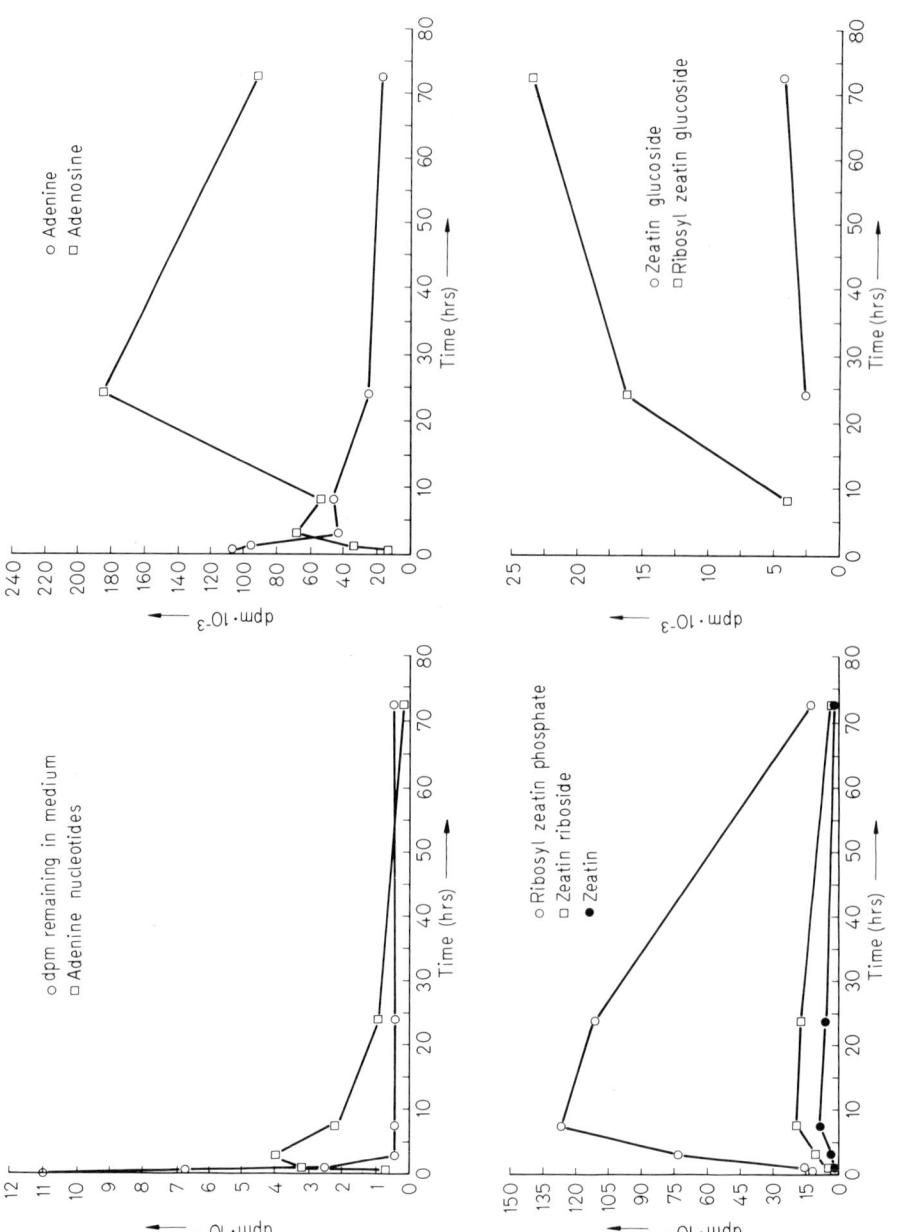

Fig. 3. Time course of cytokinin biosynthesis in *V. rosea* crown gall tissue

ribosyl-O-glucoside. The results of a time course study of zeatin metabolism are presented in Fig. 4.

In contest to zeatin and 2iP ^{14}C-zeatin-O-glucoside was taken up slowly by the tissue and very rapidly degraded so that at all times its concentration in the tissue was very low.

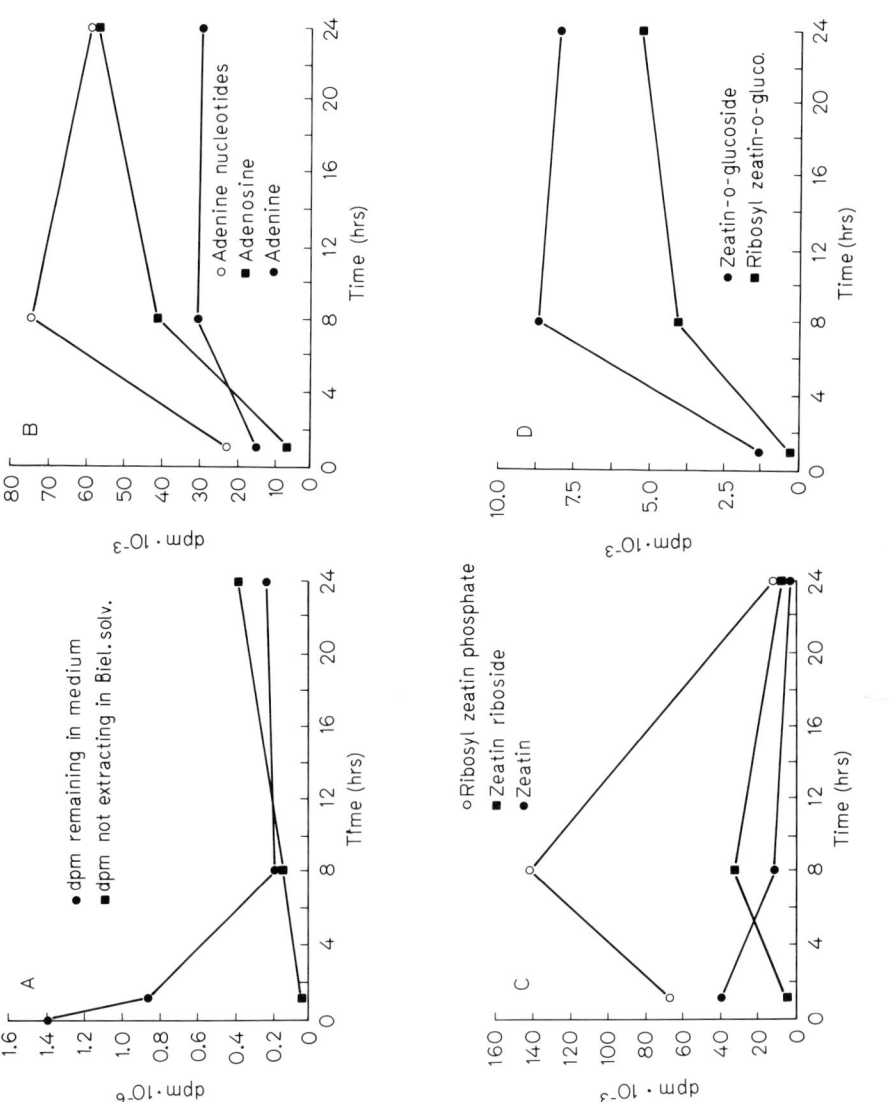

Fig. 4. Metabolism of 8-^{14}C zeatin by *V. rosea* crown gall tissue

5 Cytokinins in the tRNA of *Vinca rosea* Crown Gall and Normal Tissue

The relationship between free cytokinins and those in tRNA is still far from being understood. As an adjunct to our biosynthetic studies we have identified and made rough quantitative estimates of the cytokinins in the tRNA of crown gall and normal tissue.

Purified tRNA from both tissues was enzymatically hydrolysed and fractionated by HPLC. The identity of the UV-absorbing and biologically active peaks with the elution volumes of cis and trans zeatin riboside was confirmed by GCMS of the TMS derivatives. The relative amounts of cis and trans zeatin riboside per mg of tRNA in the two tissues were estimated by UV absorbance as:

a) *V. rosea* crown gall tRNA
   i) cis-zeatin riboside : 42 ng
   ii) trans-zeatin riboside : 18 ng
b) *V. rosea* normal tRNA
   i) cis-zeatin riboside : 40 ng
   ii) trans-zeatin riboside : not detectable

6 Conclusions

The preliminary results presented in this paper tend to confirm the suitability of *V. rosea* crown gall tissue for studies into cytokinin biosynthesis. The results of adenine feeding experiments suggest that biosynthesis is taking place at the nucleotide level and thus confirms the published work using cell-free systems from different sources (Chen and Melitz, 1979; Taya et al., 1978). In addition, the relatively rapid rate of incorporation of adenine (as $5'$-AMP) into zeatin riboside-$5'$-monophosphate suggests a route independent of tRNA synthesis. However, further studies of the rate of tRNA turnover in this tissue are required before firm conclusions can be drawn. This is particularly necessary in the light of the findings of Borek et al. (1977) on the turnover of tRNA in certain animal tumours.

Although the metabolism experiments described in this paper clearly show that externally supplied cytokinins can be converted into compounds known to occur endogenously in the tissue, considerable further research needs to be undertaken to demonstrate that this metabolism is indeed related to the metabolism of the endogenous compounds. One of the striking features of the metabolism studies is the high rate of breakdown of the applied cytokinins. Superficially this suggests that the high levels of cytokinins found in *V. rosea* crown gall tissue do not arise from the absence of enzymes for their catabolism. However, attempts to isolate "cytokinin oxidase" type enzymes from *Vinca* crown gall tissues have so far been unsuccessful (T. Stuchbury, pers. comm.).

The results of feeding experiments with $[^{14}C]$-zeatin-O-glucoside strongly suggest that the fate of this compound when supplied externally does not parallel that of the endogenous compound, which tends to accumulate in the tissue. This highlights the sort of ambiguity that can exist in the interpretation of plant hormone metabolism experiments.

We are at present investigating in situ labelling techniques with heavy isotopes to attempt to overcome this problem. As stated in the introduction the work presented in this paper is essentially preliminary in nature. In many instances our initial investigations have demonstrated to us the shortcomings of many of the methods used in cytokinins studies. This applies particularly to identifications and estimations of incorporations based on low resolution chromatographic techniques. Nevertheless we consider that *V. rosea* crown gall tissue is an excellent experimental material for studies of cytokinin biosynthesis and metabolism.

We hope to extend the work presented here to obtain a picture of cytokinin turnover in this tissue which will lead eventually to an understanding of the way in which this process is controlled.

Acknowledgments. We wish to thank Professor P.F. Wareing for his interest and encouragement throughout this work. We also wish to thank the ARC for financial support.

References

Borek E, Baliga BS, Gehrke CW, Kuo CW, Belman S, Troll W, Waalkes TP (1977) High turnover rate of transfer RNA in tumor tissue. Cancer Res 37: 3362-3366

Chen C-M, Melitz DK (1979) Cytokinin biosynthesis in a cell-free system from cytokinin-autotrophic tobacco tissue cultures. FEBS Lett 107: 15-20

Horgan R, Kramers MR (1979) High performance liquid chromatography of cytokinins. J Chromatogr 173: 263-270

NcCloskey JA, Hashizume T, Basile B, Ohno Y, Sonoki S (1980) Occurrence and levels of cis- and trans-zeatin ribosides in the culture medium of a virulent strain of *Agrobacterium tumefaciens.* FEBS Lett 111: 181-183

Miller CO (1974) Ribosyl-trans-zeatin, a major cytokinin produced by crown gall tumor tissue. Proc Natl Acad Sci USA 71: 334-338

Morris RO (1977) Mass spectroscopic identification of cytokinins. Plant Physiol 59: 1029-1033

Peterson JB, Miller CO (1976) Cytokinins in *Vinca rosea* L. crown gall tumor tissue as influenced by compounds containing reduced nitrogen Plant Physiol 57: 393-399

Peterson JB, Miller CO (1977) Glucosyl zeatin and glucosyl ribosyl-zeatin from *Vinca rosea* L. crown gall tumor tissue. Plant Physiol 59: 1026-1028

Scott IM, Horgan R, McGaw BA (1980) Zeatin-9-glucoside, a major endogenous cytokinin of *Vinca rosea* L crown gall tissue. Planta (in press)

Stuchbury T, Plani LMS, Horgan R, Wareing PF (1979) The biosynthesis of cytokinins in crown gall tissue of *Vinca rosea.* Planta 147: 97-102

Summons RE, Duke CC, Eichholzer JV, Entsch B, Letham DS, MacLeod JK, Parker CW (1979) Mass spectrometric analysis of cytokinins in plant tissues. Biomed Mass Spectro 6: 407-413

Taya Y, Tanaka Y, Nishimura S (1978) 5'-AMP is a direct precursor of cytokinin in *Dictyostelium discoideum.* Nature (London) 271: 545-547

Wood HN (1970) Revised identification of the chromophore of a cell division factor from crown gall tumor cells of *Vinca rosea* L. Proc Natl Acad Sci USA 67: 1283-1287

Wood HN, Braun AC, Brandes H, Kende H (1969) Studies on the distribution and properties of a new class of cell division-promoting substance from a higher plant species. Proc Natl Acad Sci USA 62: 349-356

Section 2
Cytokinin Metabolism and Physiological Responses in Plant or Bacterial Systems

Cytokinin Translocation and Metabolism in Species of the Leguminoseae: Studies in Relation to Shoot and Nodule Development

R.E. Summons, D.S. Letham, B.I. Gollnow, C.W. Parker, B. Entsch,
L.P. Johnson, J.K. MacLeod, and B.G. Rolfe[1]

1 Introduction

Legumes, and lupin species in particular, are of considerable interest to workers in the field of cytokinin metabolism. Some lupin species exhibit sequential leaf senescence and abscission and cytokinins have been implicated in the regulation of these phenomena [1]. White lupin (*Lupinus albus*) is one of the few annual plants which readily yields phloem exudate [12, 13] and hence would be invaluable for studies of cytokinin translocation phenomena. Bioassay studies indicate that young lupin fruits (pods) contain high levels of cytokinins when compared with the level in the primary inflorescence of flowering plants [3, 4, 5] and consequently such pods may constitute useful organs for studies of cytokinin biosynthesis or accumulation from other plant organs. Finally, but nonetheless of great importance, *Rhizobium* species induce nitrogen-fixing root nodules on the roots of legumes. Both effective and ineffective isogenic *Rhizobium* mutants are now available for study [15] and these may provide definitive information concerning the proposal that *Rhizobium*-derived cytokinin plays a key role in nodule development [14]. Thus in one group of plants, the study of many facets of cytokinin-dependent physiology may be facilitated.

Cytokinin metabolism in lupins appears to be extremely complex and involves most of the known modifications to the basic zeatin moiety. For example when tritiated zeatin was supplied to de-rooted seedlings of *L. angustifolius* and *L. luteus*, at least 16 labelled metabolites were formed and either fully or partly characterised [11]. The metabolic modifications included the removal of the side chain to give adenine, adenosine, AMP and other purines. Reduction of the side chain, ribosylation and glucosylation also occurred yielding dihydrozeatin, zeatin riboside and/or dihydrozeatin riboside, *O*-β-D-glucopyranosylzeatin (OGZ), *O*-β-D-glucopyranosyldihydrozeatin (OGDZ) and the corresponding 9-ribosides (OGZR and OGDZR, respectively). Also identified were zeatin and dihydrozeatin nucleotides, 7-glucosyl zeatin and the two unusual alanine conjugates lupinic acid and dihydrolupinic acid [7, 11].

Hence with lupin tissues we have considerable diversity of metabolic modifications and also a solid background of data on which to base physiological studies. In this communication, we report studies of cytokinin metabolism in lupin species in relation to plant development. A preliminary study concerning pea nodules is also reported.

---

[1] Research School of Biological Sciences, and Research School of Chemistry, Australian National University, Canberra, ACT 2601, Australia

2 Materials and Methods

2.1 Chemicals

[^{3}H]Zeatin riboside for studies of cytokinin translocation was synthesized as detailed previously [17]. Lupinic acid and cytokinin glucosides were prepared by published methods [2, 6, 7] as were the penta-[^{2}H]-labelled compounds [16] used for internal standards in the mass spectrometric quantitation of cytokinins. Three dyes used as markers during TLC were: Meldola Blue (Gurr, Searle Diagnostic, High Wycombe, England) which was purified by TLC, Drimarene Brilliant Blue K-BL (Polysciennces Inc., Warrington, P.A., USA), and Rhodamine B (Hopkins and Williams, England).

2.2 Mass Spectrometric Quantitation

Tissue extraction and HPLC purification of plant extracts were according to procedures described previously [16, 17, 18], deuterium-labelled cytokinins being added to the extracting solvent. HPLC fractions corresponding to the O-glucosides and lupinic acid were analyzed using the direct inlet probe [16, 18] and the ratios of endogenous cytokinin to penta-^{2}H analogue (D_0/D_5) computed from the appropriate ions in selected peak groups [16, 18]. HPLC fractions containing zeatin, dihydrozeatin and their ribosides were analyzed after TMS-derivatization using multiple-ion-detection GC-MS. GC was performed on a column (2 m x 2 mm internal diameter) of 2% OV-17 on Gas Chrom Q, with He flow of 25 ml/min and temperature programmed from 200° to 280°C at 10°C/min. The mass spectrometer operating in the EI mode (70 eV) was programmed to function in two MID modes. The first, to measure zeatin and dihydrozeatin, scanned ions at m/z 348, 350, 353 and 355 ($M^+ - CH_3$ for D_0 and D_5 compounds) using the column bleed ion at m/z 281 as a lock mass. The second mode, to measure zeatin riboside and dihydrozeatin riboside, scanned ions at m/z 624, 626, 629 and 631 using column bleed at m/z 529 as lock mass. The compounds could be analyzed separately or in mixture. Single ion current traces were recorded on a multi-pen recorder and D_0/D_5 ratios computed from these using appropriate scaling factors.

2.3 Chromatographic Methods

For TLC, layers of Merck silica gel 60 PF_{254} and Serva cellulose (Woelm green fluorescent indicator added to latter) were used. Solvents for TLC were as follows (proportions by volume):

- A, butan-1-ol-14N ammonia-water (6:1:2, upper phase)
- B, butan-l-ol-acetic acid-water (12:3:5)
- C, ethyl methyl ketone-propan-l-ol-14 N ammonia-water (10:1.1:1:1)
- D, methyl acetate-ethanol-2,2-dimethoxypropane (90:10:1) with 5 drops of formic acid per 100 ml (see ref. ll)
- E, chloroform-methanol (9:1)
- F, ethyl methyl ketone-acetic acid-water (16:1:4)

2.4 Extraction of Radioactive Metabolites from Tissue

The tissue was dropped into methanol-water-acetic acid (70:30:3) chilled to $-20°C$. After 2-3 days at this temperature to inactivate enzymes the mixture was homogenized at $4°C$ and then centrifuged. The pellet was re-extracted with 70% methanol at $23°C$. The combined extracts were evaporated under reduced pressure and the residue dissolved in 50% ethanol for TLC studies.

2.5 Translocation of Zeatin Riboside in Lupin Shoots

The roots of 11-weeks-old *Lupinus angustifolius* plants were excised and the stem bases placed in [^{3}H]zeatin riboside solution (4 μM; 400 mCi/mmol) for 20 h. The various parts of the shoot were then extracted. The portion of the stem which contacted the zeatin riboside solution was discarded.

In accordance with the procedure devised by Pate and coworkers [13], phloem sap was collected by excising the tips of the pods of *Lupinus albus* plants (cv. Ultra). The roots had been placed in [^{3}H]zeatin riboside solution (30 μM; 80 mCi/mmol) for 42 h before sap was collected.

2.6 Uptake of [^{3}H]Adenosine by Lupin Seeds

Under sterile conditions, the seeds were excised from the pods of *Lupinus luteus* plants 5-6 days after anthesis and placed on filter paper in Petri dishes. The paper was wetted with a sterile nutrient solution containing [^{3}H]adenosine (0.1 mCi/ml; 22 Ci/mmol); the nutrient solution contained the inorganic components of the revised medium of Murashige and Skoog [10] modified by replacing NH_4NO_3 and KNO_3 with L-asparagine (350 mg/l). After 75 h, the seeds were rinsed with water and extracted.

2.7 Metabolism of [^{3}H]Zeatin Riboside in Nodulated Pea Roots

Under sterile conditions, *Pisum sativum* seeds were germinated and placed on a N-free nutrient agar in large plastic Petri dishes (diameter, 13.5 cm; 1 seed per dish). The seedling shoots were pulled through holes cut in the tops of the Petri dishes. A culture of *Rhizobium leguminosarum* was streaked across the surface of the agar in close proximity to the growing roots. After initiation of nodules was apparent, and later when they were well developed, root tips were excised and replaced with agar (0.9%) in a small vial containing [^{3}H]zeatin riboside (50 μM). After 2-7 days, the nodules were excised and extracted.

3 Results and Discussion

3.1 Cytokinin Quantitation Using Mass Spectrometric Methods

The most rigorously accurate mass spectrometric quantitation methods utilise an appropriate isotope labelled internal standard for each compound to be measured. We have accomplished a facile synthesis of 16 penta-deuterium-labelled cytokinins using a set of reaction sequences beginning with $^2H_6$-acetone [16]. Calibrated aliquots of solutions of deuterium-labelled metabolites were added to the extraction solvents so that all subsequent manipulative losses of endogenous cytokinin would be accompanied by a corresponding loss of labelled standard. Plant extracts were purified using a cellulose-phosphate ion-exchange column and reverse-phase HPLC. Pure individual metabolites or simple mixtures were converted to TMS-derivatives and examined by GC-MS using multiple ion detection or by direct inlet probe using selected short mass range scans.

Use of GC-MS to resolve a fraction containing zeatin, dihydrozeatin, zeatin riboside, and dihydrozeatin riboside is illustrated in Fig. 1 which shows an elution profile for these compounds produced with short integrating scans. Figure 2 shows a MID trace for a fraction containing dihydrozeatin riboside to obtain the ratio of M^+-15 ions (m/z 631 and 626) for quantitation.

3.2 Cytokinin Levels in Lupin Pods

The careful bioassay work of Davey and Van Staden [3, 4, 5] has provided useful information regarding cytokinin distribution in *Lupinus albus* plants. Developing seed appeared to contain particularly high levels of cytokinin activity. Using the above-mentioned mass spectrometric methods, we have quantitated cytokinin levels in seeds (Table 1) and pod walls of *Lupinus luteus*. Levels of cytokinins in developing seed (ca. 14 days from petal fall) were considerably higher than in seed approaching maturity (ca. 24 days from petal fall). The dominant cytokinins in the former were zeatin riboside (level similar to that found in sweet corn kernels [16], dihydrozeatin riboside,

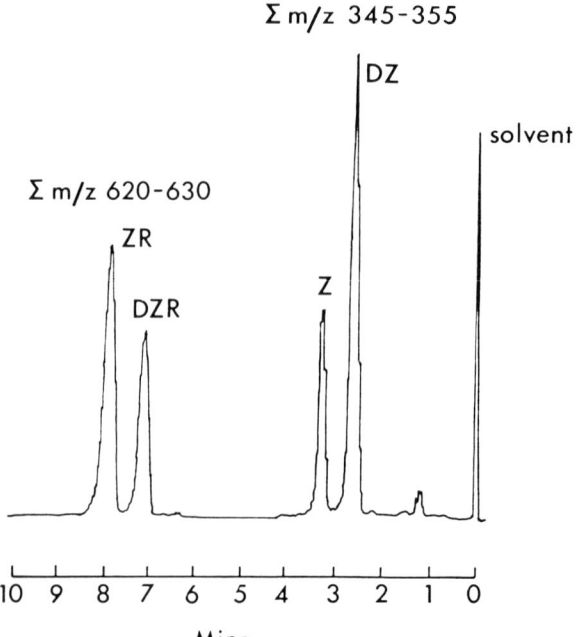

Fig. 1. The elution profile of a mixture of zeatin (*Z*), dihydrozeatin (*DZ*), zeatin riboside (*ZR*) and dihydrozeatin riboside (*DZR*) during GC-MS analysis. From 0 to 5 min, the mass spectrometer monitored the total ion current in the m/z range 345-355; from 5 to 10 min, the m/z range 620-630 was monitored

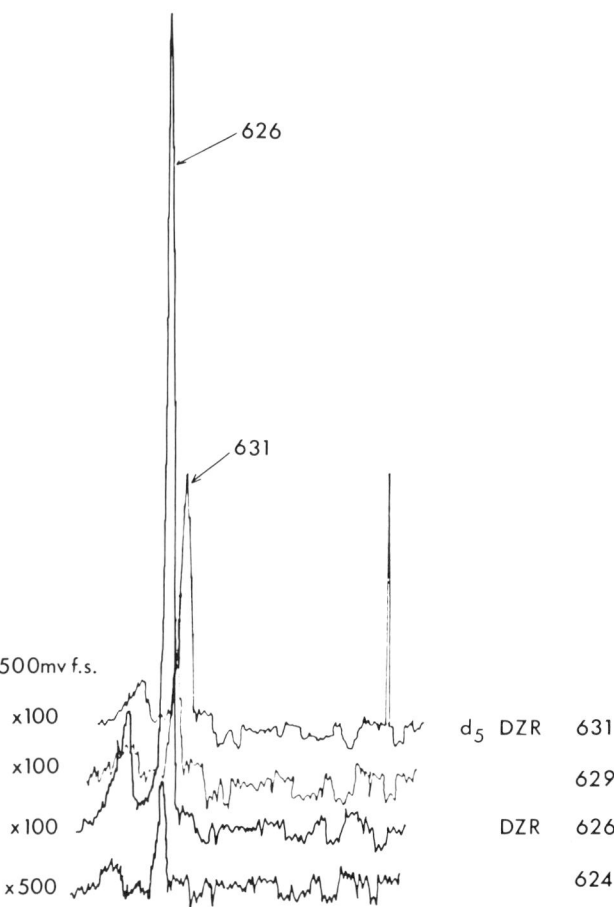

Fig. 2. A multiple-ion-detection tracing for a fraction containing [$^2H_5$]dihydrozeatin riboside (d_5DZR) and endogenous dihydrozeatin riboside (*DZR*). The fraction was obtained from extract of lupin seed by procedures outlined under Section 2. The ion of m/z 629 is due to an isotopic impurity in the $d_5 DZR$. The ions of m/z 629 and 624 were used to quantitate zeatin riboside which elutes after the *DZR* and OGDZR. OGDZR was the dominant cytokinin in seed approaching maturity and in pod walls, values for the latter tissue having been published previously [18]. Although lupinic acid could not be detected in the lupin seed, it was unambiguously identified and quantitated in immature pod walls (1.6 µg/100 g fresh weight; cf. zeatin level, 2.0 µg/100 g). This is the first report of the natural occurrence of lupinic acid. Recently, we have identified the related compound, dihydrolupinic acid, in *L. luteus* root nodules.

Our results establish that *Lupinus luteus* seeds and pod walls contain a complex of cytokinins, 10 of which have been quantitated. Seed development is accompanied by marked changes in the relative proportions of cytokinins and OGDZR is the dominant cytokinin in seed approaching maturity. The *O*-glucosyl moiety and saturated side-chain are structural features which appear to confer resistance to enzymic degradation [11, 20, 23]. Hence OGDZR is probably a very stable metabolite of zeatin, and it is relevant to note that OGDZ can under some circumstances accumulate markedly in bean leaves [22]. Experiments are in progress to determine whether OGDZR persists

Table 1. The cytokinin content of *Lupinus luteus* seed determined by mass spectrometry using deuterium-labelled (D_5) standards

	D_5 standard added (μg/100 g)	Cytokinin level (μg/100 g)	
		Developing seed	Seed near maturity
Zeatin	12.3	1.85	1.35
Dihydrozeatin	16	0.11	0.03
Zeatin riboside	18.9	39.3	3.8
Dihydrozeatin riboside	20.3	66.6	15.0
OGZ	11	0.18	0.17
OGDZ	20	1.76	0.42
OGZR	18.6	5.39	3.0
OGDZR	19.6	39.2	20.4
Lupinic acid	19.8	ND[a]	ND[a]
Dihydrolupinic acid	20.2	ND[a]	ND[a]

[a] ND = not detected

in dry lupin seed, and, if present, to determine its metabolic fate and significance during seed germination.

3.3 Lupin Seed as a Site of Cytokinin Biosynthesis

Developing seed has long been recognised as a site of cytokinin accumulation and lupin seed is clearly no exception. The degree to which this accumulation results from synthesis and/or importation from neighbouring tissue is not well understood. Hahn et al. [8] detected increasing cytokinin levels in developing pea seed in vitro using bioassay techniques and suggested this seed is a site of biosynthesis. However other workers contradict this finding [9, 19].

It was therefore relevant to carry out the following experiment with immature *L. luteus* seeds. Seeds from pods collected about 5 days after anthesis were placed on sterile media with [^{3}H]adenosine. After an incubation of 75 h, the seeds were extracted and the extract was subjected to TLC on silica gel (plates developed twice with solvent A). A histogram of the separation of [^{3}H]metabolites is shown as Fig. 3. Apart from adenosine and nucleotide, which together comprise the bulk of extractable radioactivity, there were two radioactive zones which chromatographed in the region of OGZ and of OGZR. Excision and extraction of zone E (see Fig. 3) and sequential chromatography on silica gel (solvent B) and then on cellulose with solvent C yielded at Rf 0.27 (see Fig. 4) a radioactive metabolite which accounted for 0.3% of the total [^{3}H]extracted. This could not be distinguished from OGDZ by TLC in several solvent systems. Treatment of the radioactivity in the OGDZ zone with β-glucosidase [11] yielded a [^{3}H]metabolite which co-chromatographed with dihydrozeatin on silica gel (solvents D and E). This confirms that the seed is capable of cytokinin biosynthesis, at least in this in vitro situation. Other zones from the original TLC have not been as critically examined, but OGZR, if

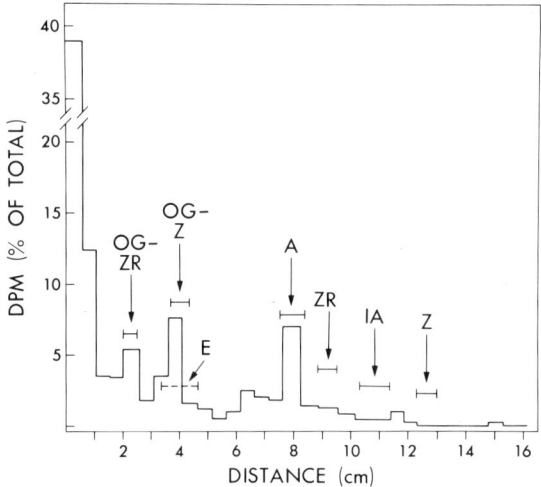

Fig. 3. The distribution of radioactivity over a thinlayer chromatogram (silica gel, solvent A) of the complete extract of immature *Lupinus luteus* seed supplied with [^{3}H]adenosine. The chromatogram was developed twice. The positions of cochromatographed markers are indicated by the *barred lines; Z* zeatin; *IA* N^6-isopentenyl-adenosine; *ZR* zeatin riboside; *A* adenosine; *OGZ* O-glucosylzeatin; *OGZR* O-glucosylzeatin riboside. The zone denoted as E was eluted for furhter TLC

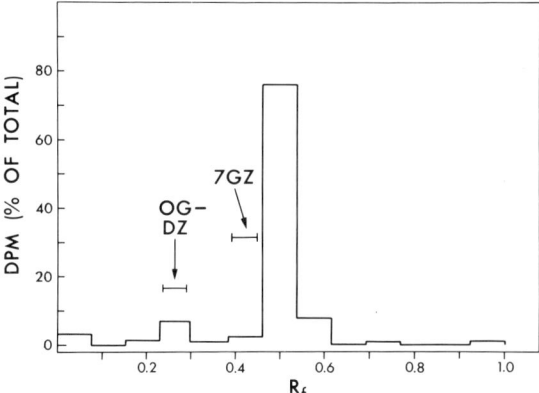

Fig. 4. The distribution of radioactivity over a thin-layer chromatogram (cellulose, solvent C) of a fraction derived from the extract of lupin seed supplied with [^{3}H]-adenosine. The fraction was obtained from zone E in Fig. 3 by TLC on silica gel (solvent B). *OGDZ* and *7GZ* denote the locations of cochromatographed O-glucosyldihydrozeatin and zeatin 7-glucoside respectively

present, is only a minor contributor to the zone of radioactivity which chromatographed with it in Fig. 3.

3.4 Cytokinin Translocation in Lupin Shoots

Zeatin riboside is thought to be the major form in which cytokinin moves from root to shoot in the xylem sap of many species. A study of the translocation of [^{3}H]zeatin riboside in lupin shoots is described in the following experiment. The roots of *Lupinus angustifolius* plants were excised and the stem bases were immersed in a solution of [^{3}H]zeatin riboside for 20 h. About 4 µg of the riboside was taken up by each shoot which weighed about 60 g. Each shoot had one set of partly developed pods and a second inflorescence with flowers and very immature pods at the top of a trained lateral. The shoots were dissected, the various parts extracted, the extracts chromatographed in the presence of suitable marker dyes, and histograms showing the distribution of [^{3}H] on the TLC plates were constructed (Fig. 5, Table 2; note that the TLC zones mentioned in Table 2 are defined in the legend to Fig. 5).

The above experiment is commented on briefly below.

1. A large proportion (64%) of the zeatin riboside taken up remained in the stem, and about one third of this was in the bark. It has not been established whether this was due to transverse movement from the xylem, or to phloem transport from leaves. However the former seems more probable. Direct upward movement through the bark from the zeatin riboside solution could not account for the radioactivity in this tissue.
2. The ^{3}H content of the laminae increases and that of the petioles decreases, along the stem from base to apex.
3. In the parts extracted, exluding stem, the ^{3}H level was least in the seeds and greatest in the developing lateral shoots.

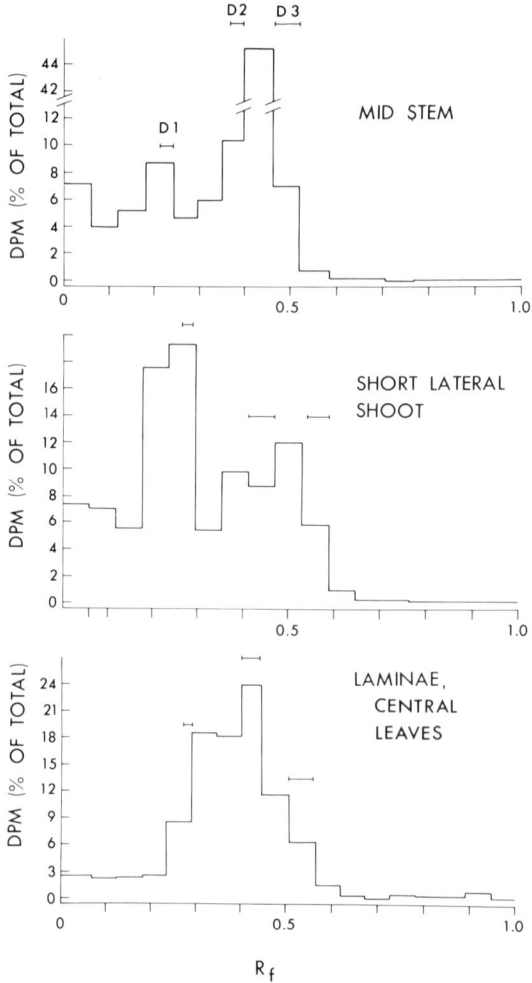

Fig. 5. The distribution of radioactivity over thin-layer chromatograms (silica gel, solvent B) of extracts of selected regions of *Lupinus angustifolius* shoots. The shoots had been supplied with [^{3}H]zeatin riboside via the transpiration stream. The locations of three marker dyes are denoted by barred lines; *D1* Meldola Blue; *D2* Drimarene Brilliant Blue; *D3* Rhodamine B. In Table 2, the region from the origin to the top of D1 is termed the "lupinic acid/nucleotide region"; the zone immediately above D1 is the "OG-Z region" and covers half the distance between D1 and D2; the region of D2 plus the remaining zone immediately below it, is the "adenosine/Z9G region"; the zone between D2 and D3 is the "Z/ZR region"

Table 2. The levels of radioactivity extracted from various regions of de-rooted *Lupinus angustifolius* plants. The shoots had been supplied with [^{3}H]zeatin riboside through the transpiration stream (see Sect. 2). The percentage of total cpm in particular regions of thin-layer chromatograms (silica gel, solvent B) is also listed. These regions are defined relative to three marker dyes in the legend to Fig. 5. Known cytokinin metabolites which chromatograph in these regions are given in the footnote to this table

Region of shoot	Extracted Radioactivity (cpm/g fresh wt x 10^{-3})	% of total cpm in TLC zones[a]			
		Z/ZR region	Adenosine/ Z9G region	OG-Z region	Lupinic acid /nucleotide region
Seeds, first inflorescence	14.0	11.5	24.9	8.6	43.1
Laminae, basal leaves	25.3	11.9	37.1	18.4	24.5
Laminae, central leaves	31.0	11.6	42.2	18.6	18.0
Flowers and small pods of second inflorescence plus associated small leaves	36.7	16.0	26.9	8.5	40.7
Laminae, upper leaves	37.5	11.6	43.4	17.9	20.3
Petioles, upper leaves	53.2	27.3	25.2	10.2	29.3
Pod walls, first inflorescence	57.3	9.6	26.6	7.1	48.4
Petioles, central leaves	76.5	23.0	26.4	9.9	31.1
Petioles, basal leaves	121.8				
Small lateral shoots (on basal stem)	232.8	12.0	18.5	5.4	56.7
Mid stem	273.2	45.2	16.4	4.7	25.2
Small lateral shoots (on mid stem)	288.3	8.4	18.8	5.6	49.4
Upper stem	293.8	45.2	15.1	3.4	28.7
Basal stem	430.0	40.1	15.3	3.9	30.5

[a] Metabolites which chromatograph in these regions are as follows:
Z/ZR: zeatin, zeatin riboside and their dihydro derivatives
Adenosine/Z9G: adenine, adenosine, zeatin 9-glucopyranoside and its dihydro derivative.
OG-Z: OGZ, OGDZ, OGZR, OGDZR, zeatin 7-glucopyranoside
Lupinic acid/nucleotide: Lupinic acid, dihydrolupinic acid, zeatin nucleotide, AMP

4. In actively growing parts, (seeds and pod walls of first inflorescence, second inflorescence, lateral shoots) a high percentage (>40%) of the [^{3}H] was in very polar metabolites, i.e., in compounds which chromatograph in the lupinic acid/nucleotide region during TLC.
5. The laminae of upper, central and basal leaves do not appear to differ in their metabolism of zeatin riboside. However in lower leaves, a much greater proportion of the [^{3}H] is located in the petiole.
6. The percentages of [^{3}H] in the *O*-glucoside and the adenosine/9-glucoside regions are much greater in leaf laminae than in other tissues.

The identities of specific metabolites accumulating in various regions of the shoot are now being examined by further chromatography. When this is completed, it may be possible to associate aspects of shoot development and senescence with specific metabolites. However the above translocation studies are consistent with the view that xylem cytokinins are important in lateral shoot growth. An interesting observation was that cytokinin did not move preferentially to developing seed.

Little is known of the identities of the cytokinins which move in phloem sap (see e.g., [21]). [^{3}H]Zeatin riboside was supplied to the roots of *Lupinus albus* plants and phloem sap was collected from the top of the pods. The distribution of [^{3}H] when this sap was chromatographed on silica gel (solvent B) is presented as Fig. 6. Critical chromatography of the zone of Rf 0.40-0.50 established that the major labelled component was dihydrozeatin which was the principal [^{3}H]-metabolite in the exudate. Hence dihydrozeatin, derived from zeatin riboside taken up by the root, moved through the phloem of the pod wall.

3.5 Cytokinins in Root Nodules

It is generally assumed that the cytokinins in developing root nodules of legumes are derived from the *Rhizobium* bacteria present. However the following experiment suggests movement of cytokinin from the root tip to developing nodules.

Tips were excised from nodulated root systems of pea seedlings growing on sterile agar. The tips were replaced by agar blocks containing [^{3}H]zeatin riboside so as to mimic the natural supply of root-tip-derived cytokinin. Extracts of developing nodules gave higher radioactivity counts than did the adjacent root tissue, suggesting accumulation in the nodules. Radioactivity also moved into mature nodules and sufficient radioactivity was present in this tissue for tentative chromatographic analysis.

TLC studies (silica gel; solvents A, B) indicated that OGZR and lupinic acid or dihydrolupinic acid were the major metabolites present. Hence the root tip in vivo may supply cytokinin to root nodules of pea seedlings.

We have outlined work in one group of plants concerning determination of cytokinin levels in tissues, biosynthesis, translocation and metabolic modification. While considerable progress has been made in the areas of cytokinin quantitation and metabolic modification in recent years, our knowledge of the remaining areas is meagre indeed. However our understanding of metabolic modification may facilitate advances in cytokinin biosynthesis. Precursor incorporation into the stable metabolites now identified may be more readily detected than incorporation into the basic cytokinins, zeatin and zeatin riboside.

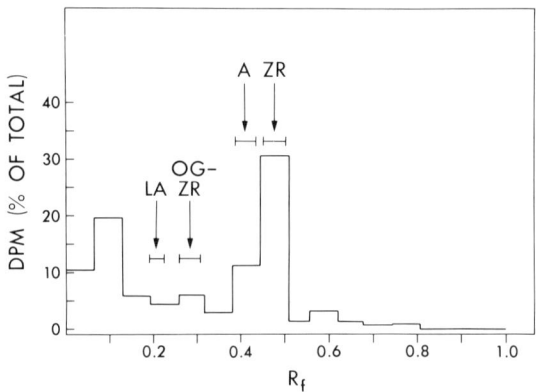

Fig. 6. The distribution of radioactivity over a thin-layer chromatogram (silica gel, solvent B) of phloem sap exuded by the pods of *Lupinus albus* plants supplied with [^{3}H]zeatin riboside through the roots. *Barred lines* denote the position of cochromatographed markers; *ZR* zeatin riboside; *A* adenosine; *OGZR* O-glucosylzeatin riboside; *LA* lupinic acid

References

1. Burrows WJ, Carr DJ (1967) Planta 73: 369
2. Cowley DE, Duke CC, Liepa AJ, MacLeod JK, Letham DS (1975) Aust J Chem 31: 1095
3. Davey JE, Van Staden J (1977) Physiol Plant 39: 221
4. Davey JE, Van Staden J (1978) Physiol Plant 43: 77
5. Davey JE, Van Staden J (1978) Physiol Plant 43: 82
6. Duke CC, Letham DS, Parker CW, MacLeod JK, Summons RE (1979) Phytochemistry 18: 819
7. Duke CC, MacLeod JK, Summons RE, Letham DS, Parker CW (1978) Aust J Chem 31: 1291
8. Hahn H, de Zacks R, Kende H (1974) Naturwissenschaften 61: 170
9. Krechting HCJM, Varga A, Bruinsma J (1978) Z Pflanzenphysiol 87: 91
10. Murashige T, Skoog F (1962) Physiol Plant 15: 473
11. Parker CW, Letham DS, Gollnow BI, Summons RE, Duke CC, MacLeod JK (1978) Planta 142: 239
12. Pate JS, Atkins CA, Hamel K, McNeil DL, Layzell DB (1979) Plant Physiol 63: 1082
13. Pate JS, Sharkey PJ, Lewis OAM (1974) Planta 120: 229; (1975) Planta 122: 11
14. Phillips DA, Torrey JG (1972) Plant Physiol 49: 11
15. Rolfe B, Gresshoff PM (1980) Aust J Biol Sci 33: No 4 (in press)
16. Summons RE, Duke CC, Eickholzer JV, Entsch B, Letham DS (1979) Biomed Mass Spect 6: 407
17. Summons RE, Entsch B, Letham DS, Gollnow BI, MacLeod JK (1980) Planta 147: 422
18. Summons RE, Entsch B, Parker CW, Letham DS (1979) FEBS Lett 107: 21
19. Van Staden J, Button J (1978) Z Pflanzenphysiol 87: 129
20. Van Staden J, Papaphilippou AP (1977) Plant Physiol 60: 649
21. Vonk CR (1979) Physiol Plant 46: 235
22. Wang TL, Thompson AG, Horgan R (1977) Planta 135: 285
23. Whitty CD, Hall RH (1974) Can J Biochem 52: 789

Uptake and Metabolism of Cytokinins in Tobacco Cells: Studies in Relation to the Expression of Their Biological Activities

M. Laloue, C. Pethe-Terrine, and J. Guern[1]

1 Introduction

The mechanisms of action of cytokinins remain to be elucidated. In particular, we do not know how these active substances, as a signal, are recognized by plant cells and how this recognition process initiates the sequence of biochemical events which are involved in the expression of their biological activities in these cells.

In order to get some insight into this recognition process, we think it is necessary to study the quantitative relationships between the intensity of the biological response which results of the application of a cytokinin and the intensity of the signal, that is to say the concentration of active molecules at their site(s) of action. Obviously, this will be attained only when we understand the uptake and metabolic processes which necessarily regulate the internal levels of these active molecules.

The processes involved in the uptake of cytokinins by plant cells have received only little attention. We are aware of only one report in which the absorption process was considered (Doree et al., 1972). These authors showed that in *Acer* cells supplied with kinetin, there is an exchangeable pool of kinetin which is built up rapidly and that kinetin uptake is closely related to its metabolism.

Concerning the nature of the metabolites formed, it is well established now that exogenous cytokinins are metabolized to their respective ribonucleoside and ribonucleoside-5'-phosphates and to various conjugates, the nature of which depends on the nature of the biological system (see the review by Letham, 1978). In addition to this, cytokinins can be degraded by removal of the N^6-substituant which is revealed by the formation of adenine and its derivatives adenosine and adenylic nucleotides.

With this respect, natural cytokinins also occur under different metabolic forms. After the purification and identification of zeatin from immature corn kernels (Letham et al., 1964), zeatin riboside and zeatin riboside-5'-phosphate were soon identified from the same source (Letham, 1968).

More recently, numerous glucosides of zeatin and its derivatives ribosylzeatin, dihydrozeatin and ribosyldihydrozeatin have been shown to be naturally occurring (Peterson and Miller, 1977; Summons et al., 1977, 1979, 1980; relevant articles in this vol). Without doubt, the isolation and characterization of these conjugates were greatly helped by the fact that they had been previously identified as zeatin metabolites.

---

[1] Laboratoire de Génétique et Physiologie du Développement des Plantes, C.N.R.S., 91190 Gif-sur-Yvette, France

Such a diversity of metabolic forms of cytokinins raises two questions. First, is there one or several metabolic forms specifically responsible for their biological activity? Second, what is the significance of these metabolic transformations in term of regulation of the levels of this or these active cytokinin metabolites in plant cells?

With respect to the first question, Hecht et al. (1971, 1975) concluded from the relative activities of cytokinin bases, cytokinin ribosides and some structural analogs that the active form of cytokinins might be the cytokinin base itself. This is indeed possible, but not demonstrated since structure-activity relationships may also reflect differences in uptake or in metabolism of different molecules rather than differences in their intrinsic activities. Consequently, to establish unequivocally that in order to stimulate the division of cytokinin-requiring cells, cytokinin bases either do not need to be metabolically "activated" or need to be transformed into a specific metabolite which should be in this case a metabolite formed from cytokinin ribosides, it is necessary to take into account their respective uptake and metabolism.

Therefore, to ascertain the relative importance of these mechanisms in the expression of the biological activity of cytokinin bases and cytokinin ribosides, we have studied the uptake and the metabolic transformations of benzyladenine, $N^6(\Delta^2$-isopentenyl)adenine and their corresponding ribosides by tobacco cells cultivated in liquid medium. The metabolic transformations of these cytokinins in tobacco cells have been previously described (Laloue et al., 1974, 1975, 1977). They are summarized in Fig. 1 with mention of the possible enzymatic activities involved. (For this aspect, see the article by Chen, this vol.). In this paper, we present some aspects of the permeability of tobacco cells to these cytokinins and their metabolites and discuss their role in the regulation of their respective internal levels in relation with their metabolic transformations.

2 Materials and Methods

2.1 Tobacco Cell Suspensions

Three cell lines of tobacco cells (*Nicotiana tabacum* Wisconsin 38) were used in these studies. Cell line no. 19, which is cytokinin-dependent, and cell line no. 13, which is cytokinin-autonomous, were originally cloned by Tandeau de Marsac and Jouanneau (1972). Strain no. 19-3 which is also cytokinin-requiring was derived from cell line no. 19 (Laloue, 1980). Cells of these strains are grown as cell suspensions as previously described (Laloue et al., 1975, 1977).

Cultures are initiated at a cell population density of 20 to 25,000 cells per ml of medium. In the case of cell strain no. 19-3, maximal cell population density at stationary phase of growth is about 235 000 ± 15 000 cells per ml and the fresh weight of 10^5 cells is 38 mg ± 2 mg. This value, over-estimated by about 10% due to contamination by residual culture medium in cell aggregates, is not corrected.

Growth stimulation of suspensions of cytokinin-requiring tobacco cells (cell line 19-3) was assayed as described previously (Gregorini and Laloue, 1980).

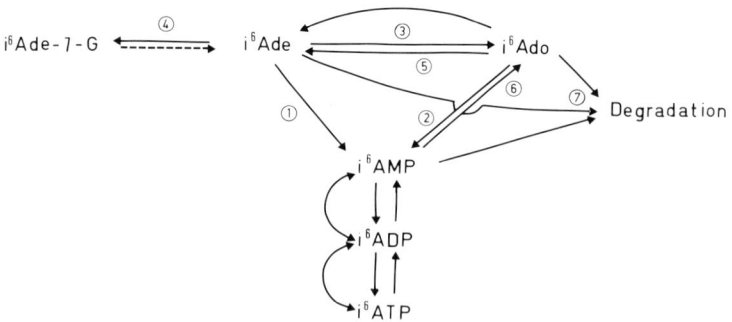

Fig. 1. Metabolic fate of cytokinins in tobacco cells. This scheme which summarizes the metabolic transformations of i^6 Ade and i^6 Ado applies also to bz^6 Ade and bz^6 Ado. $i^6 AMP$, formed either from i^6 Ade (Laloue et al., 1975) by an adenine phosphoribosyltransferase-type enzyme (*1*) (Pethe-Sadorge et al., 1972), or from i^6 ado (Laloue et al. (1974) by an adenosine-kinase-type enzyme (*2*) (Doree and Terrine, 1973). Alternatively it can be formed from i^6 Ade via the formation of i^6 Ado (*3*) see below. $i^6 ADP$ and $i^6 ATP$ are also formed (Laloue et al., 1974). Enzymes involved unknown i^6 *Ade-7-glucoside* is formed from i^6 Ade (Laloue et al., 1975). Cytokinin 7-glucosyltransferase (*4*) (Entsch et al., 1979). i^6 *Ade,* formed in small amount when i^6 ado is supplied (Laloue et al., 1975). (*5*) cytokinin ribonucleoside hydrolase (Pacès, 1976) or nucleoside ribohydrolase (Whitty and Hall, 1974). i^6 Ado, formed when i^6 Ade is supplied (Laloue et al., 1975) possibly by an adenosine phosphorylase (*3*) from i^6 Ade (Chen and Petschow, 1978) and by a 5'-nucleotidase (*6*). Degradation products are adenine, adenosine and adenylic nucleotides (Laloue et al., 1975, 1977). The metabolic level (base, riboside or ribonucleotides, at which side-chain removal occurs is not known. Cytokinin oxidase (*7*) (Whitty and Hall, 1974)

2.2 Cytokinin Uptake and Metabolism

Synthesis of [^{14}C]-labeled cytokinins has been described (Laloue et al., 1974, 1975). N^6(para-^{3}H-benzyl)adenine with a specific activity of 26 Ci xmmol^{-1} was prepared in collaboration with J.P. Jouanneau (Marseille, France) according to Sussman and Firn (1976). Tritiation of N^6-(parabromobenzyl)adenine was performed by CEA (France).

Uptake was measured by the radioactivity absorbed by the cells separated from the culture medium by Millipore filtration. It is essential to note that *cells were not washed*. Cytokinin metabolites in cells were analyzed after cold perchloric acid extraction as described previously (Laloue et al., 1974, 1975, 1977).

When [^{14}C]-labeled cytokinins were used, analysis of cytokinin metabolites in culture media was performed on 10 ml aliquots of media from which cells were removed by centrifugation (5 min, 700 g). The media were extracted three times with 10 ml of water saturated *n*-butanol. The butanolic fractions were pooled, and reduced to amost dryness at 25°C under reduced pressure. The residue was dissolved in 1 to 2 ml of 20% ethanol for Sephadex LH-20 column chromatography as described (Laloue et al., 1974, 1975). When ^{3}H-bz^6 Ade was used, analysis was performed on 3 ml aliquots of media, the *n*-butanol extraction step being omitted.

3 Results

3.1 Cellular Accumulation of Exogenous Cytokinins

The main feature of the uptake and the metabolism of an exogenous cytokinin by tobacco cells is that these processes result in its cellular accumulation, as shown in Table 1. After 48 h of incubation, 88% of the supplied benzyladenine have been absorbed by the cells. This corresponds to an overall accumulation of 470 times since the volume of the cells estimated from their fresh weight, on the basis of a density equal to 1, represents only 1.5% of the volume of the suspension.

Obviously, bz^6Ado-5'-phosphates and bz^6Ado-7-glucoside are the major metabolites accumulated while they are barely present in the medium. This suggests that cells are impermeable to these metabolites. This conclusion is further supported by the observation that they cannot be washed out of the cells by replacing the culture medium by a new medium without any cytokinin. The alternative hypothesis that such concentration gradients of these metabolites between cells and the medium could be maintained by active transport processes, the rate of the influx being much greater than the rate of the efflux, can be eliminated on the basis that bz^6Ade-7-glucoside is only very slowly absorbed by the cells (Table 2). In this respect the fact that the bz^6Ade-7-glucoside, which is metabolically very stable (Gawer et al., 1977), is present in small amounts in the washing medium is particularly important because it indicates that cells were not damaged during the rinsing and washing procedure.

On the contrary, benzyladenine itself does not accumulate to a great extent in the cells and this seems to be due to the fact that bz^6Ade leaks out of the cells as shown by the results of the washing experiment. Incidently, the amount of bz^6Ade released from the cells in 6 h represents 18 times the amount of bz^6Ade initially present in the cells. In fact, this bz^6Ade originates from the bz^6Ado-5'-phosphates pool(s). This result is interesting because it demonstrates a metabolic interconversion between these metabolites and the base, via the riboside very likely (cf. Fig. 1).

The same phenomenon is observed when the cytokinin supplied is bz^6Ado (cf. Table 3). In this case, about 90% of the bz^6Ade formed is released in the culture medium.

3.2 Cell Permeability to Lipophilic Cytokinins

The observations that bz^6Ade does not accumulate as such in cells when supplied in the medium and that it leaks out in the culture medium when formed from bz^6Ado in cells led us to investigate the mechanisms of its uptake. This study was performed with cells from cultures at stationary phase of growth for the reason that their metabolically driven accumulation of cytokinins is less intense than that of cells from exponentially growing cultures.

3.2.1 Kinetics of bz^6Ade Uptake

As shown in Fig. 2 A, the uptake of bz^6Ade by tobacco cells as a function of time presents two different phases. First, during an initial phase which lasts less than 3 min after the addition of the cytokinin, a significant amount of bz^6Ade rapidly enters the

Table 1. Accumulation of benzyladenosine-5'-phosphates and benzyladenine-7-glucoside in tobacco cells and "washing-induced" specific leakage of benzyladenine. Tobacco cells (strain 19-3) were cultivated aseptically in presence of N^6 (para-^{3}-H-benzyl)adenine 0.057 µM (200 ml culture, initial cell population density: 25,000 cells x ml^{-1}).

After 48 h of culture, aliquots of the cells and of the medium were analyzed. Then cells were washed as follows:
1 – Cells were separated from the medium by filtration on a nylon net (mesh 30 µm). The medium was discarded.
2 – Cells were rinsed in situ on the nylon net with 100 ml of fresh medium without cytokinin (rinsing medium).
3 – Cells were resuspended in 300 ml of fresh culture medium by cuting the nylon net.

Aliquots of the cells and the washing medium were analyzed after 1 h and 6 h of incubation respectively. Results are expressed in pmol

Cytokinin and cytokinin metabolites	Before washing			Rinsing medium	1 h			After washing 6 h			
	Cells 1.5 g	Medium 97 ml	$\frac{C_i^a}{C_e}$	100 ml	Cells 1.5 g	Medium 300 ml	C_i/C_e	Cells 1.5 g	Medium 300 ml	$\frac{C_i}{C_e}$
bz^6 Ado-5'-phosphates	3,220	21	9,900	n.d.^{c}	3,080	25	24,600	2,500	25	20,000
bz^6 Ade-7-glucoside	1,250	37	2,200	0.5	1,220	35	7,000	1,230	30	8,200
bz^6 Ado	70	74	60	n.d.	50	50	200	90	110	160
bz^6 Ade	26	550	3	9	20	97	41	27	470	11
Unidentified^b	480	–								

$^a \frac{C_i}{C_e} = \frac{Q_i}{Q_e}$, Qi = picomoles x g^{-1} of cells, Qe = picomol x ml^{-1} of medium.

^{b} Likely derivatives of the benzyl substituent

^{c} Non-detected

Table 2. Rate of uptake of bz^6 Ade-7-glucoside by tobacco cells (cell line no. 19), as compared to that of bz^6 Ado. Experiments 1 and 2 are from Laloue (1977), experiment 3 is from Gawer et al. (1977)

bz^6 Ade Metabolites	Concentration (μM)	Incubation time (h)	Cell population density (cells/ml)	Amount absorbed (pmol x 10^5 cells^{-1})	Rate of uptake pmol x h^{-1} x 10^5 cells^{-1}
bz^6 Ade-7-glucoside					
Exp. 1	0.6	24	40,000	14.3[a]	0.6
Exp. 2	0.57	24	12,500	30[a]	1.25
Exp. 3	0.5	48	400,000	41[a]	0.85
bz^6 Ado	0.5	4	100,000	160[b]	40

[a] bz^6 Ade-7-glucoside absorbed is not significantly metabolized [b] Essentially converted to bz^6 Ado-5'-phosphates

Table 3. Leakage of bz^6 Ade from cells supplied with bz^6 Ado-8-^{14}C. Bz^6 Ado-8-^{14}C was supplied at the final concentration of 1 μM to a suspension of tobacco cells (cell line 19-3) from a 6 days old culture in exponential phase of growth. This suspension was diluted 2.3 times with conditioned culture medium in order to reduce the cell population density to 40,000 cells x ml^{-1} (16 mg fresh weight per ml) in order to increase the ratio of the volume of the medium to the volume of the cells. Results are expressed in pmol per ml of cell suspension

bz^6 Ado and bz^6 Ado metabolites[a]	Time of incubation			
	1 h 30	3 h	5 h	7 h
Medium				
bz^6 Ado	873	741	645	584
bz^6 Ade	4.8	12.3	27.5	56.7
Cells				
bz^6 Ado-5'-phosphates	113	235	320	334
bz^6 Ado	4.7	5.9	9	14
bz^6 Ade	0.5	1.9	3	7.4

[a] bz^6 Ade-7-glucoside is only detectable in small amounts in cells after 5 h of incubation.

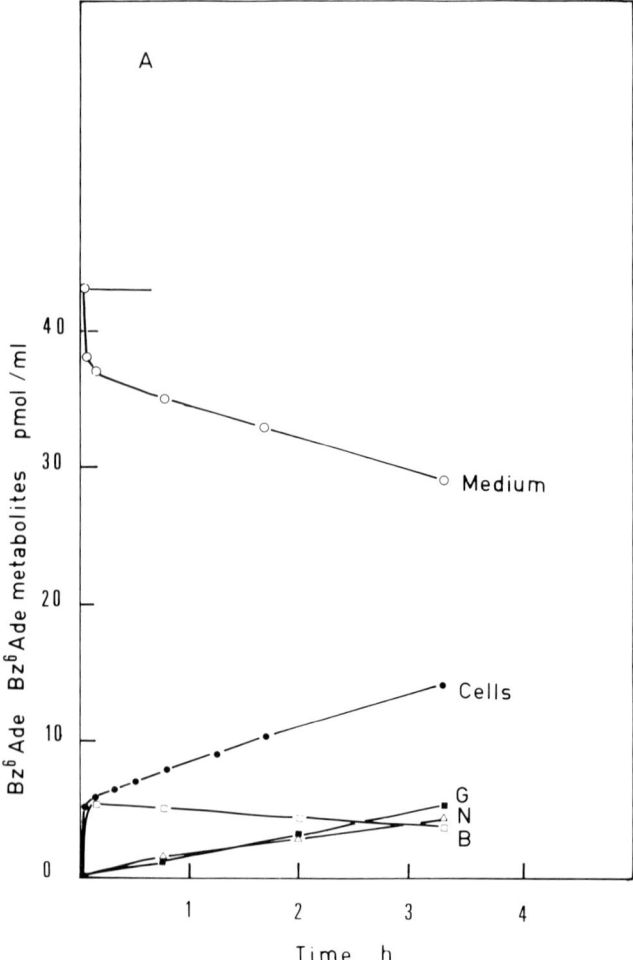

Fig. 2A. Short-term kinetics of bz^6 Ade uptake and metabolism by tobacco cells. Bz^6 Ade-8-^{14}C was supplied at a concentration of 0.043 µM to a 100 ml cell suspension (cell line 19-3) from a culture at stationary phase of growth (250,000 cells × ml^{-1}). Results are expressed in picomoles per ml of suspension (*medium*) or per 100 mg of cells (*cells*)

cells without being metabolized and without being accumulated since its "concentration" in the cells is only 1.2 times the concentration at which it remains in the medium. During this initial phase, the average rate of uptake is about 40 pmol/h/10^5 cells if we assume it is constant during these three first minutes. Then bz^6 Ade is absorbed at a much slower rate than initially. This rate is constant (2.5 pmol/h/10^5 cells) and corresponds to the rate at which it is metabolized, and hence accumulated in the cells.

During this second phase, however, the size of the pool of bz^6 Ade remains subconstant, decreasing slowly in relation with the decrease of its concentration in the medium due to its metabolic conversion and accumulation in the cells.

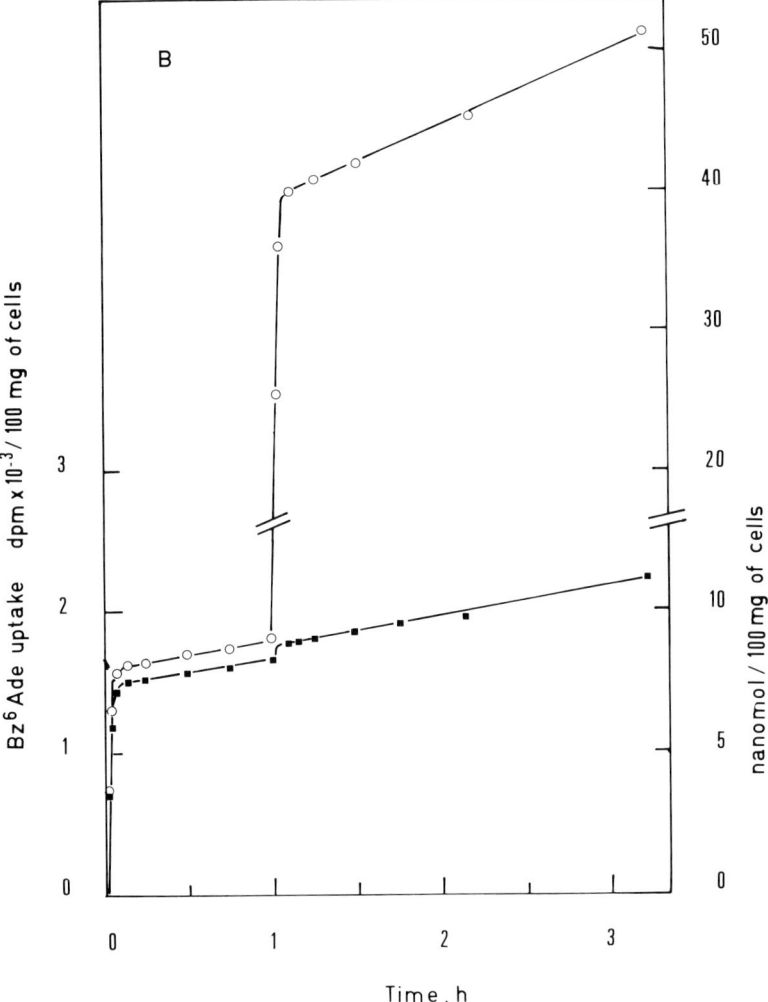

Fig. 2B. Bz^{6}Ade-8-^{14}C with a specifix activity of 2 mCi x mmol^{-1} was supplied at a concentration of 10 μM to 50 ml of a tobacco cell suspension (same nature as in **A**). After 1 h of incubation, 50 nmol of bz^{6}Ade-8-[^{14}C] (with a specific activity of 50 m Ci x mmol^{-1}) were added to the remaining 35 ml of cell suspension. This increased the specific acticity of the external bz^{6}Ade to 8 mCi x mmol^{-1} without increasing significantly its concentration which in fact increased by 12%. Results are expressed in dpm per 100 mg of cells, converted to pmol per 100 mg of cells

That such a relation between the level of cellular bz^{6}Ade and its concentration in the medium exists is demonstrated by the isotopic equilibrium experiment described in Fig. 2B. Indeed, the rapid increase of radioactivity in the cells shows the presence of an inward flux of bz^{6}Ade which must be equilibrated by an outward flux of the same intensity since the size of the cellular pool of bz^{6}Ade remains constant.

3.2.2 Influence of bz^6 Ade Concentration on the Size of its Cellular Pool

As shown by the results of the experiment described in Table 4, the size of the cellular pool of bz^6Ade measured after 3 h of incubation is linearly proportionnal to the concentration at which it remains present in the incubation medium. Furthermore, the ratio of its concentration in the cells to its concentration in the medium indicates that it does not accumulate in the cells as compared to the metabolites which are formed.

Table 4. Influence of the concentration of supplied bz^6Ade on the size of its cellular pool. Bz^6-Ade-8-^{14}C was added at the concentrations indicated (final concentration in the suspensions) to 50 ml aliquots of a tobacco cell suspension pooled from three 100 ml cultures at stationary phase of growth (250,000 cells x ml^{-1} or 100 mg of cells, fresh weight). Remaining amounts of bz^6Ade in the medium (pmol per ml of suspension) and sizes of cellular pools of bz^6Ade (pmol per 100 mg of cells) were measured after 3 h of incubation

	Benzyladenine, μM				
Benzyladenine	0.043	0.1	1	10	40
Cellular pool after 3 h of incubation	3.8	9	106	1,010	4,000
Remaining amount in the medium of 1 ml of suspension	29	70	740	8.250	33,700
$\dfrac{Ci}{Ce}$	1.18	1.15	1.29	1.10	1.06

3.2.3 Diffusion: a Possible Uptake Mechanism of Lipophilic Cytokinins by Plant Cells

The data presented above show that bz^6Ade is readily exchanged in and out of tobacco cells. This property is well explained on the basis of simple diffusion. Indeed, this behavior appears to be specific for lipophilic cytokinin bases only. Furthermore, the fact that in relatively short-term experiments bz^6Ade is present in the cells at a concentration which is only slightly higher than the concentration at which it is present in the culture medium can be considered as evidence that it is distributed throughout the cells, vacuoles included.

3.3 Dynamics of Cytokinin Metabolites Pools

3.3.1 Benzyladenine and Its Metabolites

We have studied the evolution of the cellular levels of bz^6Ade and its metabolites in cytokinin-requiring tobacco cells during a culture cycle. Cells were subcultured in presence of tritiated bz^6Ade at the usual concentration of 0.057 μM, and at an initial cell population density of 25,000 cells (about 10 mg fresh weight of cells) per ml of suspension.

After one day of culture, there are 4.8 pmol of bz^6Ade per 10^5 cells; at two days and thereafter, its level is approximatively constant at 0.7 pmol per 10^5 cells. Hence, the level of bz^6Ade in the cells which is cold perchloric acid-extractible is maximal

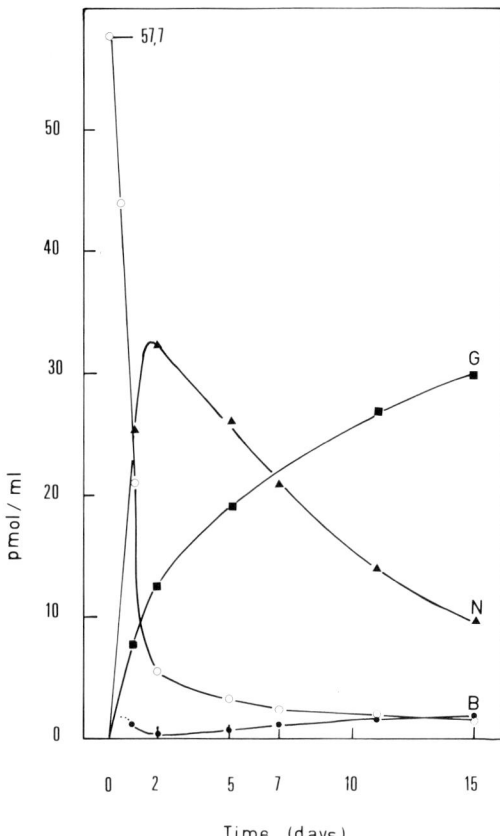

Fig. 3. Long-term kinetics of bz^6 Ade uptake and metabolism by tobacco cells. Cytokinin-requiring tobacco cells (line 19-3) were cultivated in presence of bz^6-Ade-^{3}H (same experiment as in Table 1). Disparition of bz^6 Ade from the culture is represented by B (o—o). With respect to the cells, only bz^6 Ado-5'-phosphates (N), bz^6 Ade-7-glucoside (G) and bz^6 Ade (B •—•) are represented. After 15 days, there are 9.3 pmol of unidentified metabolites per ml resulting from presumed degradation by removal of the benzyl group

within the first 24 h of incubation in relation to its high level in the medium, in agreement with what is observed on a short-term basis (cf. Fig. 2). However, on a longer-term basis and already after 48 h, the cellular level of bz^6 Ade is higher than the level which can be predicted from a simple diffusion equilibrium. Hence, bz^6 Ade appears to be slightly "accumulated" in cells, relatively to the culture medium.

The value of the concentration factor represented by the ratio Ci/Ce is 3 after two days (cf. Table 1), 5.5 after five days and about 10 after fifteen days. The significance of this cellular "accumulation" of bz^6 Ade is not clear. In this respect, however, the results of the "washing" experiment described in Table 1 indicate that about 80% of the cold perchloric acid extractible bz^6 Ade molecules present in the cells after two days of culture are in fact not exchangeable with the bz^6 Ade present in the culture medium, since they remain associated with the cells upon washing. At the present time, we interpret this phenomenon by considering that these nonexchangeable bz^6 Ade molecules are irreversibly bound to adsorption sites in the cells.

Nevertheless, it must be emphasized that the cellular level of bz^6 Ade is always low as compared to the level of bz^6 Ado-5'-phosphates. The same observation applies to the level of bz^6 Ado which represents in general only 1 to 2% of the level of its corresponding 5'-phosphates.

These, and the 7-glucoside, are the major metabolites of bz^{6}Ade which account for its accumulation in tobacco cells, as shown already in Table 1. But, as illustrated in Fig. 3, bz^{6}Ade-7-glucoside accumulates continuously while bz^{6}Ado-5'-phosphates accumulate only transiently. Indeed, the level of the latter increases until about 90% of the supplied bz^{6}Ade are absorbed. Then their accumulation stops and finally their internal level decreases, while the level of bz^{6}Ade-7-glucoside increases correlatively.

Although we have no direct evidence to support it, this is probably due to the fact that at a certain level of accumulation of the nucleotides, their formation becomes equilibrated by their conversion to bz^{6}Ado and to bz^{6}Ade. Then this equilibrium is constantly displaced by the formation of the bz^{6}Ade-7-glucoside which is essentially irreversible.

As a result, bz^{6}Ado-5'-phosphates appear relatively stable as shown by their metabolic half-life which is about 8 days.

3.3.2 i^{6}Ado and Its Metabolites

As shown in Fig. 4, i^{6}Ado is also accumulated in tobacco cells. This corresponds essentially to the formation of i^{6}Ado-5'-phosphates, since very little i^{6}Ade is formed and as a consequence only a small amount of i^{6}Ade-7-glucoside is formed (Table 5). But at the difference of bz^{6}Ade, i^{6}Ado is rapidly degraded by removal of the Δ^2-isopentenyl side chain, probably at the nucleotide level. As a result, the cellular level of i^{6}Ado-5'-phosphates decreases rapidly if i^{6}Ado is removed from the incubation medium (see Laloue et al., 1977). In the experiment which is described, the half-life of the pool of i^{6}Ado-5'-phosphates formed from the exogenous i^{6}Ado is only 2 h, as compared to 8 days in the case of bz^{6}Ado-5'-phosphates.

Table 5. Metabolites of i^{6}Ado-8-^{14}C formed and accumulated by cytokinin-requiring tobacco cells. From experiment described in Fig. 4. Cells were extracted by cold perchloric acid after 3 h of incubation. Results are expressed in pmol per 10^5 cells

Uptake: 328 pmol		Overall cellular accumulation Ci/Ce = 18	
i^{6}Ado and i^{6}Ado metabolites accumulated			
i^{6}Ado, i^{6}Ado metabolites		i^{6}Ado catabolites	
i^{6}ATP + i^{6}ADP	135	AMP, ADP and ATP	53
i^{6}AMP	112	Ado	16
io^{6}Ado-5'-phosphates	2	Ade	5
i^{6}Ade-7-G	< 1		
i^{6}Ade	1		
i^{6}Ado	2		

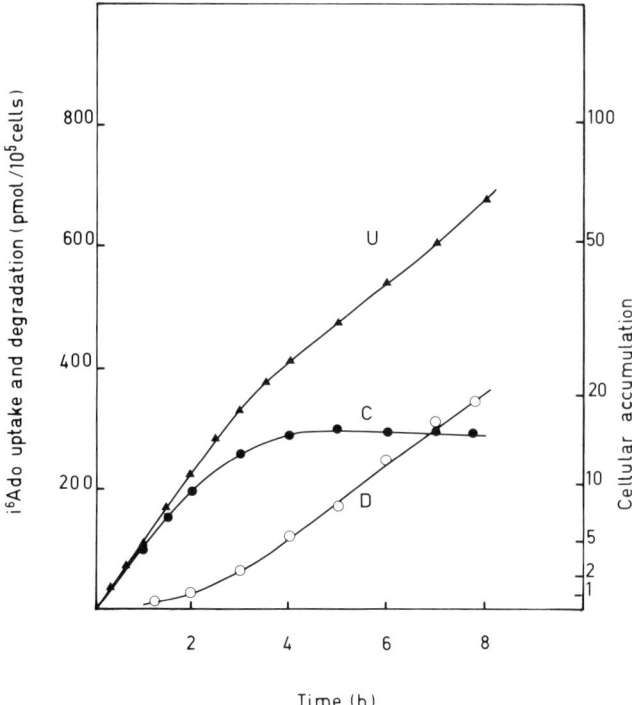

Fig. 4. Time course of $N^6(\Delta^2$-isopentenyl)adenosine accumulation (U) and degradation (D) in cytokinin-requiring tobacco cells ($8[^{14}C]$) i^6 Ado was supplied at a concentration of 1 μM to a tobacco cell suspension (cell line no. 19, 0.85 x 10^5 cells/ml from a 5-days-old culture). Results are expressed in pmol/10^5 cells. Ci/Ce represents the overall accumulation of i^6 Ado (*curve U*). *Curve C* corresponds to i^6 Ado metabolites which retained the Δ^2-isopentenyl chain (i^6 Ade, i^6 Ade-7-glucoside, i^6 AMP, i^6 ADP and i^6 ATP). See Terrine and Laloue (1980) for analytical details

4 Discussion

4.1 Factors which Regulate Cellular Levels of Cytokinin and Cytokinin Metabolites

The data presented point out that cellular accumulation of lipophilic cytokinin bases bz^6 Ade and i^6 Ade and their respective ribosides supplied to tobacco cells results in their uptake and subsequent transformation to cytokinin-riboside-5'-phosphates and cytokinin-7-glucosides to which cells appear to be impermeable. Hence, cell permeability to cytokinins and to their metabolic derivatives appears to play a crucial role in the control of their cellular levels, in relation with their metabolic transformation as illustrated in the model presented in Fig. 5.

4.1.1 Cell Permeability

This factor may be considered in two ways.

First, cells are essentially not permeable to cytokinin 7-glucosides and cytokinin riboside-5'-phosphates, as demonstrated by the fact that they remain accumulated in cells against high concentration gradients and they are not, as such, washable out of cells.

However, bz^6-Ade-7-glucoside is absorbed by tobacco cells, but at a very slow rate rate as compared to rates at which cytokinin ribosides are absorbed.

Second, cells are obviously very permeable to cytokinin bases and ribosides. Permeation of bz^6Ade and i^6Ade is simply interpreted on the basis of their diffusion through cell membranes and consequently, their internal levels are essentially determined by the levels at which they are present in the culture medium. Mechanisms involved in the permeation of cytokinin ribosides are definitely different, but remain to be elucidated.

4.1.2 Metabolic Transformations

Basically, it is possible to distinguish two types of metabolic pathways: interconversion pathways and inactivation pathways which are degradation and 7-glucosylation.

Interconversion pathways are responsible for the formation of the base, the riboside and the riboside-5'-phosphates (cf. Fig 1) and their main feature is that their overall equilibrium is very much in favor of the formation of the nucleotides, as shown by the fact that the level of cytokinin riboside represents always only a few percent of the level of the corresponding riboside-5'-phosphates. Whether this feature is general in plant cells or specific to the strains of the tobacco cells we are using is not known.

We can only observe that most of the isolations and identifications of free cytokinins in plants or plant tissues have been made at the base or nucleoside levels, while in most metabolic studies, cytokinin riboside-5'-monophosphate is a major metabolite. For example, ribosylzeatin has been shown to be the major cytokinin present in *Vinca rosea* crown-gall tissue (Miller, 1974), but Stuchbury et al. (1979) recently reported that labeled adenine is incorporated into zeatin, zeatin robiside and zeatin riboside-5'-phosphate(s) in the ratio of 1:2:20 respectively (see also the article by Horgan, this vol.). Therefore, it is our feeling that in general, not enough attention has been paid to cytokinin ribotides as naturally occurring cytokinins.

Degradation is particularly intense in the case of isopentenylated cytokinins, as shown by the half-life of i^6Ado-5'-phosphates in cytokinin-requiring tobacco cells, which is only a few hours. In this respect, degradation of exogenously supplied i^6Ado is also very rapid in cytokinin-autonomous cells and appears to be under the control of cytokinins themselves (results not shown here, see Terrine and Laloue, 1980). These observations are of particular interest because they imply in these tissues higher rates of cytokinin biosynthesis than those which may be assumed from the observed levels of endogenous cytokinins.

In contrast, bz^6Ade is not degraded to a great extent and this explains its accumulation as bz^6Ade-7-glucoside in long-term experiments we consider now as inactivation. First, we have previously shown that 7-glucosylation is not a prerequisite of the expression of the biological activity of cytokinins on the basis that (1) bz^6-Ade-7-glucoside itself is not active per se, (2) bz^6Ado and i^6Ado stimulate division of cytokinin-requiring tobacco cells without being significantly glucosylated, especially in the case of i^6Ado (Laloue, 1977). Second, formation of cytokinin-7-glucoside in tobacco cells is practically irreversible in the sense that its rate of reutilization is extremely low, in the range of 0.02 to 0.03 picomol/h x 10^5 cells for an internal level of 70 pmol/10^5 cells (Gawer et al., 1977).

4.2 Metabolic Activation of Cytokinins

If it is clear that 7-glucosylation of cytokinin bases does not correspond to a metabolic activation, as discussed above, it remains that after 24 h to 36 h, division of cytokinin-requiring tobacco cells can be correlated in general with high levels of cytokinin ribotides and low levels of cytokinin base and riboside, whichever form, base or riboside, is supplied. Therefore, it is not surprising to observe that bz^6Ado and i^6Ado are as active as bz^6Ade and i^6Ade, respectively, to stimulate the growth of suspensions of cytokinin-requiring tobacco cells as shown in Fig. 5.

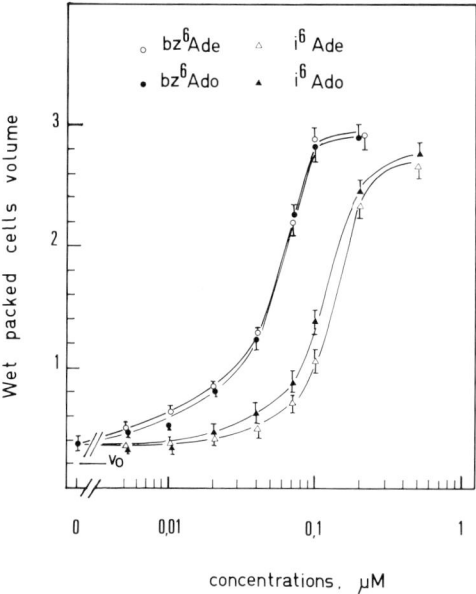

Fig. 5. Yields, after 9 days, of 10 ml suspension cultures of cytokinin-requiring tobacco cells (line 19-3) grown in presence of serial concentrations of bz^6Ade, i^6Ade and their respective ribosides. Each value is the mean (± SE) of four cultures

To evaluate the role of the riboside moiety in the expression of the biological activtiy of cytokinins, it is therefore necessary to compare the biological activities of cytokinin bases and cytokinin ribosides under experimental conditions which limit their metabolic utilization in order to obtain the greatest possible difference in the internal levels of cytokinin base between the situation where the cytokinin supplied is the base and the situation where the cytokinin supplied is the riboside, as predicted from the model of regulation of their internal levels presented in Fig. 6.

Indeed, if metabolic utilization of supplied cytokinins is limited by using very low initial cell population densities, in the situation where the base is supplied its cellular level should remain at its highest level possible in relation to the high level which is maintained in the medium, while in the situation where the riboside is supplied, the cellular level of base should remain as low as possible, since most of the base formed would diffuse out of the cells in the medium.

We are presently developing such experimental conditions with low cell densities in order to verify this model.

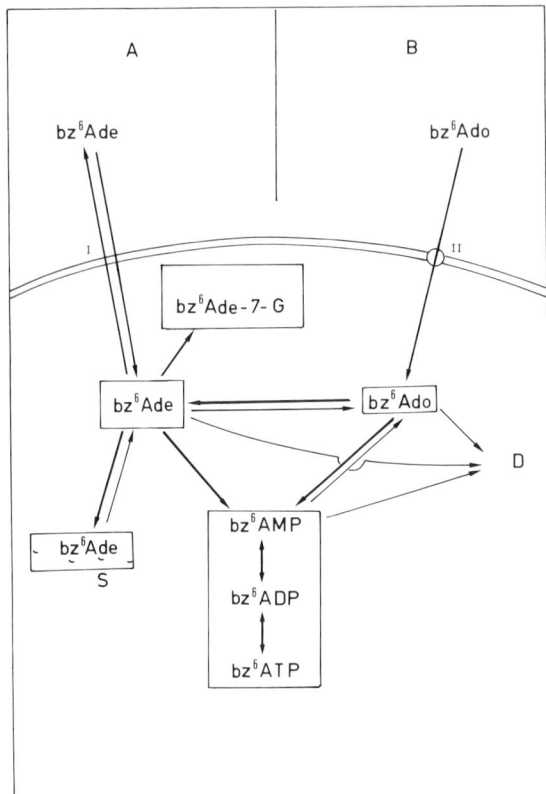

Fig. 6 A, B. Schematic model accounting for the relationships between uptake and metabolism of lipophilic cytokinin bases (A) and their ribosides (B) by tobacco cells. *I* Diffusion of the cytokinin base through the plasmalemma, *II* Uptake of the cytokinin riboside by a presumed transport system which remains to be elucidated. For the different metabolic pathways see legend of Fig. 1. *S* represents adsorption sites.

Nevertheless, independently of the problem of the active form of cytokinins, these results point out a central role of cytokinin riboside-5'-phosphates in the regulation of internal levels of the various metabolic forms of cytokinins. Such a central role is further supported by the recent report by Chen and Melitz (1979) that cytokinin biosynthesis occurs at the monophosphate level, i^6AMP being synthesized by direct isopentenylation of AMP.

References

Chen CM, Melitz DK (1979) Cytokinin biosynthesis in a cell-free system from cytokinin autotrophic tobacco tissue cultures. FEBS Lett 107: 15-20

Chen CM, Petschow B (1978) Metabolism of cytokinin. Ribosylation of cytokinin bases by adenosine phosphorylase from wheat germ. Plant Physiol 62: 871-874

Dorée, M., Terrine C (1973) Enzymatic synthesis of ribonucleoside-5'-monophosphates from N^6-substituted adenosines. Phytochemistry 12: 1017-1023

Dorée M, Terrine C, Guern J (1972) Plant cell permeability to kinetin. In: Kaldewey Vardar H (ed) Modern aspects of hormonal regulation of plant growth and development. Proc Adv Study Inst Izmir, 1971. Chemie, Weinheim, p 221

Entsch B, Parker CW, Letham DS, Summons RE (1979) Preparation and characterization, using high-performance liquid chromatography, of an enzyme forming glucosides of cytokinins. Biochim Biophys Acta 570: 124-139

Gawer M, Laloue M, Terrine C, Guern J (1977) Metabolism and biological significance of benzyl-adenine-7-glucoside. Plant Sci Lett 8: 267-274

Gregorini G, Laloue M (1980) Biological effects of cytokinin antagonists 7-(pentylamino) and 7-(benzylamino)-β-methylpyrazolo(4, 3-d)pyrimidines on suspension cultured tobacco cells. Plant Physiol 65: 1090-1095

Hecht SM, Bock RM, Schmitz RY, Skoog F, Leonard NJ, Occolowitz J (1971) Question of the ribosyl moiety in the promotion of callus growth by exogenously added cytokinins. Biochemistry 10: 4224-4228

Hecht SM, Frye RB, Werner D, Hawrelak SD, Skoog F, Schmitz RY (1975) On the "activation" of cytokinins. J Biol Chem 250: 7343-7351

Laloue M (1977) Cytokinins: 7-glucosylation is not a prerequisite of the expression of their biological activity. Planta 134: 273-275

Laloue M (1980) Etude de l'importance fonctionnelle des processus d'absorption et des transformations métaboliques des cytokinines dans l'expression de leur activité biologique chez les cellules de *Nicotiana tabacum*. Thèse de Doctorat ès Sciences, Paris, CNRS (in press)

Laloue M, Terrine C, Gawer M (1974) Cytokinins: formation of the nucleoside-5-'-triphosphate in tobacco and *Acer* cells. FEBS Lett 46: 45-50

Laloue M, Gawer M, Terrine C (1975) Modalités de l'utilisation des cytokinines exogènes par les cellules de tabac cultivées en milieu liquide agité. Physiol Veg 13: 781-796

Laloue M, Terrine C, Guern J (1977) Cytokinins: metabolism and biological activity of N^6 (Δ^2 isopentenyl)adenosine and N^6-(Δ^2 isopentenyl)adenine in tobacco cells and callus. Plant Physiol 59: 478-483

Letham DS (1968) A new cytokinin bioassay and the naturally occuring cytokinin complex. In: Wightman, F., Setterfield, G. (eds) Biochemistry and Physiology of Plant Growth Substances. Runge Press, Ottawa, p 19

Letham DS (1973) Regulators of cell division in plant tissues. XV. Cytokinins from *Zea mays*. Phytochemistry 12: 2445-2455

Letham DS (1978) Cytokinins. In:Letham DS, Goodwin PB, Higgins TJV (eds) Phytohormones and related compounds: a comprehensive treatise, vol I. Elsevier North-Holland, Amsterdam Oxford New York, p 205

Letham DS, Shannon JS, McDonald TR (1964) The structure of zeatin, a kinetin-like factor inducing cell division. Proc Chem Soc London, p 30

Miller CO (1974) Ribosyl-*trans*-zeatin, a major cytokinin produced by crown-gall tumor tissue. Proc Natl Acad Sci USA 71: 334-338

Pacès, V (1976) Metabolism of cytokinins in barley leaves. Biochem Biophys Res Commun 49: 460-466

Peterson JB, Miller CO (1977) Glucosyl zeatin and glucosyl ribosyl-zeatin from *Vinca rosea* L. grown-gall tumor tissue. Plant Physiol 59: 1026-1028

Pethe-Sadorge P, Signor Y, Guern J (1972) Sur la synthèse des nucléosides-5'-monophosphates de cytokinines par les cellules d'*Acer pseudoplatanus*. CR Acad Sci Paris 275: 2493-2496

Stuchbury T, Palni LM, Horgan R, Wareing PF (1979) The biosynthesis of cytokinins in crown-gall tissue of *Vinca rosea*. Planta 147: 97-102

Summons RE, MacLeod JK, Parker CW, Letham DS (1977) The occurence of raphanatin as an endogenous cytokinin in radish seed. Identification and quantitation by gas chromatographic-mass spectrometric analysis using deuterium-labelled standards. FEBS Lett 82: 211-214

Summons RE, Entsch B, Parker CW, Letham DS (1979) Mass spectrometric analysis of cytokinins in plant tissues. III. Quantitation of the cytokinin glycoside complex of lupin pods by stable isotope dilution. FEBS Lett 107: 21-25

Summons RE, Entsch B, Letham DS, Gollnow BI, MacLeod JK (1980) Regulators of cell division in plant tissues. XXVIII. Metabolites of zeatin in sweet-corn kernels: purifications and identifications using high performance liquid chromatography and chemical-ionization mass spectrometry. Planta 147: 422-435

Sussman MR, Firn R (1976) The synthesis of a radioactive cytokinin with high specific activity. Phytochemistry 15: 153-155

Tandeau de Marsac N, Jouanneau JP (1972) Variation de l'exigence en cytokinine de lignées clonales de cellules de tabac. Physiol Veg 10: 369-380

Terrine C, Laloue M (1980) Kinetics of N^6-(Δ^2isopentenyl)adenosine degradation in tobacco cells. Evidence of a regulatory mechanism under the control of cytokinins. Plant Physiol 65: 1090-1095

Genetic Regulation of Cytokinin Metabolism in *Phaseolus* Tissue Cultures

D.J. Armstrong[1], S.-G. Kim[1], M.C. Mok[2], and D.W.S. Mok[2]

1 Introduction

The regulatory mechanisms controlling hormonal metabolism play a central role in plant development. For normal development to occur, hormone levels appropriate to each tissue at each stage of development must be maintained. Mutations affecting the metabolism of particular plant hormones and resulting in various developmental abnormalities have been described, but the genetic controls operating on hormonal metabolism have not been extensively investigated and are likely to be complex.

We are attempting to identify the genetic factors controlling cytokinin metabolism in *Phaseolus*. The selection of *Phaseolus* as the plant material to be investigated was based on the ready availability of suitable genetic materials, the development in our laboratories of methods of culturing rapidly growing *Phaseolus* callus tissues (Mok and Mok, 1977), and the existence of a number of significant problems in both interspecific and intraspecific gene transfer in this plant group. The present paper summarizes the general approach and principal results obtained in this investigation to date.

2 Screening of *Phaseolus* Genotypes for Evidence of Genetic Variations in Cytokinin Metabolism

Our initial investigations of the genetic regulation of cytokinin metabolism in *Phaseolus* have centered on the identification of intrinsic genetic differences in the cytokinin metabolism of selected *Phaseolus* genotypes. The success of this approach has depended on the presence of genetic diversity in cytokinin metabolism in *Phaseolus* and on our ability to recognize phenotypic expressions of variations in cytokinin metabolism. Phenotypic traits of this type are more easily defined for tissue culture systems than intact plants. Therefore, tissue culture methods have been used to screen for genotypes of interest.

Our initial screening procedures have been based on two assumptions: (1) Genetic variations in cytokinin metabolism involving differences in metabolic transformations, uptake, or catabolism, should be reflected in differences in cytokinin structure-activity relationships in cytokinin-dependent *Phaseolus* tissue cultures. (2) Genetic variations in the regulation of cytokinin biosynthesis should be reflected (at least in some cases)

---

[1] Department of Botany and Plant Pathology, Oregon State University, Corvallis, OR 97331, USA
[2] Department of Horticulture, Genetics Program, Oregon State University, Corvallis, OR 97331, USA

in differences in the ability of *Phaseolus* tissue cultures to exhibit cytokinin-autonomous growth (growth on cytokinin-free medium). Examples of both cases are now available.

Seeds of plant introduction lines (P.I. lines) have been provided by the Regional Plant Introduction Station, Washington State University (Pullman, WA). Seeds of commercial cultivars of *Phaseolus* have been provided by Asgrow Seed Company. For screening purposes, *Phaseolus* callus cultures were established from the hypocotyls of five-day-old seedlings using the standard *Phaseolus* tissue culture medium (Mok and Mok, 1977) containing 2.5 μM picloram as an auxin source and 5 μM kinetin. The callus tissue that formed on the initial explants was transferred once (first passage) on the same medium and tested in the second passage for various traits indicative of possible differences in cytokinin metabolism. All cultures were grown in the dark at 27°C and harvested after 5 weeks growth. The use of picloram as an auxin source is essential to obtain vigorous cultures of *P. vulgaris* and has the additional advantage that this auxin promotes optimal growth of *Phaseolus* callus cultures over a wide range of concentrations (Mok and Mok, 1977). Thus, the effects of possible genetic variations in auxin metabolism are minimized under these test conditions.

3 Interspecific Differences in the Metabolism of Cytokinin Bearing Isoprenoid Side Chains

As shown in Table 1, tests of tissue cultures derived from *P. vulgaris* cv. Great Northern and *P. lunatus* cv. Kingston revealed dramatic differences in the responses of these callus tissues to cytokinins bearing unsaturated isoprenoid side chains [zeatin and N^6-(Δ^2 isopentenyl)adenine] (Mok et al., 1978). The corresponding saturated analogs (dihydrozeatin and N^6-isoamyladenine) exhibited more nearly equivalent activity in the two types of cultures. In *P. lunatus* callus cultures, there was little difference in the activity of cytokinins bearing saturated and unsaturated side chains. However, in *P. vulgaris* cultures, the presence of a double bond in the cytokinin side chain resulted in a marked reduction in cytokinin activity. Thus, zeatin and N^6-(Δ^2-isopentenyl)adenine were only weakly active in supporting the growth of cytokinin-dependent *P. vulgaris* callus tissue.

A possible explanation for the observed differences in cytokinin structure-activity relationships of *P. lunatus* and *P. vulgaris* tissue cultures is suggested by the report by Whitty and Hall (1974) that cytokinins with saturated side chains are resistant to attack by a cytokinin oxidase from maize that readily cleaves the unsaturated side chains of zeatin and N^6-(Δ^2-isopentenyl)adenine. Thus, the weak activity of the latter two compounds in promoting the growth of *P. vulgaris* cv. Great Northern callus tissue may be related to elevated levels of cytokinin oxidase activity in the tissue cultures derived from this genotype. Preliminary results (unpublished) comparing the metabolism of N^6-(Δ^2 isopentenyl)adenosine-8-^{14}C in callus tissues derived from *P. vulgaris* cv. Great Northern and *P. lunatus* cv. Kingston tend to support this hypothesis. In Great Northern callus tissues, this cytokinin is rapidly metabolized to a compound with chromatographic properties similar to adenosine. In Kingston callus tissue, the labeled N^6-(Δ^2-isopentenyl)adenosine persists for longer periods of time and the major metabolic products appear to be compounds other than adenosine.

Table 1. Response of *Phaseolus vulgaris* cv. Great Northern and *Phaseolus lunatus* cv. Kingston callus tissues to cytokinins with isoprenoid side chains

Cytokinin	Cytokinin concentration (μM) required for one half maximum yield of tissue[a]	
	Great Northern	Kingston
Zeatin	0.400	0.008
Dihydrozeatin	0.006	0.020
N^6-(Δ^2-isopentenyl)adenine	4.000	0.079
N^6-isoamyladenine	0.020	0.100

[a] Calculated from the data of Mok et al. (1978)

In view of the weak activity of zeatin in promoting the growth of *P. vulgaris* callus tissue, it is of interest that dihydrozeatin has been reported to be one of the major cytokinins produced by *P. vulgaris* seedlings (Wang et al., 1977; Sondheimer and Tzou, 1971). Reduction of the N^6-side chain of zeatin may serve as one mechanism to protect against excessive cytokinin destruction. The cytokinins produced by *P. lunatus* have not been examined, but it is possible that *P. lunatus* and *P. vulgaris* may differ in their ability to reduce the zeatin side chain as well as in cytokinin oxidase activity.

Genetic analysis of interspecific differences is obviously difficult. In the case of *P. vulgaris* and *P. lunatus* crosses, it is possible to use embryo culture to obtain progeny from the cross *P. vulgaris* (female) x *P. lunatus* (male), but the reciprocal of this cross has yet to be obtained (Mok et al., 1978). Embryos of the *P. lunatus* (female) x *P. vulgaris* (male) cross typically do not develop beyond the four cell stage (Rabakoarihanta et al., 1979). The phenotype of callus tissues derived from the hybrid *P. vulgaris* x *P. lunatus* has been examined and found to more closely resemble that of *P. vulgaris* (weak response to zeatin) than *P. lunatus* (unpublished). However, until progeny from the reciprocal cross are obtained, we cannot be certain whether this result is due to nuclear interactions or the influence of the cytoplasm.

Genetic variation in the regulation of hormonal metabolism might be expected to influence the recovery of progeny from wide crosses, and the observed differences in cytokinin structure-activity relationships in *P. vulgaris* and *P. lunatus* tissue cultures suggest that this may be a factor in regulating interspecific crosses in *Phaseolus*. If this is the case, biochemical or genetic manipulation of hormonal metabolism may be an important tool in obtaining wide crosses in this plant group.

4 Inheritance of Cytokinin Autonomy in *Phaseolus vulgaris* Callus Tissues

Both interspecific and intraspecific differences in ability of *Phaseolus* callus tissues to exhibit cytokinin autonomous growth (growth on cytokinin-free media) were detectable by our screening procedures, and the inheritance of this trait in selected *P. vulgaris* genotypes has been examined in some detail (Mok et al., 1980). Of ten genotypes of *P. vulgaris* tested in this study, one required cytokinin for callus growth, six exhibited some to moderate growth on cytokinin-free media, and the remaining three grew vigorously and uniformly in the absence of cytokinin. Of ten *P. lunatus* genotypes tested, six were strictly cytokinin-dependent in tissue culture and four displayed irregular callus growth on cytokinin-free media. The behavior of the callus tissues was independent of the tissue of origin (hypocotyl, petiole, cotyledon, or root) and the time in culture prior to testing for the autonomous growth habit. Callus tissues from all genotypes tested grew well on cytokinin-containing media.

The inheritance of the cytokinin requirement of *P. vulgaris* callus tissues has been examined in crosses between the strictly cytokinin-dependent genotype P.I. 200960 and two cytokinin-independent genotypes (cv. Gallatin 50 and P.I. 286303). The growth of hybrid tissues on cytokinin-free media was intermediate between and significantly different from the parental tissues. No reciprocal cross differences were evident in the F_1 populations. The F_2 progeny exhibited both parental and intermediate phenotypes. Again, no reciprocal cross differences were observed. Among 170 F_2 progeny derived from the crosses P.I. 200960 x cv. Gallatin 50 and reciprocal, 46 exhibited the cytokinin-dependent phenotype. The remaining progeny gave rise to callus tissues exhibiting either the intermediate or completely cytokinin-autonomous phenotype. Distinguishing between these latter two groups was difficult, but at least 24 of the F_2 progeny gave rise to callus tissues exhibiting vigorous cytokinin-autonomous growth equivalent to that of the original Gallatin 50 phenotype. Similar results were obtained for the F_2 progeny derived from reciprocal crosses between P.I. 200960 and P.I. 286303. Among 125 F_2 progeny, 36 exhibited the cytokinin-dependent phenotype and 15 were classified as completely cytokinin-autonomous. In backcrosses of the F_1 progeny of P.I. 200960 x P.I. 286303 (and reciprocal) to P.I. 200960, only cytokinin-dependent and intermediate phenotypes were observed. Among 44 backcross progeny examined, 17 were strictly cytokinin-dependent.

The frequency distribution of strictly cytokinin-dependent progeny in the F_2 and backcross populations from the crosses described above indicates that the cytokinin requirement of *P. vulgaris* callus tissues is regulated by one set of alleles. Recovery of the completely cytokinin-autonomous parental phenotype in the F_2 population was lower than expected, and this may reflect the presence of additional genes influencing the level of cytokinin autonomy or genetic variation in other factors that affect the rate of growth of the callus tissues. The biochemical basis of the observed differences in the cytokinin requirement of *P. vulgaris* callus tissues has yet to be investigated, and we cannot be certain whether the genetic factors we have identified are controlling rates of cytokinin biosynthesis or influencing the conversion of endogenous cytokinins to inactive metabolites. However, the rather simple mode of inheritance of the cytokinin-autonomous and cytokinin-dependent growth habits in this system suggests that we should be able to distinguish between these possibilities.

5 Genetic Variation in the Response of *Phaseolus lunatus* Callus Tissues to Diphenylurea

N,N'-Diphenylurea and a number of structurally related compounds exhibit cytokinin activity similar to that of N^6-substituted adenine derivatives in various bioassay systems. Diphenylurea stimulates cell division in carrot tissue cultures (Shantz and Steward, 1955) and exhibits cytokinin activity in the tobacco callus bioassay (Bruce and Zwar, 1966), but it is ineffective in promoting the growth of cytokinin-dependent soybean callus tissue (Dyson et al., 1972). Therefore, it was of interest to determine whether genetic variation in ability to respond to diphenylurea was present in *Phaseolus*. Callus tissues from sixteen *P. lunatus* genotypes have been tested for their ability to respond to diphenylurea. The result of these tests are summarized in Table 2. It should be noted that the diphenylurea concentration used here (32 μM) was near the limits of solubility under these test conditions. The response of the *P. lunatus* callus tissues varied with genotype. Genotypes such as P.I. 256845 and P.I. 260415 did not respond to diphenylurea. Genotypes such as cv. Jackson Wonder and cv. Henderson Bush exhibited relatively vigorous callus growth on diphenylurea-containing media, although in no case did the yield of callus tissue approach that obtained on kinetin-containing media.

The growth of *P. lunatus* callus tissues on diphenylurea-containing media was noticeably irregular when compared with either the growth of the same tissues on kinetin-containing media or the response of tobacco (Wisconsin 38) callus tissue to diphenylurea. When transferred to cytokinin-free media, *P. lunatus* callus tissues cultured for one transfer on diphenylurea were found to have acquired the capacity for continued and indefinite growth in the absence of an exogenous source of cytokinin, while control cultures grown on 5μM kinetin remained cytokinin-dependent (Mok et al., 1979). All genotypes that responded to diphenylurea behaved in a similar fashion when transferred to cytokinin-free media (Table 2). Preliminary results with other cytokinin-active urea derivatives indicate that the development of cytokinin autonomy may be a general response of these genotypes to this class of compounds. Genotypes such as P.I. 260415, that do not respond to diphenylurea, exhibit little tendency to display cytokinin-autonomous growth under any culture conditions tested to date. The growth of such genotypes is promoted by some of the more active urea derivatives, but transformation to the cytokinin-autonomous growth habit has not been observed.

The ability of diphenylurea to promote the development of cytokinin-autonomous growth in *P. lunatus* callus tissues has now been examined in more detail in selected *P. lunatus* genotypes. The interaction of kinetin and diphenylurea in promoting the growth of callus tissue derived from one such genotype, P.I. 257422(w), is shown in Fig. 1. In this and in other genotypes, the presence of diphenylurea in the tissue culture media appears to have little effect on the concentration of kinetin required for optimal growth of callus tissue. Table 3 shows the results obtained when callus tissues of P.I. 257422(w) were transferred to cytokinin-free media. As expected, tissues grown on diphenylurea in the previous passage exhibited cytokinin-autonomous growth. However, tissues grown on suboptimal concentrations of kinetin also exhibited growth on cytokinin-free media. When both compounds were included in

Table 2. Response of *Phaseolus lunatus* genotypes to N,N'-diphenylurea

Genotype	Average fresh weight (g/flask)[a]			
	Second passage (Cytokinin media)		Third passage (Cytokinin-free media)	
	Diphenylurea (32 μM)	Kinetin (5 μM)	From diphenylurea	From kinetin
Henderson Bush	4.66	24.34	7.21	0.54
Jackson Wonder	6.58	24.64	11.68	0.07
257409	2.45	23.36	7.61	0.33
257422(w)	6.59	22.24	5.17	0.32
264239	4.83	17.38	5.09	0.68
Early Thorogreen	3.98	24.56	1.13	0.10
Nema Green	1.30	22.88	7.80	0.08
Kingston	2.80	23.10	5.40	0.08
180324	1.64	21.90	12.18	0.10
256812	0.74	18.18	11.86	0.09
257560	2.58	21.64	1.84	0.11
195339	0.07	15.35		
202830	0.08	14.75		
256845	0.13	15.84		
257547	0.06	17.74		
260415	0.06	15.66		

[a] Average of eight flask. Tissues were harvested after 35 days growth.

Table 3. Interaction of kinetin and N,N'-diphenylurea in inducing cytokinin autonomy in callus cultures of *Phaseolus lunatus* P.I. 257422(w)

Kinetin concentration in previous passage	Diphenylurea concentration in previous passage	
	0 μM	32 μM
	Avg. Fresh Weight (g/flask)[a] on Cytokinin-Free Media	
0.00	–	8.64
0.03 μM	–	4.70
0.1 μM	5.86	5.46
0.3 μM	0.54	2.58
1.0 μM	0.64	1.29
3.0 μM	0.20	0.52
10.0 μM	0.31	0.22

[a] Average of eight flask. Tissue were harvested after 35 days growth

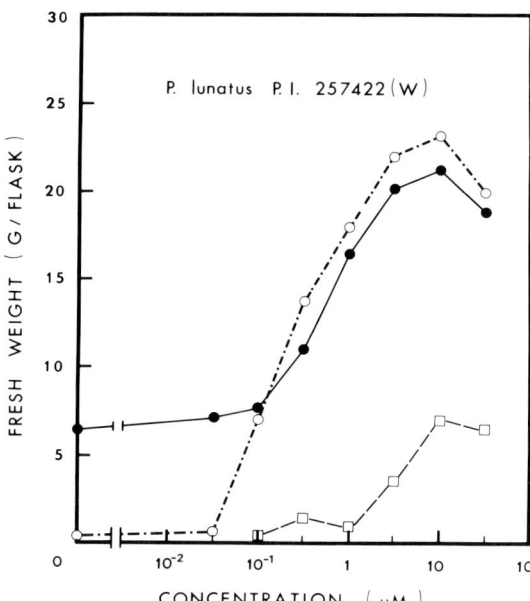

Fig. 1. Interaction between kinetin and N,N'-diphenylurea in promoting callus growth of *Phaseolus lunatus* P.I. 257422 (w). Callus tissues were planted on media containing 2.5 μM picloram and various concentrations of kinetin alone (o—·—·—·—o), kinetin plus 32 μM diphenylurea (●-----●), and diphenylurea alone (□-----□). Tissues were harvested and weighed after 35 days. The average fresh weight of four replicate flask was determined

the same tissue culture media, the subsequent yield of tissue on cytokinin-free media was less than that obtained after exposure to diphenylurea alone, and the development of cytokinin autonomy was completely suppressed at optimal kinetin concentrations. At intermediate kinetin concentrations, however, the presence of diphenylurea in the tissue culture media appears to increase the tendency of the tissue to display cytokinin-autonomous growth when transferred to cytokinin-free media. Similar results have been obtained with other *P. lunatus* genotypes that respond to diphenylurea.

Our initial observations of the effect of diphenylurea on the growth of *P. lunatus* callus cultures led us to suggest that the apparent cytokinin activity of this compound and other substituted urea derivatives might be due to indirect effects on the metabolism of endogenous cytokinin (Mok et al., 1979). While this hypothesis may be correct, the control of cytokinin autonomy in *P. lunatus* callus cultures appears to be complex. The development of cytokinin autonomy in these cultures is dependent upon genotype and is influenced by the exogenous supply of cytokinin-active adenine derivatives as well as by exposure to cytokinin-active urea derivatives. Therefore, we cannot be certain as yet that the effects of these two classes of compounds involve distinct mechanisms.

6 Discussion

The results of our investigations indicate that both intraspecific and interspecific diversity in cytokinin metabolism are present in *Phaseolus*. The identification of intrinsic genetic differences in cytokinin metabolism in existing *Phaseolus* genotypes provides an opportunity to bring together techniques of both biochemical and genetic analysis in investigations of the regulatory mechanisms controlling cytokinin metabolism. In a given species, the range of genetic diversity in cytokinin metabolism is likely to be limited, but comparative studies of cytokinin metabolism in more widely separated genetic materials also show promise of providing useful insights into the mechanisms by which the levels of these hormones are regulated in plant tissues. Thus, intrinsic as well as induced genetic variations in hormonal metabolism may be useful tools in studies of the metabolism and mechanism of action of these compounds.

Acknowledgments. This research has been supported by the Science and Education Administration of the U.S. Department of Agriculture under Grant 78-59-2411-0-1-028-0 from the Competitive Research Grants Office, by the Research Council of Oregon State University (NIH Biomedical Research Support Grant RR07079) and by the Oregon Agricultural Experiment Station. This is technical paper No. 5616 of the Oregon Agricultural Experiment Station.

References

Bruce MI, Zwar JA (1966) Cytokinin activity of some substituted ureas and thioureas. Proc R Soc London Ser B 165: 245-265

Dyson WH, Fox JE, McChesney JD (1972) Short term metabolism of urea and purine cytokinins. Plant Physiol 49: 506-513

Mok DWS, Mok MC, Rabakoarihanta A (1978) Interspecific hybridization of *Phaseolus vulgaris* with *P. lunatus* and *P. acutifolius.* Theor Appl Genet 52: 209-215

Mok MC, Kim S-G, Armstrong DJ, Mok DWS (1979) Induction of cytokinin autonomy by N,N'-diphenylurea in tissue cultures of *Phaseolus lunatus* L. Proc. Natl Acad Sci USA 76: 3880-3884

Mok MC, Mok DWS (1977) Genotypic responses to auxins in tissue cultures of *Phaseolus.* Physiol Plant 40: 261-264

Mok MC, Mok DWS, Armstrong DJ (1978) Differential cytokinin structure-activity relationships in *Phaseolus.* Plant Physiol 61: 72-75

Mok MC, Mok DWS, Armstrong DJ, Rabakoarihanta A, Kim S-G (1980) Cytokinin autonomy in tissue cultures of *Phaseolus:* A genotype-specific and heritable trait. Genetics 94: 675-686

Rabakoarihanta A, Mok DWS, Mok MC (1979) Fertilization and early embryo development in reciprocal interspecific crosses of *Phaseolus.* Theor Appl Genet 54: 55-59

Shantz EM, Steward FC (1955) The identification of compound A from coconut milk as 1,3-diphenylurea. J Am Chem Soc 77: 6351-6353

Sondheimer E, Tzou DS (1971) The metabolism of hormones during seed germination and dormancy. II. The metabolism of 8-^{14}C-zeatin in bean axes. Plant Physiol 47: 516-520

Wang TL, Thompson AG, Horgan R (1977) A cytokinin glucoside from the leaves of *Phaseolus vulgaris* L. Planta 135: 285-288

Whitty CD, Hall RH (1974) A cytokinin oxidase in *Zea mays.* Can J Biochem 52: 789-799

Metabolism of the Cytokinins in Mosses

M. Bopp and U. Erichsen[1]

Introduction

1.1 Effect of Cytokinins in Mosses

One of the most pronounced effects of cytokinin is the induction of buds on the protonema of mosses (Hahn and Bopp, 1968; Szweykowska, 1974; Nehlsen, 1978). Most of the moss species treated with cytokinin between 10 nM and 100 μM show a similar reaction. Certain protonemal filaments which are target cells for bud formation quickly change their morphogenetic behavior and form buds instead of filamentous side branches. In some mosses a short side branch which has already been formed can also change its shape under the influence of cytokinin. The primary steps of this reaction can be seen as early as 6 to 10 h after application as a modification of the tip region of the cells and shortly later in an increase of the RNA content in these cells. These reactions of the moss protonema are very specific, because only N-6-substituded adenine derivatives – true cytokinins – are active (Hahn and Bopp, 1968), whereas other substances like substituted ureas, which have a strong effect in other test systems (Takahashi et al., 1978), do not induce buds in the mosses. Also the ribosides of benzyladenine and other cytokinins are much less active than the purine bases themselves when they are applied from outside (Whitacker and Kende, 1974; Spiess, 1975). But this lack of an effect seems to be more a consequence of a restricted uptake than of an inefficiency of the substance. It can be shown that the application of ribose-1-phosphate together with the kinetin enhances the bud formation by about 100%, whereas ribose-5-phosphate does not change the number of buds induced by kinetin (Fig. 1). From this one may conclude that for the effect of cytokinin in the cells the ribosides and not the bases are necessary, but no further experiments or arguments are available at the moment to support this assumption.

It should be mentioned that cytokinin can also stimulate the cell division in moss protonema, first shown by Szweykowska et al. (1972). Under the influence of cytokinins one finds in all types of protonema cells a higher rate of cell division, but this cell division is much less specific to cytokinin than the bud induction. Other substances like urea also have a stimulating effect on cell division, mostly in the tip cells because these are the places of growth of the filaments. Also the intercalary divisions, induced by separation of intact cells, as a sign of regeneration, are enhanced by cytokinin treatment.

---

[1] Botanisches Institut der Universität, Im Neuenheimer Feld 360, D-6900 Heidelberg, FRG

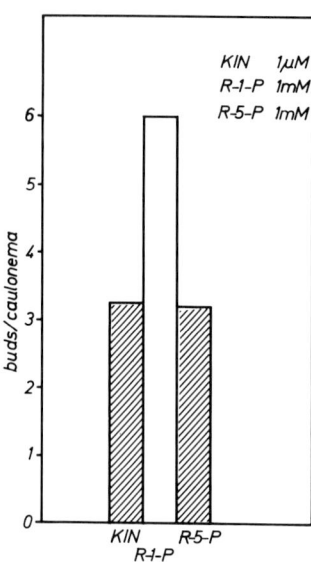

Fig. 1. Bud induction on isolated caulonema filaments by kinetin and combination of kinetin and ribose-1-phosphate or kinetin and ribose-5-phosphate. Only the first can enhance the effect of kinetin

For our purposes, however, the effect of bud formation in comparison to the cytokinin metabolism may stay in the foreground.

1.2 Cytokinin in Mosses

The development of mosses is quite different from that of higher plants and also the regulation shows distinct differences between both groups of plants. In respect to the hormonal control of regulation we know clear effects only of two phytohormones in mosses: auxin and cytokinin, the other hormones, abscisic acid, gibberellins and ethylene seem to be less important, perhaps, they are without specific actions. Therefore it is of great interest to ask whether the first two hormones are found in mosses. For auxin this is not completely clear but for cytokinin the presence was demonstrated as early as 1966 (Bauer, 1966). Later on Beutelmann and Bauer (1977) have identified this "bryokinin" as N^6-isopentenyladenine. It is present in a stable callus derived from the moss hybrid *Funaria* x *Physcomitrium*. This callus contains high amounts of the cytokinin, but it was not detected in normal growing moss protonema or moss plants so far. However in recent times Cove and his coworkers (Ashton et al., 1979) working with mutants of *Physcomitrella patens* found mutants which show the same shape as mosses treated with higher concentrations of cytokinin. They call these mutants cytokinin overproducers. It was possible to isolate from tissue of these plants as well as from the culture medium by chromatography, UV-spectroscopy and mass spectroscopy N^6-(Δ^2-isopentenyl)adenine. Also zeatin was present as a minor component (Wang et al., 1979).

From these results it is very probable that the normal protonema contains a certain amount of cytokinin, but the level must be very low. It can be assumed that this low level in wild-type plants is regulated by a balance between synthesis of cytokinin on one side and degradation, metabolism or inactivation (Laloue, 1977) on the other side. Therefore the aim of the following paper is to show whether a metabolism of cytokinin in mosses exists, although Wang et al. (1979) were not able to demonstrate a significant metabolism of exogenous [^{14}C]-benzyladenine or isopentenyladenine either in the over-producers or in the wild-type protonemas.

2 Material and Methods

To study the metabolism of cytokinin in mosses we use the protonema of the moss *Funaria hygrometrica* developed from spores. This protonema exists in two forms: chloronema and caulonema, the second form appears after about 10 days of growth. About 10 days later buds are formed on the caulonema filaments. The protonema was treated shortly before bud formation with kinetin-8-C^{14} with a radioactivity of 10 μCi/ml in small droplets of water giving a final concentration of 40 mM kinetin in the substrate.

For the extraction the protonema were washed several times to remove the outer kinetin solution and then extracted with Tris buffer pH 8.8 in the cold. After centrifugation the clear supernatant was chromatographed on a microscale thin layer chromatography, in two dimensions (Erichsen et al., 1978). First dimension: n-butanol:acetone: acetic acid:H_2O:5% ammonia (3,5:2,5:1,5:1:1,5). Second dimension: n-butanol:acetone:chloroform:25% ammonia (4:3:3:1).

The chromatography was performed on "DC-Alufolie" (Merck) or "Nanoplates" Sil UV 254 (Macherey and Nagel) 4 x 5 cm.

The chromatograms were compared with a standard two-dimensional TLC and the spots localized by UV light. Then the radioactivity was measured in a scintillation-counter 5 min after finishing the chromatogram to avoid changes in the scintillation liquid by the Kieselgel. The sum of activity on the plate was compared with the total activity in the extract as 100%.

For the treatment with enzymes, the extract was incubated 24 h at room temperature with the enzyme and then analyzed in the usual manner. As a control an extract was compared which stood at room temperature without enzymes. To stop the activity of endogenous phosphatases 0.01 M NaF was added.

The concentrations of the enzymes were: Phosphatase 10 mg/1 ml; Trypsin 4 mg/1 ml; RN-ase 0.1 mg/1 ml.

3 Experimental Results and Discussion

3.1 Distribution of Radioactivity After ^{14}C-Kinetin Application

The application of labeled kinetin results in an uptake which is somewhat faster in the first 4 h and then nearly linear for at least 24 h. In this case the total activity in the cells was measured. A separation of the buffer extract by thin layer chromatography after 2 h gives only a very low radioactivity on the kinetin spot itself, which is less than 5% of the total activity in the extract. But 75% is found in the front of the chromatogram. Other pronounced parts of the chromatogram with a high labeling rate are the starting point with 8% and the adenine spot, which contains also about 8%. The adenine spot is clearly separated from kinetin and one finds a region without radioactivity between both substances. The distribution pattern, which we found after 2 h, changes only slightly in the following time, therefore after 24 h the relative activity in the different spots is nearly the same (Fig. 2). This means that the kinetin is metabolized very quickly and is not at all stable for a longer time. In comparison to the results with kinetin the content of BAP also labeled with ^{14}C remains much higher after 24 h, which means that the half-life of this cytokinin is longer than that of kinetin (Fig. 6). BAP therefore is more stable and less metabolized than kinetin.

The main products of the metabolism besides adenine are in both cases different nucleotides which are present in the front fraction. We found there about 9% of the activity as ATP, 11.5% as ADP, 39% as AMP and around 40% other nucleotides — perhaps of kinetin. The radioactivity in the start fraction belongs to RNA and substances associated with proteins — we do not know exactly whether this is kinetin, kinetinriboside, or other products.

These results are supported by fluorography of a two dimensional microgel-electrophoresis of a Tris buffer extract — the first dimension was separated by an electrophoresis in a gradient gel from 2% to 40%, the second dimension on a SDS-gel. Also under these conditions, in which the ^{14}C-kinetin was applied for 20 h, it is clearly seen that the activity is distributed on several different spots. The first group of spots is associated with ribosomes in the first dimension — which means ribosomal RNA and proteins, the second group is found in a fraction which contains the t-RNA. This part is also labeled when radioactive uridine or adenine are applied, the third group is in the region of the large subunit of fraction I protein, and the last is a very clear spot, which coincides with one ribosomal protein, perhaps poly (A) of m-RNA.

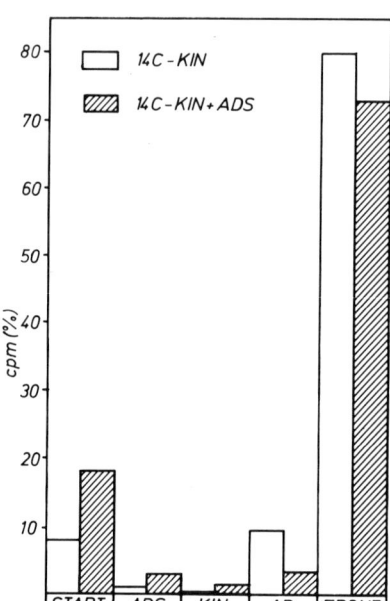

Fig. 2. Distribution of radioactivity in the different metabolites after separation by thin layer chromatography. Extracted from protonema of *Funaria hygrometrica* by Tris buffer pH 8.8. Treatment with labeled kinetin 8-C-14 for 20 h, with and without simultaneous application of unlabeled adenosine. 100% = total activity on the chromatogram

The main result of this observation is that kinetin which is incorporated in the cells is metabolized to a large part to adenine so that only a very small amount remains presents as free kinetin or derivatives of kinetin. In comparison to higher plants (Parker et al., 1978) the degradation to adenine and adenine compounds seems to be very high and the main pathway of cytokinin inactivation. Also BAP is metabolized in the same way but not as fast as kinetin.

But as long as cytokinin is applied from outside, this can always saturate the system which needs cytokinin for bud formation. As soon as the cytokinin is removed from the nutritional medium, the internal cytokinin falls under the level necessary for saturation. The result is that buds do not grow any longer as such but change to filamentous growth (Brandes and Kende, 1968). Also from this observation it is clear that the cytokinin cannot be stable in the moss cells.

3.2 Enzyme Treatment of Cell Extracts

To answer the question what kind of incorporation is found in the different fractions of the thin layer chromatography, we treated the extract with different enzymes and compared the distribution of activity before and after the treatment. For better comparison the distribution is always given in percent of the total activity.

Under the influence of phosphatase (Fig. 3) the activity in the front fraction decreases to about 25%-35%.

The activity which was lost from the front appears as adenine, adenosine and kinetinriboside on the chromatogram. This confirms that in the front fraction, beside other substances, nucleotides are present which must be adenine-derivatives such as AMP, ADP and ATP and kinetin-nucleotides, namely kinetinriboside-phosphate.

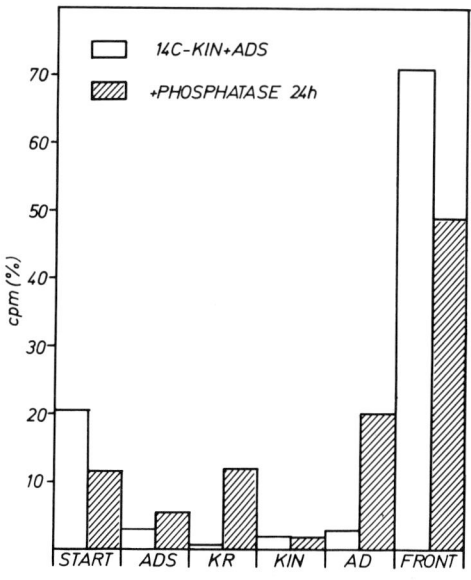

Fig. 3. Distribution of radioactivity after treatment of the buffer extract with phosphatase for 24 h

The treatment with trypsin to remove proteins (Fig. 4) from the extract gives a distribution which is partly similar, partly different from that with phosphatase. Because the extract itself contains active phosphatases, the similarity is due to these endogenous enzymes.

But one clear and important difference was seen with trypsin: the radioactivity in the kinetin spot increased threefold, which means that in contrast to the phosphatase experiments kinetin is released when proteins are destroyed. We do not know at the moment in which form kinetin is bound to the protein and experiments are in progress to demonstrate which protein, separated in the electrophoresis, contains cytokinin. There are some arguments that superstructure proteins which can be centrifuged together with ribosomes contain the kinetin; these proteins can then be separated in SDS-gel in the different bands and one was found to contain radioactivity from kinetin.

A third treatment was carried out with RNase (Fig. 5). In this case the total activity was reduced by about 25%. But the strongest reduction was found in the start fraction. The treatment with RNase results only in a significant increase of adenine and adenosine, which is to be expected as these substances are very quickly incorporated in ribosomal RNA, which runs in the start fraction. Our method is not precise enough to find out whether cytokinin itself is incorporated in the RNA as an intact molecule, because we cannot show unequivocally differences smaller than 3%.

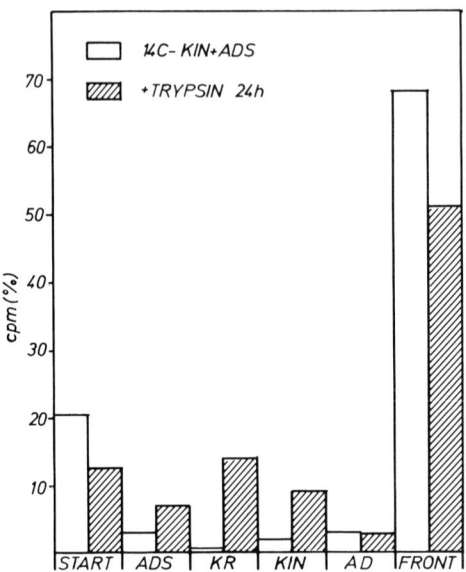

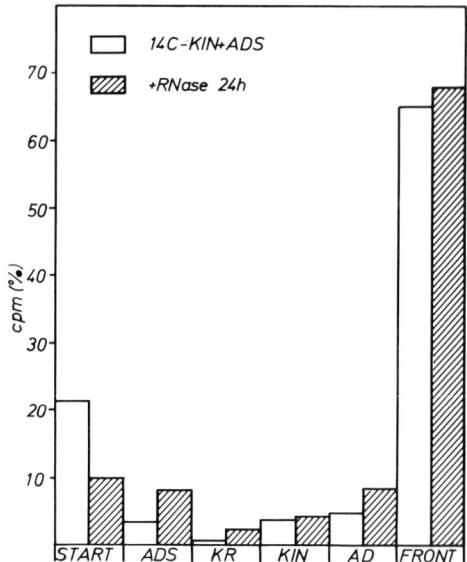

Fig. 4. Distribution of radioactivity after treatment of the buffer extract with trypsin for 24 h

Fig. 5. Distribution of radioactivity after treatment of the buffer extract with RN-ase for 24 h. In all cases protonemata are treated for 20 h with labeled kinetin and unlabeled adenosine and then extracted

The material is also insufficient to demonstrate the metabolism of kinetin or BAP to glucosides or other derivatives (Laloue, 1977), nevertheless all the results fit the assumption that cytokinins are metabolized in different ways in the moss protonema, and that the products of the metabolism are incorporated in RNA and DNA.

3.3 Effect of Adenosine

Under the assumption that a great deal of the metabolic products of exogenous kinetin are adenosine and adenine, the level of these substances in the cells must be enhanced by application of cytokinins. If this is correct one can expect that a simultaneous application of adenosine and kinetin can change the activity of the enzymes which remove the side chain in a freed-back reaction (cf. Chen, 1980) and this should result in a different distribution of radioactivity.

We found a remarkable change in the pattern when adenosine was given together with ^{14}C kinetin (Fig. 1) or BAP (Fig. 6). The activity in the start fraction was nearly twice as high as without adenosine, on the other hand the incorporation in the adenine fraction was drastically reduced and in the front fraction the portion of labeled adenine nucleotides is only 47%, whereas without adenosine this part is about 60% of the activity in this area. The incorporation in adenosine was, with and without an external adenosine source, too low to find any change. With ^{14}C-BAP (Fig. 6) we also found an enhanced incorporation of radioactivity in the start fraction, whereas the incorporation in adenine was not significantly changed, however in the nucleotides the activity was lower. But the content on unchanged BAP was clearly higher than with kinetin.

In general the results of these experiments are in agreement with the assumption that the start fraction contains one part of cytokinins or cytokinin derivatives in which

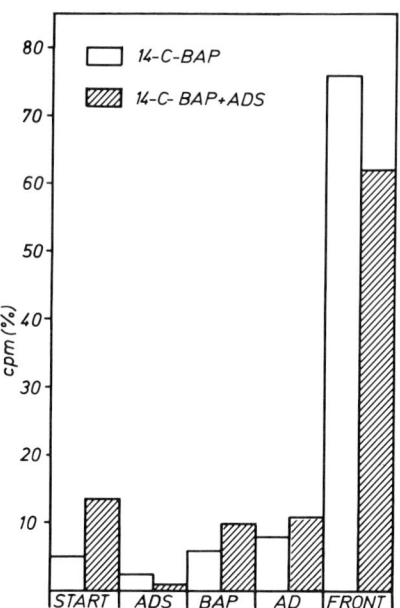

Fig. 6. Distribution of radioactivity in the different metabolites after separation by thin layer chromatography. Treatment with benzylaminopurin 8-C-14 for 20 h with and without simultaneous application of unlabeled adenosine

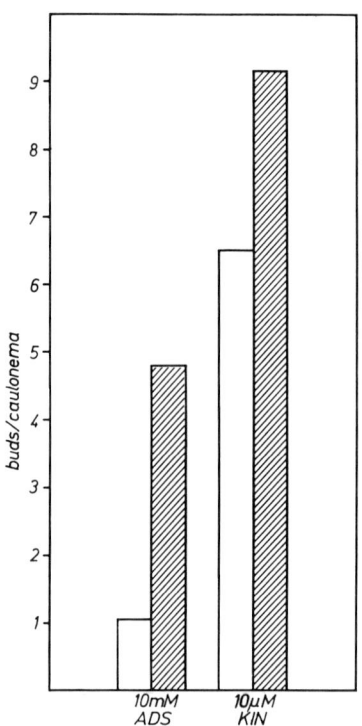

Fig. 7. Bud formation on isolated caulonema filaments (*open bars*) and on protonemas which are divided into halves (*striped bars*). The caulonema grow 72 h on adenosine or kinetin

the side-chain is not removed, and we conclude that the external adenosine application enhances the internal level of cytokinin available for morphogenetic regulation, because it is not degraded to adenine and adenosine. If this is correct, one should expect that adenosine can enhance the bud induction effect of kinetin. The following picture (Fig. 7) shows that this is the case (Bopp, 1977; Erichsen et al., 1978). Depending on the concentration of cytokinin and adenosine the bud formation is much higher than with adenosine alone or cytokinin alone. Adenosine has nearly no effect, applied to isolated filaments — because, as we know from other experiments, such filaments must have a very low internal cytokinin level — but in a whole protonema also adenosine alone can enhance the "natural" bud formation by about 300%. It looks as if adenosine alone may induce the buds. But because it is quite sure that adenosine has an influence on the metabolism of the endogenous kinetin, cytokinin can be accumulated to a higher extent in the cells, high enough to produce buds as in the cytokinin overproducers.

3.4 Incorporation of Radioactivity in the Cells

A further approach to the question of cytokinin metabolism in the cells gives the distribution of radioactivity in intact cells after labeling with ringlabeled cytokinins. Autoradiographic pictures of cells which are treated for a longer time, 24 or 48 h with [^{14}C]-kinetin show a characteristic distribution of the silver grains (Erichsen et al.,

1978). Altogether the apical cells of the main filament and of the side branches contain the highest incorporation rate. This distribution is independent of whether the cells can take up the labeled cytokinin directly from the substrate, with which they are in contact, or if the cytokinin enters other cells of the filament and is transported to the place of incorporation. In both cases the autoradiographic picture is identical. Another remarkable place of incorporation which is not seen in cells labeled only for 12 h is the nucleus. After 24 h and even more strongly after 48 h all nuclei in the filament are labeled. This is possible because DNA-synthesis continues for a long time in the caulonema filaments (Knoop, 1978). The strong labeling of the nucleus can only be explained under the assumption that adenine is incorporated into the DNA. Therefore also these experiments demonstrate the direction of the metabolism of cytokinin.

As I have just mentioned, the pattern of labeling is the same also when the substances are transported in the filaments. However from the autoradiography we cannot recognize whether the cytokinin itself is transported, or only the derivatives. But other experiments can discriminate between the two possibilities. For this purpose kinetin was applied on one end of a filament and the bud formation was observed on the other end. With this simple method it was precisely demonstrated that the active principle, which must be kinetin or kinetinriboside, is transported through the filaments and not only adenine.

4 Conclusions

Our experiments have clearly shown that cytokinins (kinetin and also BAP) are metabolized very quickly in the cells mainly by removing the side chain and the greater part is then present as adenine derivatives, which can be incorporated in all kinds of RNA and DNA. BAP is more stable than kinetin and therefore a higher content of free benzyladenine remains present in the cells.

Some fractions which are labeled are associated with proteins and it is possible that this association is responsible for the cytokinin effect on bud formation. No evidence has been found for the assumption that cytokinin is stable for a long time in the cells. The question remains to answer whether this is due to the kind of cytokinin used, namely kinetin which is metabolized, or natural cytokinins like zeatin or isopentenyladenine which are stable (Wang et al., 1979).

In any case metabolization of cytokinin must be a part of the normal hormonal balance in the wild-type protonema.

Acknowledgements. We thank W. Beltle and S. Zimmermann very much for valuable cooperation. Supported by a grant of the Deutsche Forschungsgemeinschaft.

References

Ashton NW, Cove DJ, Featherstone DR (1979) The isolation and physiological analysis of mutants of the moss, *Physcomitrella patens,* which overproduce gametophores. Planta 144: 437-442

Bauer L (1966) Isolierung und Testung einer kinetinartigen Substanz aus Kalluszellen von Laubmoossporophyten. Z. Pflanzenphysiol 54: 241-253

Beutelmann P, Bauer L (1977) Purification and identification of a cytokinin from moss callus cells. Planta 133: 215-218

Bopp M (1977) Interactions in the regulation of development in lower plants. In: Schütte HR, Gross D (eds) Regulation of developmental processes in plants. Proc Halle, pp 278-297

Brandes H, Kende H (1968) Studies on cytokinin-controlled bud formation in moss protenema. Plant Physiol 43: 827-837

Chen CM (1980) Biosynthesis, interconversion and metabolism of N^6-(Δ^2-isopentenyl)-adenin. In: Le metabolisme et les activités moleculaires des cytokinines. Colloq Int CNRS. Résumés des communications

Erichsen U, Knoop B, Bopp M (1978) Uptake, transport and metabolism of cytokinin in moss protonema. Plant Cell Physiol 19: 839-850

Hahn H, Bopp M (1968) A cytokinintest with high specificity. Planta 83: 115-118

Knoop B (1978) Multiple DNA contents in the haploid protonema of the moss *Funaria hygrometrica* Sibth. Protoplasma 94: 307-314

Laloue M (1977) Cytokinins: 7 Glucosylation is not a prerequisite of the expression of their biological activity. Planta 134: 273-275

Nehlsen W (1978) A new method for examining induction of moss buds by cytokinin. Am J Bot 66: 601-603

Parker CW, Letham DS, Gollnow BJ, Summons RE, Duke CC, Macleod JK (1978) Regulators of cell division in plant tissue XXV. Metabolism of zeatin by Lupin seedlings. Planta 142: 239-251

Spiess LD (1975) Comparative activity of isomers of zeatin and ribosylzeatin on Funaria hygrometrica. Plant Physiol 55: 583-585

Szweykowska A, Korcz I, Jaśkiewicz-Mroczkowska B, Metelska M (1972) The effect of various cytokinins and other factors on the protonemal cell divisions and the induction of gametophores in *Ceratodon purpureus*. Acta Soc Bot Pol 41: 403-409

Szweykowska A (1974) The role of cytokinins in the control of cell growth and differentiation in culture. Tissue Culture Plant Sci 461-475 (1974)

Takahashi S, Shudo K, Okamoto T, Yamada K, Isogai Y (1978) Cytokinin activity of N-phenyl-N'-(4-pyridyl)urea derivatives. Phytochemistry 17: 1201-1207

Wang WL, Beutelmann P, Featherstone DR, Cove DJ (1979) Cytokinin production and metabolism in the moss *Physcomitrium patens*. 10th Int Conf Plant Growth Substances, Abstr 14

Whitacker BD, Kende H (1974) Bud formation in *Funaria hygrometrica*. A comparison of the activities of three cytokinins with their ribosides. Planta 121: 93-96

Cytokinin Activities of N-Phenyl-N'-(4-Pyridyl)ureas

Y. Isogai[1]

1 Introduction

About 17 years ago, when Okamoto et al. had been studying azaindene compounds [1] (e.g., 8-benzylamino-2-methyl-s-triazolo[1,5-a]pyridine, 4(7)-benzylaminobenzimidazole, etc.) which have no purine ring yet have apparently analogous chemical structure to N^6-benzyladenine (BA) or kinetin (K), it was suspected that they might have antagonistic effect on BA or K in the tobacco bioassay. But the result was that they showed cytokinin activity and gave ca 4 g of fr. wt of callus at 10 ppm, while BA and K gave similar callus yield at 0.01 ppm in the same bioassay. We were also interested in 1,3-diphenylurea (DPU). Because there were many papers reporting the cytokinin activity of DPU [1, 2], but perhaps owing to the difference of bioassay methods and also possibly its low solubility in H_2O, the results differed one from the other and it was generally concluded that its cytokinin activity is low and sporadic. We intended to clarify this confusion, and tested DPU in the tobacco bioassay many times carefully and concluded that DPU has a definite cytokinin activity, giving ca 4 g of callus at 10 ppm [1].

These results seemed to show that, on the one hand, purine ring is not always necessary for cytokinin activity, but on the other, that there are some compounds having a chemical structure common to the above-mentioned active compounds. Considering the chemical structure of azaindene compound and DPU, the compound N-benzyl-imidazole-4-carboxamide [3] which has no pyrimidine ring in the structure was synthesized as the analog of BA or K. However, this compound was quite inactive but imidazol-carboanilide (10 ppm)[2] which was obtained as an intermediate in the course of this synthesis was active. Therefore, aromatic amide derivatives were synthesized. Among them, nicotinoylanilide (10 ppm), isonicotinoylanilide (INA – 1 ppm) were found to be active.

Taking all these results into consideration, Shudo [3] synthesized N-phenyl-N'-(4-pyridyl)urea (4 PU – 0.1 ppm) which is analogous to INA and belongs to aromatic urea as DPU. 4PU was found to be highly active in the tobacco bioassay and though its optimum concentration is ten times higher, (i.e., the activity is 1/10 lower) than BA or K, it always gives 10%-20% higher callus yield than BA or K. Then, N-phenyl-N'-(3-

---

[1] Department of Biology, College of General Education, University of Tokyo, Komaba, Meguro-ku, Tokyo 153, Japan
Present adress: Industrial Research Institute (Kogyokaihatsu Kenkysho), 1201 Takada, Kashira 277, Japan
[2] Optimum concentration of various compounds in the tobacco bioassay is shown in parenthesis

pyridyl)urea (3PU – 10 ppm) and N-phenyl-N'-(2-pyridyl)urea (2PU – almost inactive) were also synthesized and assayed. But their activity was lower than 4PU or almost inactive.

After that, the study of synthesizing further higher active 4PU derivatives had been carried out but did not give fruitful results at first. Now, incidentally, together with these studies, the study of pyrimidine-type cytokinin was also carried out in our laboratory and it was found that though benzylaminopyrimidine (BAP) was quite inactive, benzylamino-6-Cl-pyrimidine [5], which was also an intermediate in the course of the synthesis of BAP, had weak activity. This fact appeared suggestive to Isogai who proposed that the introduction of Cl into position 2 of pyridyl ring of 4PU may also enhance the activity. This was immediately realized and the product, N-phenyl-N'-(2-chloro-4-pyridyl)urea (4PU-30 – 0.001 ppm), showed surprisingly higher activity and gave ca 5 g of fr. wt of callus. The result gave us the clue to proceed with the synthesis of many other highly active 4PU derivatives [6] and mainly by the steady and skilful work of Takahashi, more than 150 kinds of 4PU derivatives and analogous compounds were synthesized. Consequently, at present, it has become possible to establish clear-cut structure-activity relationships. The results will be reported in the present paper.

2 Results and Discussion

a) At first, compounds with both unsubstituted- or 2-CH_3 substituted-4-pyridyl and unsubstituted or substituted phenyl rings were synthesized. But, as shown in Tables 1 and 2, they all showed weak activity and seven of them were inactive.
b) While, as shown in Table 3 and Fig. 1, introduction of Cl into position 2 of the pyridyl ring enhances the activity remarkably, this phenomenon seems to be limited to 4PU derivatives, because in the case of INA, substitution of position 2 of the pyridyl ring has no effect for the activity as shown in 37 in Table 9.
c) Through all these examples and all the 4PU derivatives afterwards synthesized, it was clarified that mono or disubstitution of the phenyl ring reduces activity irrespective of the pyridyl ring being substituted or unsubstituted and, generally as shown in Tables 1, 2 and 3 and Fig. 2a in the following decreasing order: meta (3) > ortho (2) ≥ para (4), with the exception of meta(3)-F substitution (e.g., 1h and 3i), in which cases reduction of activity does not occur and gives activity similar to or sometimes higher than that of 1a(4PU) as shown in Fig. 2b.
d) Subsequently, compounds with both mono various substituted-4-pyridyl and unsubstituted or meta(3)-F substituted phenyl rings were synthesized and tested. As shown in Table 4 and Figs. 3 and 4, substitution of position 2 of the pyridyl ring with electronegative groups, i.e., F (8), Br (9), CF_3 (10a), CN (11), OCH_3 (4), $NHCOCH_3$ (5), OH (7) and $COOCH_3$ (12) gave higher activity, while, especially in the former four cases, the activity was remarkably higher. In the case of electropositive groups, i.e., CH_3 (2a) and NH_2 (6), the activity was similar or lower than 4PU. Again, the case of 10b which has meta(3)-F substituted phenyl

Table 1. List of compounds with unsubstituted-4-pyridyl and unsubstituted or mono substituted phenyl rings and their optimum concentrations for cytokinin activity in the tobacco bioassay

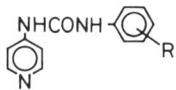

Compd. No.	R	Optimum concentration M	ppm
1a	H	4.7 x 10⁻⁷	0.1
1b	2-Cl	4.0 x 10⁻⁵	10
1c	3-Cl	4.0 x 10⁻⁶	1
1d	4-Cl	4.0 x 10⁻⁵	10
1e	3-Br	3.4 x 10⁻⁶	1
1f	4-Br	Inactive	
1g	2-F	4.3 x 10⁻⁶	1
1h	3-F	4.3 x 10⁻⁷	0.1
1i	4-F	4.3 x 10⁻⁶	1
1j	2-Me	4.4 x 10⁻⁵	10
1k	3-Me	4.4 x 10⁻⁵	10
1l	4-Me	Inactive	
1m	2-OMe	4.1 x 10⁻⁵	10
1n	3-OMe	4.1 x 10⁻⁵	10
1o	4-OMe	Inactive	
1p	3-OH	4.4 x 10⁻⁵	10
1q	4-OH	Inactive	
1r	4-COOEt	Inactive	

Table 2. List of compounds with both mono methyl substituted-4-pyridyl and unsubstituted or mono substituted phenyl rings and their optimum concentrations for cytokinin activity in the tobacco bioassay

Compd. No.	R	Optimum concentration M	ppm
2a	H	4.0 x 10⁻⁷	0.1
2b	2-Me	3.8 x 10⁻⁵	10
2c	3-Me	3.8 x 10⁻⁶	1
2d	4-Me	Inactive	
2e	2-Cl	3.6 x 10⁻⁵	10
2f	3-Cl	3.8 x 10⁻⁶	1
2g	4-Cl	3.8 x 10⁻⁵	10

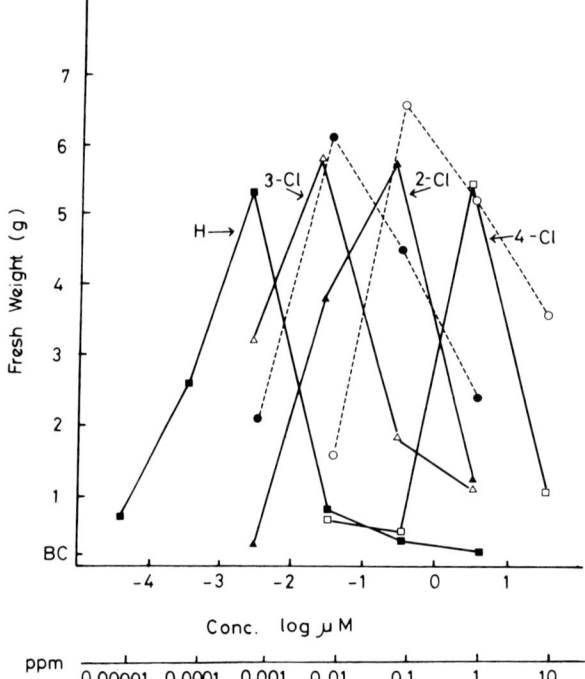

Fig. 1. Effect of compounds 3a, 3e, 3f 3g, BA and 4PU on fr. wt yield of tobacco callus

ring showed similar or higher activity to that of 10a which has unsubstituted phenyl ring.

e) In order to clarify the effect of the substitution of position 3 of pyridyl ring, compounds with both 3-CH_3 or 3-Cl substituted-4-pyridyl and unsubstituted or substituted phenyl rings were synthesized and tested. As shown in Table 5, in the case of unsubstituted phenyl ring (13a and 14a), they both showed marked reduction of activity, i.e., by 3-CH_3 substitution of pyridyl ring (13a), the activity was 1/100 lower and by 3-Cl substitution (14a), 1/10000 lower than those of the corresponding 2-substituents (2a and 3a). Therefore, position 3 of the pyridyl ring should be kept intact for giving the higher activity.

f) Concerning the disubstitution of two α-positions of the pyridyl ring, at first seven kinds of 2,6-di-CH_3 substituents with unsubstituted or substituted phenyl rings (15a, 15b, 15c, 15d, 15e, 15f and 15g) were synthesized. However, as shown in Table 6, their activity was very weak and most of them were almost or completely

Table 3. List of compounds with both mono chloro substituted-4-pyridyl and unsubstituted or mono substituted phenyl rings and their optimum concentration for cytokinin activity in the tobacco bioassay

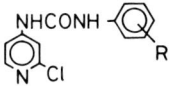

Compd. No.	R	Optimum concentration M	ppm
3a	H	4.0×10^{-9}	0.001
3b	2-Me	3.8×10^{-8}	0.01
3c	3-Me	3.7×10^{-8}	0.01
3d	4-Me	3.8×10^{-6}	1
3e	2-Cl	3.5×10^{-7}	0.1
3f	3-Cl	3.5×10^{-8}	0.01
3g	4-Cl	3.5×10^{-6}	1
3h	2-F	3.8×10^{-8}	0.01
3i	3-F	3.8×10^{-9}	0.001
3j	4-F	3.8×10^{-8}	0.01
3k	3-Br	3.1×10^{-7}	0.1
3l	4-Br	3.1×10^{-5}	10
3m	2,5-Cl$_2$	3.2×10^{-7}	0.1
3n	4-n-Pr	3.5×10^{-6}	1
3o	3-OH	3.8×10^{-7}	0.1
3p	3-OMe	3.6×10^{-7}	0.1

Table 4. List of compounds with both unsubstituted or mono substituted 4-pyridyl and unsubstituted or mono substituted phenyl rings and their optimum concentrations for cytokinin activity in the tobacco bioassay

Compd. No.	R_1	R_2	Optimum concentration M	ppm
1a	H	H	4.7×10^{-7}	0.1
2a	CH$_3$	H	4.0×10^{-7}	0.1
3a	Cl	H	4.0×10^{-9}	0.001
4	OCH$_3$	H	4.1×10^{-8}	0.01
5	NHCOCH$_3$	H	1.9×10^{-7}	0.05
6	NH$_2$	H	4.4×10^{-6}	1
7	OH	H	4.0×10^{-7}	0.1
8	F	H	4.3×10^{-9}	0.001
9	Br	H	3.4×10^{-9}	0.001
10a	CF$_3$	H	3.6×10^{-9}	0.001
10b	CF$_3$	F	3.3×10^{-9}	0.001
11	CN	H	4.2×10^{-9}	0.001
12	COOCH$_3$	H	2.8×10^{-7}	0.075

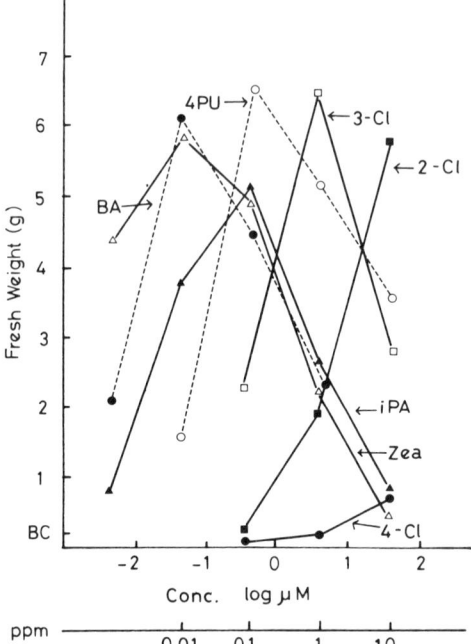

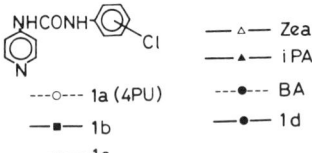

Fig. 2a. Effect of compounds 1b, 1c, 1d, 1a(4PU), BA, zeatin (Zea) and 6-(3-methyl-2-butenylamino)purine (iPA) on fr. wt yield of tobacco callus

Table 5. List of compounds with both mono 3-substituted-4-pyridyl and unsubstituted or substituted phenyl rings and their optimum concentrations for cytokinin activity in the tobacco bioassay

Compd. No.	R_1	R_2	Optimum concentration M	ppm
13a	Me	H	4.2×10^{-5}	10
13b	Me	Cl	3.8×10^{-5}	10
14a	Cl	H	4.0×10^{-5}	10
14b	Cl	Cl	3.5×10^{-5}	10

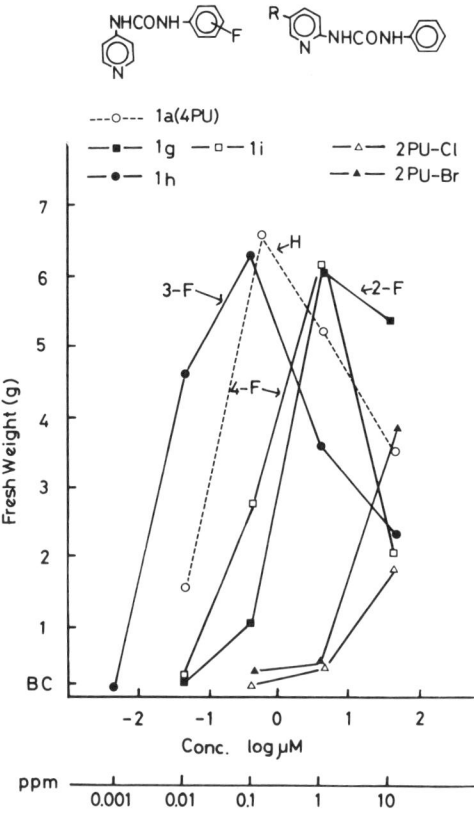

Fig. 2b. Effect of compounds 1g, 1h, 1i, 2PU-Cl, 2PU-Br and 1a(4PU) on fr. wt yield of tobacco callus

inactive. But once compounds with both 2-Cl:6-CH$_3$(16), 2,6-di-Cl (17a), 2-Cl:6-OCH$_3$(18a), 2,6-di-OCH$_3$(19), or 2,6-di-Br(20) substituted-4-pyridyl and unsubstituted phenyl rings were synthesized, they all showed remarkably higher activity. Moreover, as shown in Fig. 5, (17b), i.e., compounds with 2,6-di-Cl substituted-4-pyridyl and meta(3)-F substituted phenyl rings and (18b), i.e., compounds with both 2-Cl:6-OCH$_3$ substituted-4-pyridyl and meta(3)-F substituted phenyl rings gave higher activity than the corresponding compounds with unsubstituted phenyl rings (17a and 18a).

g) As to the cytokinin activity of the 4PU derivatives having thiourea chain, compounds with both mono-(Cl, OCH$_3$ or COOCH$_3$) substituted-4-pyridyl and unsubstituted phenyl rings were synthesized and tested. As shown in Table 7, they gave similar optimum concentrations and growth curves to the corresponding compounds with urea chain. Therefore, substitution of urea chain of 4PU derivatives with thiourea chain seems to give similar activity.

h) In addition to the above-mentioned compounds, miscellaneous types of related compounds of 4PU were synthesized and their activity was assayed. As shown in Table 8, compounds with both 2-Cl mono substituted-4-pyridyl and unsubstituted pyridyl rings (instead of phenyl ring) showed considerable activity and gave pro-

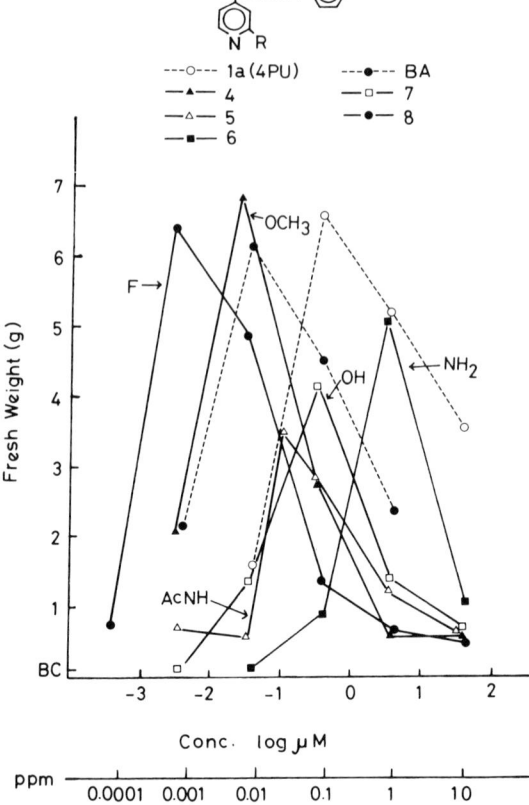

Fig. 3. Effect of compounds 4, 5, 6, 7, 8, BA and 4PU on fr. wt yield of tobacco callus

ducts with the following decreasing order of activity: 2>3>4 with respect to the position of the N atom in the pyridyl ring. But compound (24), with the highest activity, showed the optimum concentration of 0.01 ppm which is 1/10 lower than the corresponding compound with phenyl ring (4PU-30 – 0.001 ppm), while various compounds (27-36) with 4-pyridyl of 2-Cl-4-pyridyl amide derivatives shown in Table 9 showed lower activity and four of them were inactive. These results suggest that the direct combination of N-phenyl ring with urea chain and that the distance between pyridyl and phenyl rings being equal to the length of urea chain are the indispensable structural conditions for giving higher cytokinin activity of 4PU derivatives.

i) In addition to 4PU derivatives, N-phenyl-N'-(5-Cl or 5-Br-2-pyridyl)ureas (2PU-Cl or 2PU-Br), i.e., 2PU derivatives were synthesized and tested. As shown in Fig. 2b, though their optimum concentration is 10 ppm, their callus yield is very low. Considering the fact that 3PU had lower (10 ppm) activity and 2PU was almost inactive as mentioned above, it may be said that pyridyl ring combined directly with urea chain at position 4 is also a necessary structural requirement for giving higher ac-

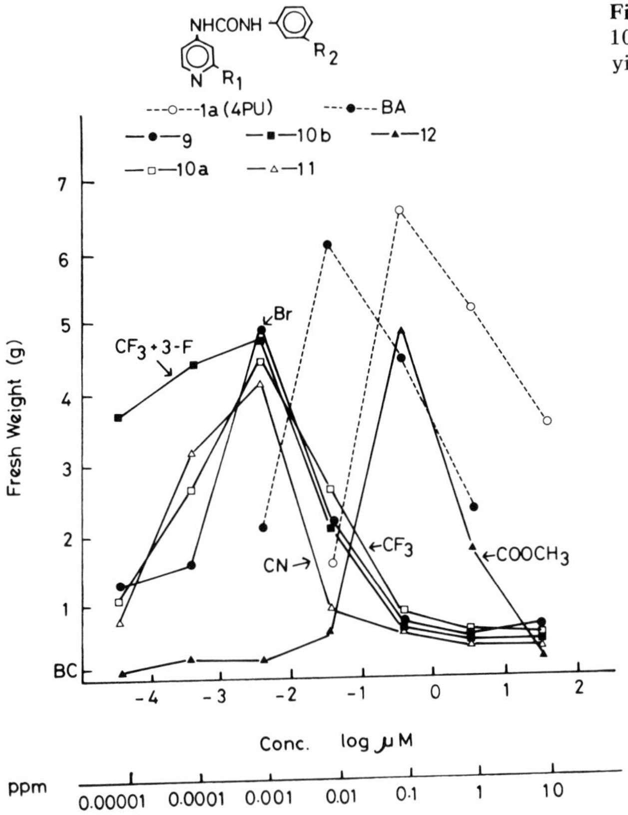

Fig. 4. Effect of compounds 9, 10a, 10b, 11, 12, BA and 4PU on fr. wt yield of tobacco callus

tivity. To sum up the conclusions of (8) and (9), the chemical structure of unsubstituted N-phenyl-N'-(4-pyridyl)urea is an essential, indispensable structure of these synthesized new type of cytokinins.

j)[3] The similarity in the biological activity of structurally unrelated classes of compounds, such as BA and 4PU derivatives, has posed the problem of their sites of action. Kurosaki tried to solve the problem using inhibitors for cytokinins. For this purpose, two inhibitors were selected; the one is N-benzyl-N'-4-bromophenylurea (BP) which was reported by Kefford et al [7] as an urea type inhibitor for DPU and BA being tested by chlorophyll retention bioassay, the other is 4-cyclobutylamino-2-methylpyrrolo[2,3-d]pyrimidine (CP)[4] which was reported by Iwamura et al. [8] as an inhibitor for the cell division activity of K. BP and CP also have inhibitory activity on the action of 4PU in the tobacco bioassay.

---

[3] This part of study was carried out by F. Kurosaki in Okamoto's laboratory and constitutes part of his doctoral thesis

[4] We are grateful to Dr. H. Iwarmura for providing the sample of 4-cyclobutylamino-2-methylpyrolo[2,3-d]pyrimidine

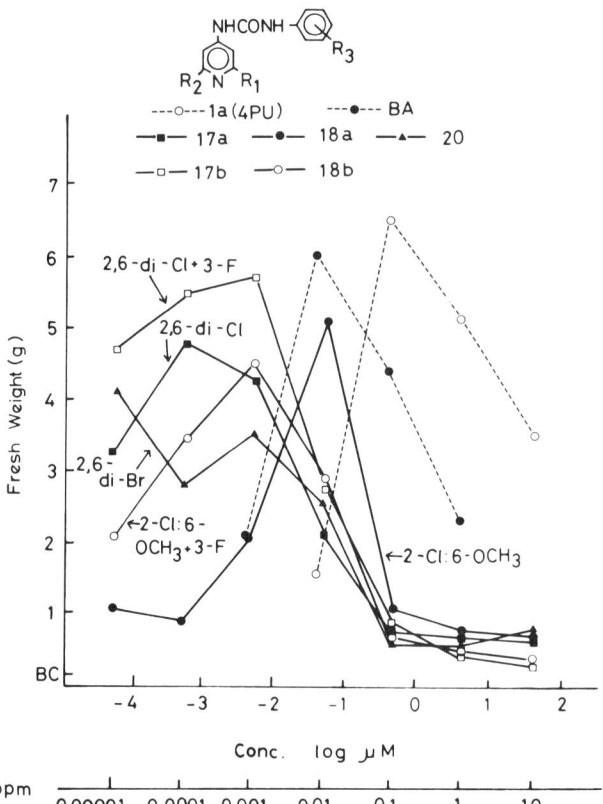

Fig. 5. Effect of compounds 17a, 17b, 18a, 18b, 20, 1a-(4PU) and BA on fr. wt yield of tobacco callus

In the tobacco bioassay, the cytokinin activity of BA and 4PU expressed as percentages of the fr. wt of callus at the optimum concentration was reduced in the presence of 33 μM of BP by its inhibitory action as shown in Fig. 6a and b. But at concentrations higher than optimum, was released up the inhibited (reduced) growth to the maximum growth. These results mean that BA and 4PU acted as competitive reversing agents. In other words, BP is a competitive inhibitor for both BA and 4PU. Similar results were obtained using CP as shown in Fig. 6c and d. In short, BP and CP exhibited "cross" competitive inhibitions on BA and 4PU. These observations suggest that 4PU and purine cytokinins have the same receptor site for cell division of tobacco callus.

k) In order to clarify whether there are some compounds which have cell-division activity in the absence of exogenously added auxin in the tobacco bioassay or not, all the 4PU compounds (ca 150 kinds) were tested for their effect on callus in the tobacco bioassay without auxin. The result was that, on the one hand, it became clear that such a compound did not exist, but on the other, it was clarified that all the 4PU derivatives which show cytokinin activity at the optimum concentration of 1 ppm or lower in the tobacco bioassay, initiate or form shoots from callus in the complete absence of exogenous auxin, and that the optimum concentration for shoot forma-

Cytokinin Activities of N-Phenyl-N'-(4-Pyridyl)ureas

Table 6. List of compounds with both disubstituted-4-pyridyl and unsubstituted or mono substituted phenyl rings and their optimum concentrations for cytokinin activity in the tobacco bioassay

Comp. No.	R_1	R_2	R_3	Optimum concentration M	ppm
15a	Me	Me	H	4.0×10^{-6}	1
15b	Me	Me	2-Me	Almost inactive	
15c	Me	Me	3-Me	Almost inactive	
15d	Me	Me	4-Me	Inactive	
15e	Me	Me	2-Cl	Inactive	
15f	Me	Me	3-Cl	3.4×10^{-5}	10
15g	Me	Me	4-Cl	Inactive	
16	Cl	Me	H	3.8×10^{-9}	0.001
17a	CE	CE	H	3.5×10^{-9}	0.001
17b	Cl	Cl	3-F	3.3×10^{-9}	0.001
18a	Cl	OMe	H	3.6×10^{-8}	0.01
18b	Cl	OMe	3-F	3.5×10^{-9}	0.001
19	OMe	OMe	H	3.7×10^{-8}	0.01
20	Br	Br	H	2.7×10^{-9}	0.001

Table 7. List of thiourea compounds with both mono substituted-4-pyridyl and phenyl rings and their optimum concentrations for cytokinin activity in the tobacco bioassay

Compd. No.	R	Optimum concentration M	ppm
21	Cl	3.8×10^{-9}	0.001
22	OCH_3	3.9×10^{-8}	0.01
23	$COOCH_3$	3.5×10^{-7}	0.1

Table 8. List of compounds with both mono substituted-4-pyridyl and 2,3 or 4-pyridyl rings and their optimum concentrations for cytokinin activity in the tobacco bioassay

Compd. No.	R	Optimum concentration M	ppm
24	2-pyridyl	4.0×10^{-8}	0.01
25	3-pyridyl	4.0×10^{-7}	0.1
26	4-pyridyl	4.0×10^{-6}	1

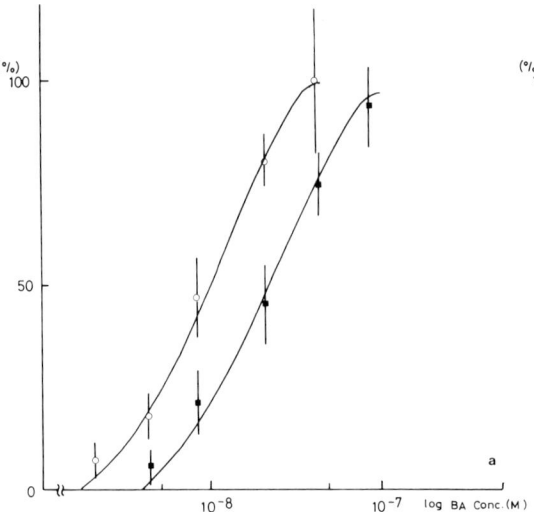

Fig. 6a. Dose response curves of BA and BA plus inhibitor. —○— BA: —■— BA + 33 µM of N-benzyl-4-bromophenyl urea (BP)

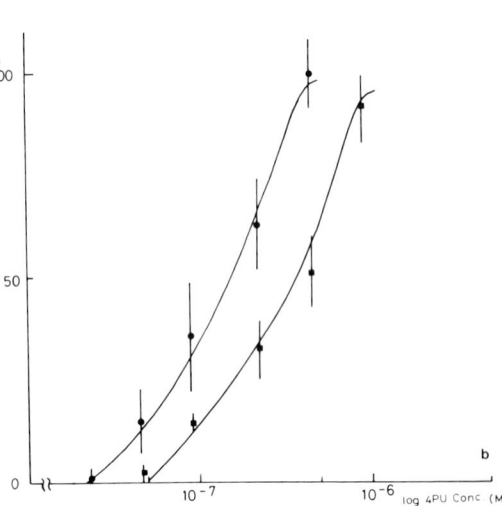

Fig. 6b. Dose response curves of 4PU and 4PU plus inhibitor. —●— 4PU: —■— 4PU + 33 µM of N-benzyl-4-bromophenyl urea (BP)

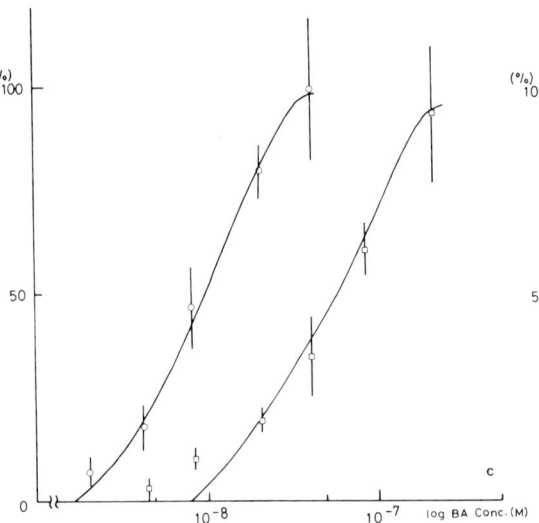

Fig. 6c. Dose response curves of BA and BA plus inhibitor. —○— BA: —□— BA + 0.5 µM of 4-cyclobutylamino-2-methylpyrrolo-[2,3-d]-pyrimidine (CP)

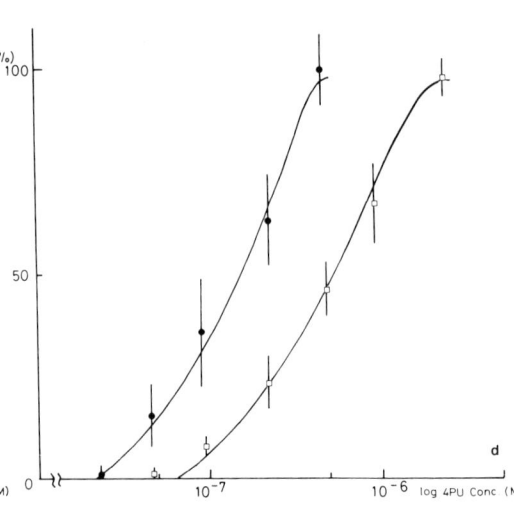

Fig. 6d. Dose response curves of 4PU and 4PU plus inhibitor. —●— 4PU: —□— 4PU + 0.4 µM of 4-cyclobutylamino-2-methylpyrrolo-[2,3-d]-pyrimidine (CP)

Table 9. List of compounds with 4-pyridyl or 2-chloro-4-pyridyl amide derivatives and 2-chloro-isonicotinoylanilide and their optimum concentrations for cytokinin activity in the tobacco bioassay

Compd. No.	R_1	R_2	Optimum concentration M(ppm)
27	H	NH_2	Inactive
28	Cl	$N(CH_3)_2$	5.9×10^{-5} (10)
29	Cl	NHCH$_2$–⟨phenyl⟩	3.8×10^{-6} (1)
30	H	$N(CH_3)$–⟨phenyl⟩	Inactive
31	H	O–⟨phenyl⟩	4.7×10^{-5} (10)
32	Cl	O–⟨phenyl⟩	4.0×10^{-6} (1)
33	Cl	S–⟨phenyl⟩	3.8×10^{-6} (1)
34	H	CH_2–⟨phenyl⟩	Inactive
35	H	⟨phenyl⟩	5.7×10^{-5} (10)
36	H	⟨furyl⟩	Inactive
37	\multicolumn{2}{|c|}{2-chloro-isonicotinoylanilide}	4.3×10^{-6} (1)

tion is ca 100 times higher than that of cytokinin activity [4, 9]. However, this phenomenon is at present limited only in the case of tobacco pith callus.

These 4PU derivatives also have chlorophyll retention activity in the bioassay using leaves of various plant species, and show promotion of the growth of lateral buds (inhibitory action of auxin in apical dominance) in tobacco plants and *Datura* sp. . Also as one of the practical agricultural uses of 4PU-30, inhibition of flower shedding and acceleration of fruit bearing in grape is now under study, and by combining application of 100 ppm of gibberellin on Delaware grape, 4PU-30 gives 175% (25 ppm), 184% (50 ppm) and 190% (100 ppm) increase of the weight of berry cluster as compared to gibberellin alone, while BA gives the values of 122% (100 ppm) and 132% (200 ppm).

3 Bioassay Procedure

Cytokinin activity was determined in the tobacco pith callus bioassay which was described in [1, 3, 4] and was based on the optimum concentration range and fr. wt yield of callus tissue. Since most of the synthetic compounds are sparingly soluble in water they were at first dissolved in a small quantity of DMSO and added to the nutrient medium. No effect of DMSO on the bioassay was observed. The same test was repeated at least three times until similar results were obtained.

References

1. Torigoe Y, Akiyama N, Hirobe M, Okamoto T, Isogai Y (1972) Cytokinin activity of azaindene, azanaphthalene, naphthalene and indole derivatives Phytochemistry 11: 1623-1630
2. Bruce MI, Zwar JA (1966) Cytokinin activity of some substituted ureas and thioureas Proc R Soc London B. 165: 245-265
3. Okamoto T, Shudo K, Isogai Y (1974) Plant growth substances. Hirokawa Publ Co Ltd, Tokyo pp 447-455
4. Isogai Y, Shudo K, Okamoto T (1976) Effect of N-phenyl-N'-(4-pyridyl)urea on shoot formation in tobacco pith disk and callus culture in vitro. Plant Cell Physiol 17: 591-600
5. Takahashi S, Yatsunami T, Shudo K, Okamoto T, Yamada K, Isogai Y (1978) Cytokinin activity of pyrimidine derivatives Chem Pharm Bull 26: 2286-2287
6. Takahashi S, Shudo K, Okamoto T, Yamada K, Isogai Y (1978) Cytokinin activities of N-phenyl-N'-(4-pyridyl)urea derivatives. Phytochemistry 17: 1201-1207
7. Kefford NP, Zwar JA, Bruce MI (1967) Antagonism of purine and urea cytokinin activities by derivatives of benzylurea. In Wightman F, Setterfield G (eds) Biochemistry and physiology of plant growth substances. Runge Press Ltd, Ottawa, pp 61-69
8. Iwamura H, Masuda N, Koshimizu K, Matsubara S (1979) Cytokinin – agonistic and antagonistic activities of 4-substituted-2-methylpyrrolo[2,3-d]pyrimidines, 7-deaza analogs of cytokinin-active adenine derivatives. Phytochemistry 18: 217-222
9. Isogai Y, Yamada K, Takahashi S, Shudo K, Okamoto T (1978) Shoot formation effect of ureas, Pyrimidines and related compounds on tobacco callus cultured in vitro. Sci Pop Coll Gen Edu Univ Tokyo 28: 93-127

Towards an Understanding of the Mechanism of Cytokinin Activity in *Mercurialis annua* L. Sex Differentiation

A. Champault[1], S. Chung[2], B. Guerin[1], G. Kahlem[1], A. Lhermitte[1], G. Teller[3], and B. Durand[1]

1 Introduction

The sexual differentiation system of *Mercurialis,* a dioecious plant, constitutes an integrated eukaryotic system. We have previously shown that the staminal and carpellate sexual differentiation programs are present in the sub-tunical cell lines of the floral primordia which are genetically male (♂) or female (♀). These cells are in fact bipotential: whether they are on ♂ or ♀ plants, they can initiate the formation of either stamens or carpels and ovules (Kahlem et al., 1975).

The pattern of the sex expression can be determined by studying the macromolecules synthesized in the stamen or carpel cells. It is not necessary to examine a large variety of these macromolecules, but rather only some of those which are specific to each sex; they constitute molecular markers or probes of differentiation. Thus, during the first cellular staminal repartitioning, a group of three isoperoxidases appears. They are synthesized in the cells of the ♂ floral primordia (except the tunic and the vascular tissues) and persist until the end of meiosis (Kahlem, 1976). Two isoesterases and two chloroplast tyrosine tRNAs (Bazin et al., 1975) are specific for ♀ flowers.

For those studying cytokinins, the interest of the system is that when these hormones are given exogenously to genetically ♂ plants, they have a feminizing effect. They induce the ♀ program in the new floral primordia formed. Then the synthesis of ♀ molecular markers occurs. These molecules are thus "cytokinin-dependent". We may add that the administration of IAA to genetically ♀ plants leads to the appearance of ♂ flowers and ♂ specific isoperoxidases (Champault, 1973). Similarly, when tissue cultures of each sex are grown, the expression of maleness or femaleness can be established by screening for two isoperoxidases, which constitute specific markers for the ♀ strain. We call these sex-related markers.

Since exogenous cytokinins feminize genetic ♂, we formulated a hypothesis that ♀ plants could contain a pool of endogenous cytokinins greater than that of ♂. This line of research could eventually lead to the determination of which of all the endogenous metabolites induce ♀ sexual differentiation, if indeed one (or several) endogenous metabolites could be shown to be specific for this differentiation. This research could also provide information for the determination of physiologically active cytokinins.

---

[1] Laboratoire de Biologie Végétale, Université d'Orléans, 45046 Orléans Cédex, France
[2] Yeungnam University, Faculty of Pharmaceutical Sciences, Taegu, Korea
[3] Laboratoire de Spectrométrie de Masse, Université Louis Pasteur, 67008 Strasbourg Cédex, France

Using a computerized gas chromatography/mass spectrometry system, we simultaneously purified, identified and assayed the currently known, frequently encountered cytokinins. The extraction and purification method was developed in our laboratory and the assay was performed in cooperation with the Mass Spectrometry Laboratory in Strasbourg (Dauphin et al., 1975, 1979, 1980). Assays were first performed on ♂ and ♀ tissue cultures, then directly on ♂ and ♀ plants and finally on autonomous tissues, obtained by infecting plants of each sex with *Agrobacterium tumefaciens*. We will see how the data gathered in these three series of comparative analyses could provide arguments for or against our hypothesis.

2 Results

2.1 Tissue Strains

Three tissue strains (♂ R, ♂ S and ♀) were selected from shoots grown *in vitro*, taken from various types of ♂ and ♀ plants. The genetic formula of the plants was determined in the laboratory by formal genetic analysis (Louis et al., 1978). Sensitive easily feminized ♂ have the formula Ab_1B_2. Others, only feminized by BAP, are termed resistant ♂ (R) and have the formula AB_1B_2. Females have the formula Ab_1b_2 or aB_1B_2. The three tissue strains obtained were grown in the presence of 10^{-4} g/l of BAP and 10^{-4} g/l of 2,4-D. As seen in Table 1, endogenous native metabolites could not be detected in ♀. Their concentration is probably too low, in spite of the lower limit of detection of the assay system, which is 10 ng. I_6 Ado and RZ could be characterized in the two male strains. It was also determined that BAP exists in ♀ (850 ng), while the two ♂ strains showed considerable differences. Resistant ♂ contain about ten times as much as sensitive ones. A portion of the BAP is transformed to RBAP which is encountered in the three strains, in the absence of a parallelism between BAP and RBAP concentrations (Fig. 1). Nucleotides, especially those of BAP, could not be detected in any of the three strains; the same was true of Z. As we will see, this is a characteristic of these tissue strains when compared to whole plants.

Table 1. Cytokinins from three callus strains (Resistant ♂, Sensitive ♂ and ♀)

Callus	♂ R AB_1B_2	♂ S Ab_1B_2	♀
Cytokinins	ng/100 g fresh weight		
I_6 Ado	7.5	28	0
Trans RZ	314	157	0
Trans Z	0	0	0
BAP	2140	233	854
R. BAP	966	428	860
BAP/RBAP Ratio	2,2	0.54	1
Z, I_6 Ade, BAP } Ribotides	0	0	0

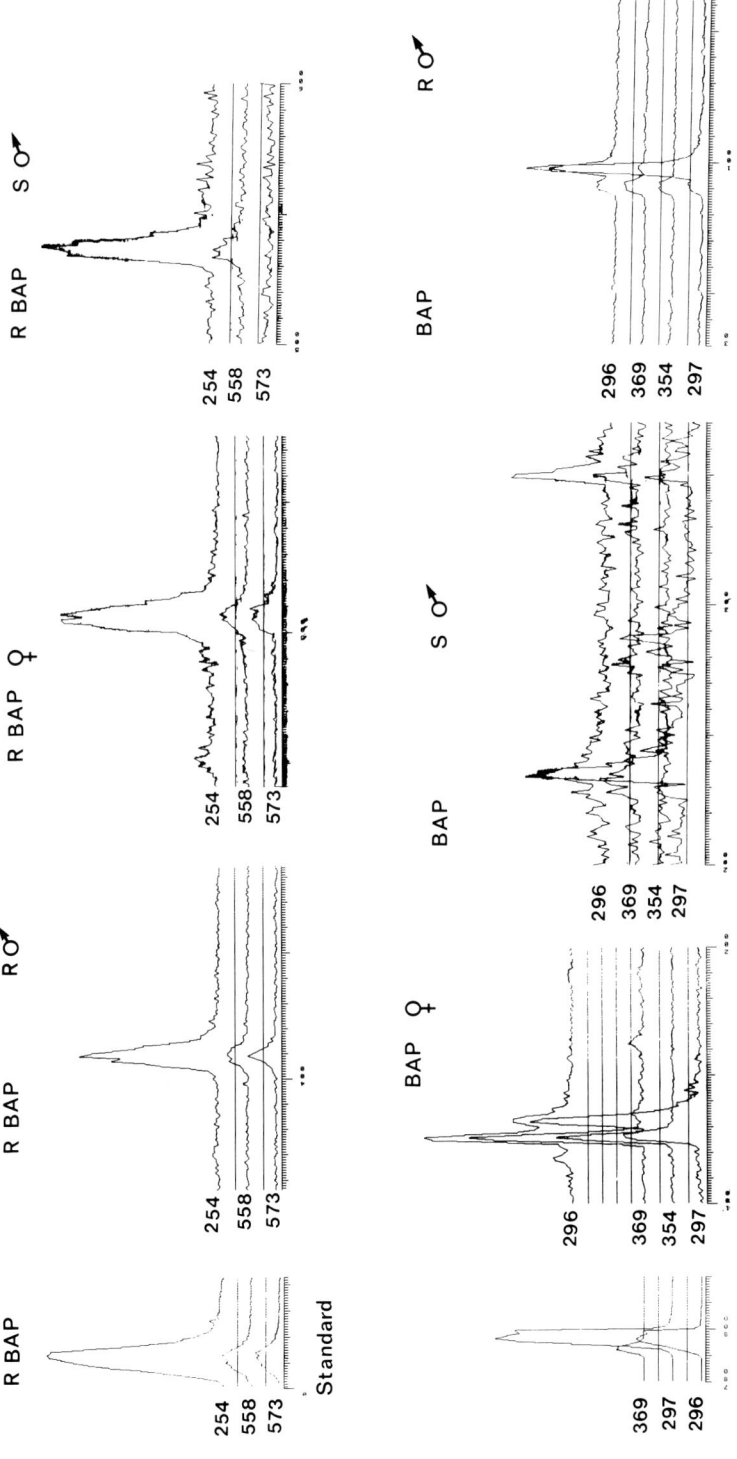

Fig. 1. RBAP and BAP detections by scanning for characteristic spectrum ions in the recorded MS spectra of three strains of *Mercurialis annua* tissue cultures. For a detailed description of the methods, see Dauphin et al. (1979)

The content of endogenous and exogenous cytokinin metabolites is clearly different in the three strains. Thus there exists a genetic basis of general cytokinin metabolism, which is related to the sex of the cultures. There is also a difference between the two ♂ with different genetic formulas.

2.2 Apices and Whole Plants

Sporophyte assays lead to analogous observations. As seen in Table 2, the common nucleosides I_6 Ado and RZ are found in both sexes. The same is not true for the free bases and ribotides: Z could be easily identified in ♀, where it was found to be the most abundant metabolite (Fig. 2). Its concentration was five times greater than that of nucleosides and it is the maximum reached by a metabolite in this species. For a comparable sample, however, that is, up to 800 g of shoot apices, Z could not be detected in ♂. This could result from either the very low concentration being below the lower detection limit of the system (10 ng) or from the fact that Z is replaced by another free base. An indirect response to this question could be given by an analysis of nucleotide contents. The I_6 Ade nucleotides are thus present in both sexes, just as the corresponding ribosides, but at lower concentrations. The Z nucleotides could be traced only in ♂. They were absent or present in very low concentrations in ♀ (Fig. 2).

Thus it may be concluded that the main differential metabolite between the sexes is trans-Z. This difference is above all quantitative, since male metabolism is oriented toward ribotide production and toward accumulation of the free base in ♀, probably resulting from a higher level of a hydrolase activity.

These results are consistent with the possibility that Z among the cytokinins, could be one of the key feminizing materials.

Table 2. Cytokinins from ♂ and ♀ sporophytes

Mercurialis plants	♂		♀	
Cytokinins	ng/100 g fresh weight			
	Apices	Whole plant	Apices	Whole plant
I_6 Ado	203	115	71	73
Cis-RZ	71	–	22	–
Trans-RZ	188	–	65	–
Trans-Z	0	0	242	positive Detection
I_6 Ade Ribotide	–	80	–	56
Z-Ribotide	–	239	–	0

– not tested

2.3 Mercurialis Crown Gall Tissue

In order to confirm these previous results, one of us inoculated the same *Agrobacterium tumefaciens* strain (15,955) in ♂ and ♀ *Mercurialis* plants. In ♀, no modification of floral morphogenesis was observed in the immediate surrounding of the tumor.

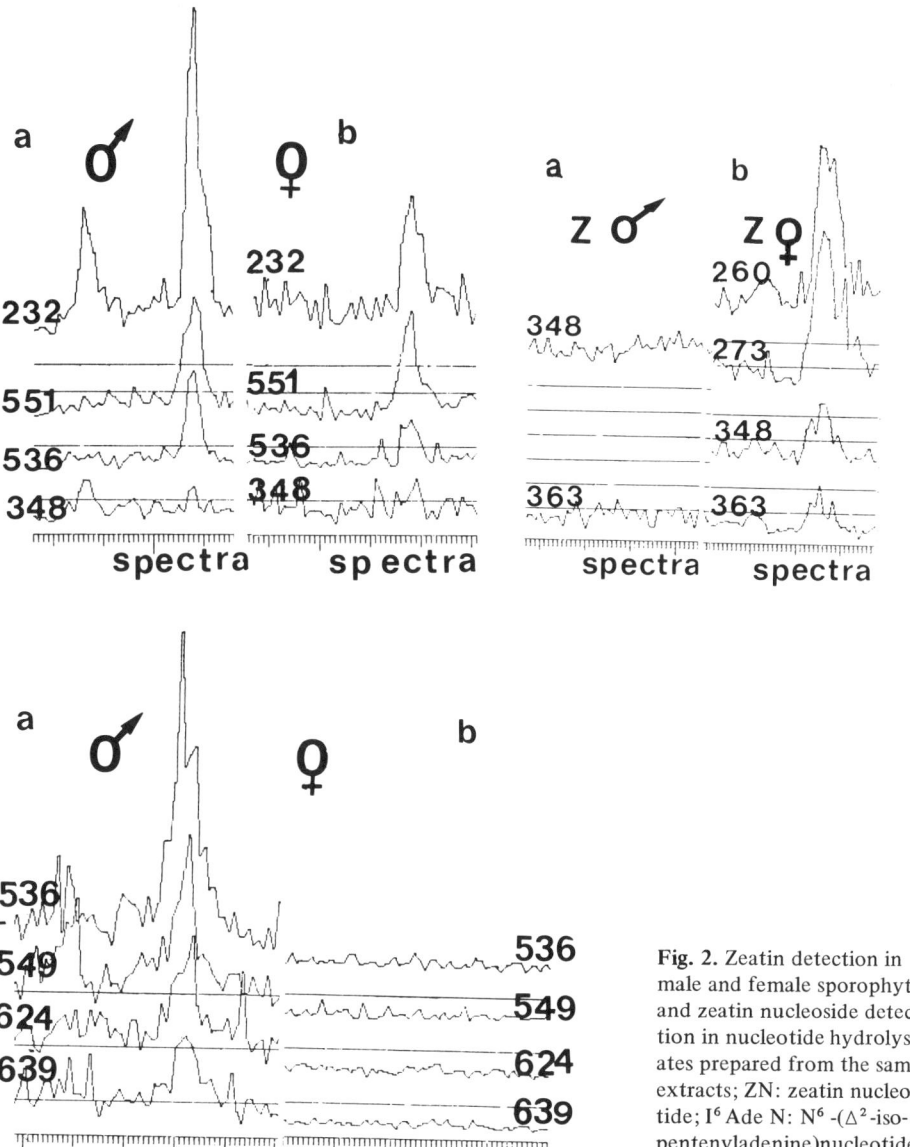

Fig. 2. Zeatin detection in male and female sporophytes and zeatin nucleoside detection in nucleotide hydrolysates prepared from the same extracts; ZN: zeatin nucleotide; I⁶Ade N: N^6-(Δ^2-isopentenyladenine)nucleotide. (Dauphin-Guerin et al., 1980)

In ♂, inoculated near a leaf next to a floral meristem, the tumor induced some anomalies. Male inflorescence was poorly developed and contained ♀ flowers and stamen-carpels. An aseptic culture of the tumor tissues was performed. It was grown near young ♂ inflorescences and feminized carpel-shaped structures were observed.

Then we analyzed the cytokinin contents of ♂ and ♀ *Mercurialis* tumor tissues. In ♂ crown-galls, there are lower quantities of RZ than in ♀ and traces of nucleotides (Table 3). Ribotides can be detected easily, since they are more abundant than I⁶Ado.

Table 3. Cytokinins from crown-gall obtained on ♂ and ♀ plants infected by *A. tumefaciens*

Crown-Gall (*A. tumefaciens* 15,955)	♂	♀
Cytokinins	ng/100 g fresh weight	
I_6 Ado	26.8	6.8
Trans-RZ	164	841
Trans-Z	936	806
Methyl-Thio Z	–	Positive Detection
I_6 Ade Ribotide	5.2	18
Z-Ribotide	–	16
I_6 Ade 9-Glucoside	–	Positive Detection

Finally, ♂ and ♀ tumor tissues produce large quantities of trans-Z 3 to 4 times more than ♀ apices.

The presence of Z in inoculated ♂ plants is thus correlated with the appearance of ♀ flowers. The presence of the *Agrobacterium* plasmid not only modified the relative quantities of metabolites by displacing the equilibrium toward Z production in ♂, but also induced clear qualitative differences:

1. RZ not longer exists in the cis- and trans-forms, as in healthy plants. The cis-form is not found in tumor tissues.

2. Another base, 2-methylthiozeatin, could be detected in ♀ crown galls. Its apparently high concentration could not be determined, since we had no standard available (Fig. 3).

Finally, as in most crown gall tissues, I_6 Ade-9-glucoside could be detected in ♀.

To summarize, ♂ *M. annua* crown gall tumor tissues produce large quantities of Z and here again the correlation between the accumulation of free base and feminization is positive. In ♀ tumor tissues, the endogenous cytokinin pool is similar, although the riboside and ribotide concentrations are higher, especially for RZ. This situation may again be interpreted as a sex-related heredity phenomenon.

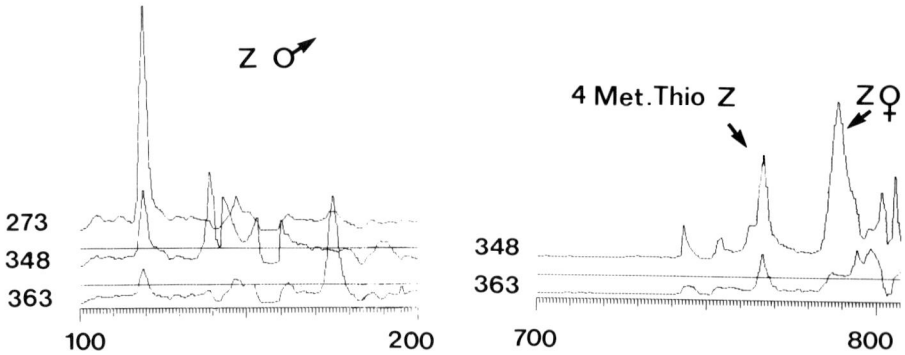

Fig. 3. Zeatin detection in male crown-gall tissue and 2-methylthiozeatin in female crown-gall tissue. For methods, see references in Figs. 1 and 2

2.4 Mercurialis annua and Glucosaminides

The results of all these analyses show that the presence of one metabolite, Z, is related to ♀ differentiation; but as we will see now, it can act in synergy with other metabolites such as methylated N-acetyl-glucosamines which have a physiological cytokinin effect but are not substituted adenines.

Thus some fractions of our extracts separated on a LH 20 Sephadex column showed a positive cytokinin bioassay that cannot be related to the detection of any known cytokinin. In the computer analysis data of these extracts, some spectra evoked glucosaminides.

Then four glucosaminides have been tried on following bioassays: Nitrate reductase, radish cotyledons, soja callus and *Mercurialis* feminization. All these glucosaminides acted in *Agrostemma* bioassay with lower concentrations than BAP and showed an equal or higher activity (Fig. 4a). On soja callus and radish cotyledons bioassay the action was weaker and intermediate in *Mercurialis* feminization, so that there is a relative specificity of the biological answer.

But all these chemical species had a common character; they acted in synergy in all bioassays with all cytokinin metabolites synthetic or native. The activity measurements reached exceptional values and inhibitor concentrations became very active (Fig. 4b).

To explain these facts a hypothesis can be proposed; some cytokinins possess a glucosaminide moiety as the glucoside moiety of some cytokinin metabolites encountered in different other species (Deleuze et al. 1972; Etham et al. 1976). To test this, we took again a ♀ crown gall extract.

The fractions issued from cation exchange Sephadex corresponding to standard glucosaminides and following those of cytokinins, were treated by N-acetyl-glucosaminidase. In the hydrolysates, RZ and Z can be easely characterized (Fig. 5); N-acetyl-glucosamine has been of course detected, but also methylated glucosaminides, sometimes linked to another sugar to constitute a disaccharide.

This suggests that, at least in *Mercurialis,* one can find molecules containing a covalent link between a cytokinin and a glucosaminide.

We have previously seen that one metabolite, Z, is related to ♀ differentiation. Nevertherless we can add that perhaps it is not alone. Other experiments of course will be necessary on this point.

2.5 Mercurialis Cytokinin Receptors

How can we imagine the role of the active cytokinin metabolite triggering the ♀ differentiation program in floral primordia? According to the hypothesis of hormone action via the participation of specific receptors, ♂ primordia would be the preferred target organs for exogenous cytokinins. The target cells would be those of the second subtunicate layer. By analogy, it may be noted that ♀ primordia are the target organs for exogenous auxins. We believed it is interesting to compare the association constants of a free base, first on a known receptor site, the ribosomes and then at the level of the chromatin of each sex.

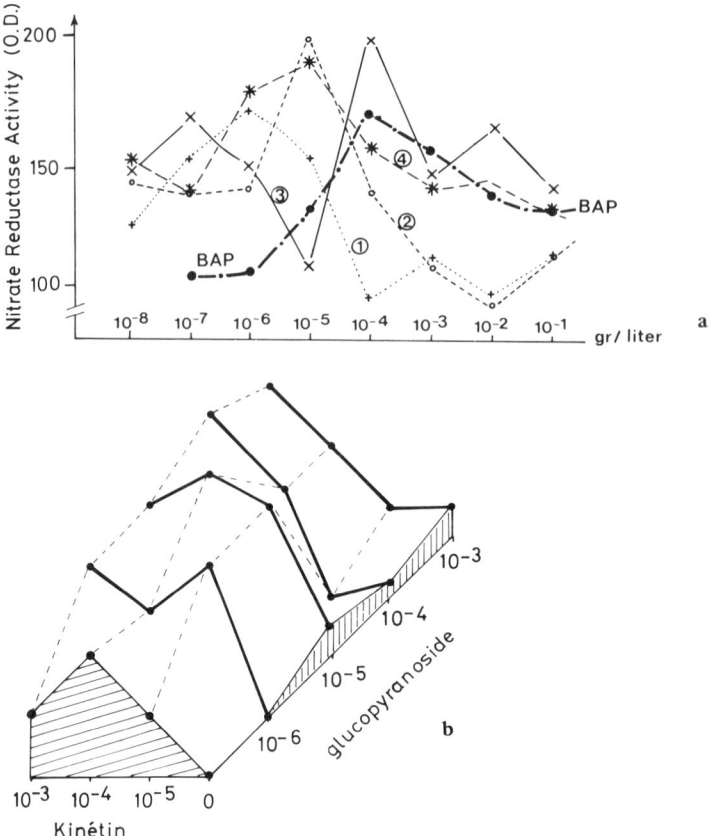

Fig. 4a, b. Nitrate reductase activity (**a**) and soybean callus tissue bioassay (**b**) for glucosaminides; a *1* O-methyl-(2-acetamido, 2-deoxy, 3-O-methyl)-α, β-D-glucopyranoside; *2* O-methyl-(2-acetamido, 2-deoxy, 3-6 di-O-methyl)-α, β-D-glucopyranoside; *3* O-methyl-(2-acetamido, 2-deoxy, 3-O-methyl)-α-D-glucopyranoside; *4* O-methyl-(2-acetamido, 2-deoxy, 3,4,6-tri-O-methyl)-α-D-glucopyranoside. **b** *ordinates*, relative weight of callus tissue

Thus, in an initial series of experiments, we compared the binding kinetics of a synthetic base – [^{14}C]-BAP – with ribosome subcellular fractions of ♂ and ♀ target organ cells, using the technique of Hertel (Chung et al., 1979). Unlabeled BAP is clearly competitive with [^{14}C]-BAP fixed on ♂ and ♀ ribosomes: ♀ ribosomes fix half as much BAP as male ribosomes and are less sensitive to competition. Using Scatchard plots, the number of sites can be calculated, as well as the association constants. Males have twice as many sites, while the association constant is 1.5 times higher in ♀. Ribosomal receptors are thus different in each sex and ♀ receptors are more avid.

It next appeared necessary to verify if Z had an increased affinity for receptors, in view of its physiological efficiency. We thus compared the binding kinetics of various bases: BAP, FAP, I_6 Ade, and Z, on one hand, and their ribosides, on the other hand.

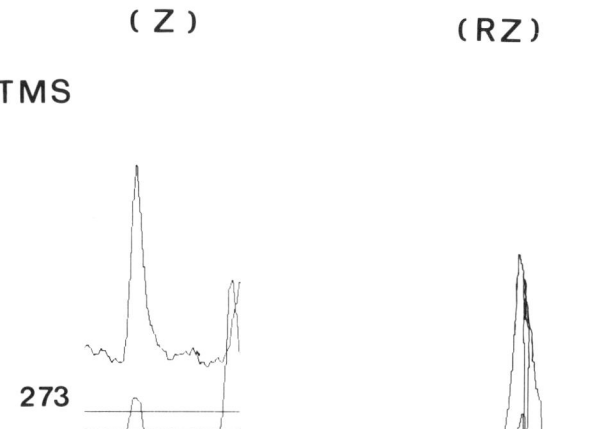

Fig. 5. RZ and Z recovery in the N-acetylglucosaminidase hydrolysates from female crown-gall extracts. Determination of characteristic M.S. ions: see references under Figs. 1 and 2

None of the ribosides were competitors of the free bases. Zeatin had the highest binding constant, which is in agreement with its physiological properties.

We then sought cytokinin ligands with chromatin. Nonhistone proteins (NHP) of ♂ and ♀ chromatin were extracted and studied without prior fractionation. [^{14}C]-BAP was added to these protein preparations and the hormone-protein complexes formed were separated by molecular sieving on Sephadex G-25. Binding kinetics were then performed by adding high concentrations of nonradioactive BAP (Fig. 6). As above, it can clearly be seen that ♂ NHP are able to bind twice as much hormone since ♀ NHP association constant to BAP is 2.4 higher than association constant of ♂ NHP. This difference is even greater than in the case of ribosomes. Finally, we may add that when ♂ plants were feminized, it was verified that if specific markers of ♀ differentiation were synthesized, then ♀ receptor sites appeared with the same characteristics (affinity, fixation, number of sites) as those obtained with genetic ♀.

3 Discussion and Conclusions

The data presented show that the *Mercurialis* system is a material well adapted for studying the mode of action of plant hormones and for studying the regulation of their metabolism. The results led us to formulate several groups of conclusions and some speculations.

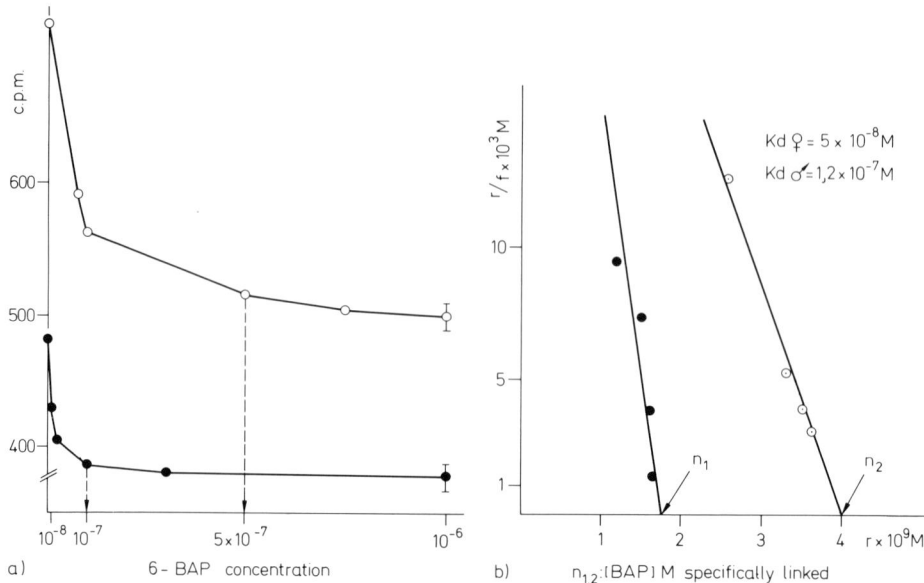

Fig. 6a. Inhibition profiles of 8-[^{14}C]-BAP binding to male and female nonhistone proteins (NHP) in the presence of increasing concentrations of unlabeled BAP; **b** Scatchard plots; *solid circles* females; *open circles* males

3.1 Cytokinin Metabolism and Sexual Differentiation

The endogenous cytokinin metabolite which appears to be active in female *Mercurialis* sexual differentiation is zeatin. Perhaps it is not the only active compound. Nevertheless the presence of Z is related to the female sex, the presence of its nucleotide to male in sporophyate.

3.2 Genetic Regulation of Cytokinin Metabolism

Zeatin in ♀ is controlled by two recessive genes: a and b_1 or b_2 with a cumulative effect. Only one is required to determine the phenomenon; perhaps it induces a higher level of hydrolase activity since the free base probably arises from RZ.

3.3 Cytokinins and Receptors in Mercurialis annua

There is generally no well-defined hormone target organ in plants. Nevertheless, the ♂ floral primordia of *Mercurialis* seem to behave as an unambiguous cytokinin target organ for exogenous cytokinins, comparable to the embryonic animal sexual cells in the case of estrogens. It is well known that this favorable situation helps to characterize hormone receptors in terms of their presence and their properties. The most active free base in the feminization, zeatin is endowed with higher binding constant.

Other cellular receptors could be detected in extracts of *Mercurialis annua* apices, including NHP, cytosol and chloroplast sites, suggesting a pleiotropic hormone effect.

We have seen that zeatin-poor and auxin-rich (Champault, 1975; Kahlem et al., 1975) ♂ primordia contain a larger quantity of cytokinin-specific receptors. Reciprocally, we also know that auxin-poor and cytokinin-rich ♀ primordia contain numerous auxin receptors. We may thus formulate the hypothesis according to which certain of the cytokinin receptors are auxin-dependent and that certain of the auxin receptors are cytokinin-dependent. The effects of the cytokinin-auxin balance in differentiation phenomena might thus be explained.

Whatever the true situation concerning the regulation of cytokinin receptor sites, the result of their interaction with the hormone in *Mercurialis* seems to be the induction of the synthesis of specific markers in all target cells of floral primordia: specific esterases, chloroplast tyrosine tRNAs, etc.

These syntheses which characterize the onset of the ♀ differentiation program seem coordinated and synchronized not only between nuclear and chloroplast compartments, but also among all the cells of the organ undergoing elaboration. Perhaps the *Mercurialis* system would be able to lead to an eventual coherent explanation for the functioning of an integrated system including nuclear regulatory genes, chemical mediators acting on a differentiation program.

References

Kahlem G, Champault A, Louis JP, Bazin M, Chabin A, Delaigue M, Dauphin B, Durand R, Durand B (1975) Détermination génétique et régulation hormonale de la différenciation sexuelle chez *Mercurialis annua* L. Physiol Vég 13: 763-779

Kahlem G (1976) Isolation and localization by histoimmunology of isoperoxydase specific for male flowers of the dioecious species *Mercurialis annua* L. Develop Biol 50: 58-67

Bazin M, Chabin A, Durand R (1975) Comparison between four isoaccepting transfer ribonucleic acids and corresponding synthetases in male and female flowers of the dioecious species *Mercurialis annua* L. Develop Biol 44: 288-297

Champault A (1973) Effects de quelques régulateurs de la croissance sur les noeuds isolés de *Mercurialis annua* L. cultivés *in vitro*. Bull Soc Bot Fr 120: 87-100

Dauphin B, Teller G, Durand B (1977) Mise au point d'une nouvelle méthode de purification, caractérisation et dosage de cytokinines endogènes extraites de bourgeons de Mercuriales annuelles. Physiol Vég 15: 747-762

Dauphin B, Teller G, Durand B (1979) Identification and quantitative analysis of cytokinins from shoot apices of *Mercurialis ambigua* by Gas Chromatography-Mass spectrometry computer system. Planta 144: 113-119

Dauphin-Guérin B, Teller G, Durand B (1980) Different endogenous cytokinins between male and female *Mercurialis annua* L. Planta 148: 124-129

Deleuze G, McChesney J, Fox J (1972) Identification of a stable cytokinin metabolite. B.B.R.C. 48: 1426-1432

Letham D, Parker C, Duke C, Summons R, MacLeod S (1976) O-glucosylzeatin and related compounds – a new group of cytokinin metabolites. Ann Bot 40: 261-263

Louis JP, Durand B (1978) Studies with the dioecious Angiosperm *Mercurialis annua* L. (2n=16): correlation between genic and cytoplasmic male sterility, sex segregation and feminizing hormones (cytokinins). Molec Gen Genet 165: 309-322

Chung SR, Durand R, Durand B (1979) Differential cytokinin binding to dioecious plant ribosomes. FEBS Lett 102: 211-215

Champault A (1975) Détection et dosage par chromatographie en phase gazeuse de l'acide indolyl acétique extrait de deux souches tissulaires de l'espèce *Mercurialis annua* L. (2n=16). CR Acad Sci Paris 280: 591-594

Relative Activities of Cytokinins and Antagonists in Releasing Lateral Buds of *Pisum* from Apical Dominance Compared with Their Relative Activities in the Regulation of Growth of Tobacco Callus

F. Skoog and A.K.B. Abdul Ghani[1]

1 Introduction

Cytokinin-antagonists which reversibly inhibit cytokinin-induced growth of callus in the tobacco bioassay at the same time, in conjunction with high cytokinin concentrations, may bring about or enhance bud formation in the callus tissue (Skoog et al., 1975).

As the potency of cytokinins in promoting bud formation is generally correlated with their activity in promoting growth of lateral buds, it became of interest to compare the relative activities of cytokinins and cytokinin antagonists in releasing lateral buds from apical dominance.

2 Materials and Methods

Twenty cytokinin-active adenine derivatives (Table 1) and 24 analogs, including 18 pyrrolo [2,3-d]pyrimidines (Table 2) mostly cytokinin-antagonists, were tested and compared for activity in releasing lateral buds of *Pisum sativum* L. var. Alaska from apical dominance (Fig. 1).

The chemicals were applied directly to the bud in the axil of the next to lowest leaf of 14-day old pea shoots severed above the upper stipule (node 2) and placed in 50 ml nutrient solution in 125 ml Erlenmeyer flasks. The lowest leaf and the bud in its axil were usually removed for ease of handling.

Bud-releasing activity was measured quantitatively in terms of the minimum dosage required to obtain significantly longer buds than the controls measured after 3 or 4 days, and relative potency was estimated by comparison with that of N^6-(Δ^2-isopentenyl) adenine (i^6Ade) as a standard.

Details of procedures are described by Abdul Ghani (1980) and Davidson (1971).

3 Results

The effect of the applied i^6Ade on the growth of bud #4 after 3 days is shown in Fig. 2. In this experiment the lengths of bud #5 and of bud #3 (the latter present in this

---

[1] Institute of Plant Development, Birge Hall, University of Wisconsin-Madison, Madison, WI 53706, USA

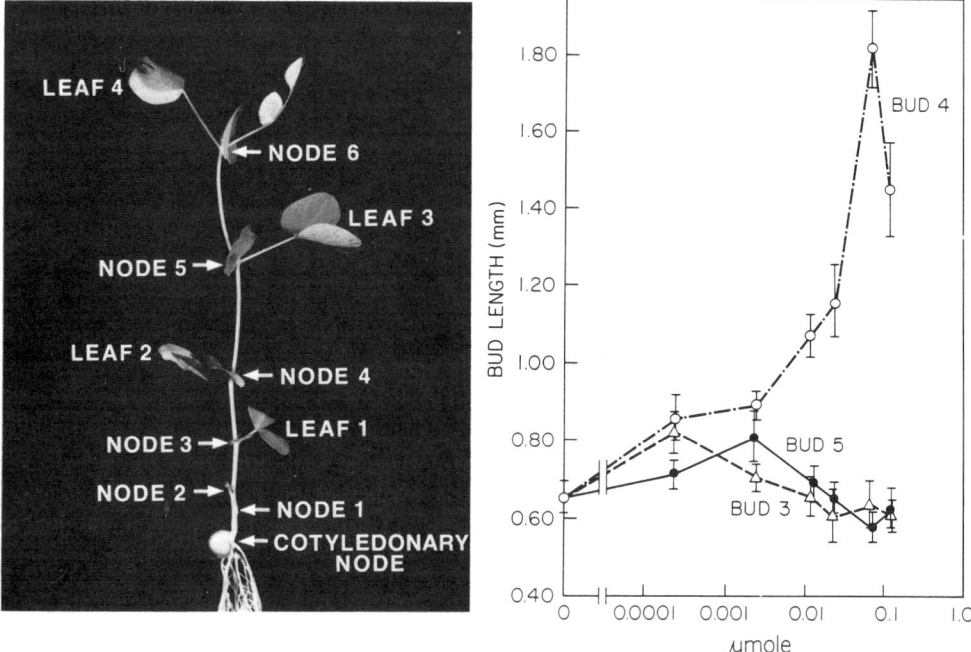

Fig. 1. A 14-day-old pea plant showing the numbering of the nodes and leaves. The stem was cut just above the second node, and in most experiments leaf 1 and the bud in its axil were removed

Fig. 2. Dose responses of lateral buds 3, 4 and 5 to i^6 Ade applied to bud 4 only. Measurements of bud length made 3 days after the treatment. Data are mean values for 9 plants per treatment

case only) were also recorded to show that only the treated bud is released. The time course of the response to an optimal dosage of i^6 Ade is shown in Fig. 3.

Increase in the mitotic index of bud #4 associated with its release from apical dominance by applied i^6 Ade or by removal of the terminal bud is shown in Fig. 4.

All tested cytokinin-active compounds released lateral buds, and in general their relative effectiveness was closely correlated with that in promoting growth of callus tissue in the tobacco bioassay (Hecht et al., 1970; Leonard et al., 1969; Schmitz et al., 1972 a and b; Skoog, 1973; Skoog et al., 1967; Skoog and Leonard, 1968; Skoog and Schmitz, 1979) (Table 1). 2-amino-trans-zeatin was the most active, followed by zeatin and i^6 Ade. N^6-benzyladenine and kinetin were 0.1 and 0.01 respectively as active as i^6 Ade. trans-Zeatin was more active than cis-zeatin (Fig. 5), but the trans- and cis-isomers of iso-zeatin were both 0.11 times as active as i^6 Ade. 9-Ribosylation generally decreased activity, especially of highly active compounds (Fig. 6). 2-Methyl-thiolation caused a five fold reduction in activity (Fig. 7). While the 2-amino substituent increased the activity of zeatin, the 2-hydroxy group reduced it to 0.11 times that of i^6 Ade. The (+) and (−) isomers of dihydrozeatin had ca. 9 and 15 times less activity, respectively, than i^6 Ade.

Cytokinins applied in the nutrient solution and taken up through the bases of the excised shoots were also effective in releasing lateral buds from apical dominance, as illustrated in Fig. 8.

Table 1. Comparisons of activities of cytokinins in promoting callus growth in the tobacco bioassay and in releasing lateral buds from apical dominance

Number	Compound Chemical name	Substituents R_1	R_2	R_3	Symbol	Minimum dosage for Cytokinin Activity (μMolar)	Releasing lateral buds (μmol/bud)
1	6-(3-Methyl-2-butenyl-amino)purine	(3-methyl-2-butenyl)	H	H	i^6 Ade	3×10^{-4}	2.5×10^{-4}
2	6-(3-Methyl-2-butenyl-amino)-9-β-D-ribofuran-osylpurine	(3-methyl-2-butenyl)	H	Ribose	i^6 A	7×10^{-3}	7.1×10^{-3}
3	6-(3-Methyl-2-butenyl-amino)-2-methylthio-purine)	(3-methyl-2-butenyl)	CH_3S	H	ms^2i^6 Ade	1.5×10^{-3}	1.3×10^{-3}
4	6-(3-Methyl-2-butenyl-amino)-2-methylthio-9-β-D-ribofuranosylpurine	(3-methyl-2-butenyl)	CH_3S	Ribose	ms^2i^6 A	2×10^{-2}	1.3×10^{-2}
5	6-(4-Hydroxy-3-methyl-trans-2-butenylamino)-purine	(4-hydroxy-3-methyl-trans-2-butenyl)	H	H	$t\text{-}io^6$ Ade	1.2×10^{-4}	2.3×10^{-4}
6	6-(4-Hydroxy-3-methyl-trans-2-butenylamino)-9-β-D-ribofuranosylpurine	(4-hydroxy-3-methyl-trans-2-butenyl)	H	Ribose	$t\text{-}io^6$ A	4.5×10^{-3}	1.4×10^{-3}
7	6-(4-Hydroxy-3-methyl-cis-2-butenylamino)purine	(4-hydroxy-3-methyl-cis-2-butenyl)	H	H	$c\text{-}io^6$ Ade	1.1×10^{-2}	1.1×10^{-2}

Relative Activities of Cytokinins and Antagonists in Releasing Lateral Buds of Pisum

#	Name	Structure			Abbreviation		
8	6-(4-Hydroxy-3-methyl-cis-2-butenylamino-9-β-D-ribofuranosylpurine		H	Ribose	c-io^{6}A	2×10^{-2}	1.3×10^{-2}
9	(+) 6-(4-Hydroxy-3-methylbutylamino)purine	HO—	H	H	(+) hio^{6}Ade	3×10^{-3}	2.5×10^{-3}
10	(−) Ditto		H	H	(−) hio^{6}Ade	3.3×10^{-4}	3.6×10^{-3}
11	6-(4-Hydroxy-2-methyl-trans-2-butenylamino)purine	HO—	H	H	iso-t-io^{6}Ade	1.0×10^{-2}	2.3×10^{-3}
12	6-(4-Hydroxy-2-methyl-trans-2-butenylamino)-9-β-D-ribofuranosylpurine	HO—	H	Ribose	iso-t-io^{6}A	2.5×10^{-2}	2.5×10^{-3}
13	6-(4-Hydroxy-2-methyl-cis-2-butenylamino)purine	HO—	H	H	iso-c-io^{6}Ade	4.0×10^{-2}	2.3×10^{-3}
14	6-(4-Hydroxy-2-methyl-cis-2-butenylamino)-9-β-D-ribofuranosylpurine		H	Ribose	iso-c-io^{6}A	4.2×10^{-2}	5.0×10^{-3}
15	6-(4-Hydroxy-3-methyl-trans-2-butenyl-amino)-2-aminopurine	HO—	NH$_2$	H	n^2-t-io^{6}Ade	2×10^{-4}	2.1×10^{-4}
16	2-Hydroxy-6-(4-hydroxy-3-methyl-trans-2-butenylamino)purine		OH	H	ho^2-t-io^{6}Ade	1.5×10^{-2}	2.1×10^{-3}
17	6-Benzylaminopurine		H	H	bzl^{6}Ade	8×10^{-4}	2×10^{-3}
18	6-Furfurylaminopurine		H	H	fr^{6}Ade	1×10^{-3}	2.3×10^{-2}
19	6-(4-Hydroxy-3-methyl-2-butenylamino)-2-methyl-thiopurine	HO—	CH$_3$S	H	ms^{2}io^{6}Ade	1×10^{-3}	8.7×10^{-4}
20	6-(4-Hydroxy-3-methyl-2-butenylamino)-2-methyl-thio-9-β-D-ribofuranosyl-purine		CH$_3$S	Ribose	ms^{2}io^{6}A	1.1×10^{-2}	6.3×10^{-2}

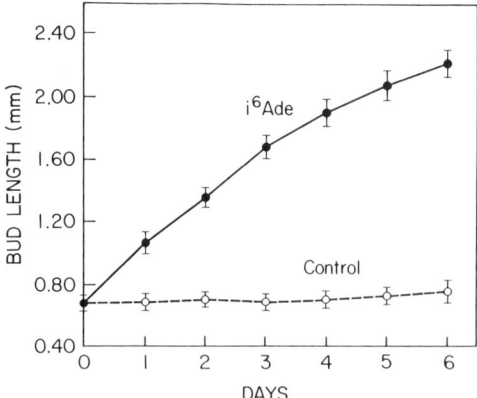

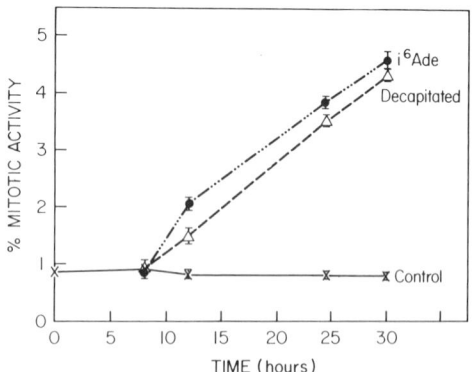

Fig. 3. Time course of response of lateral bud 4 to a single dosage of 0.05 μmol i⁶ Ade

Fig. 4. Time course of mitotic activity in lateral bud 4 of i⁶ Ade-treated and of decapitated pea shoots. The i⁶ Ade (0.05 μmol) was applied directly to bud 4 of intact shoots. In the decapitated plants the stems were cut just below node 6. Intact shoots with 10 μl carrier (50% ethanol-5% carbowax solution) applied to bud 4, were used as controls. Each *point* is the mean value for 9 tissue squashes

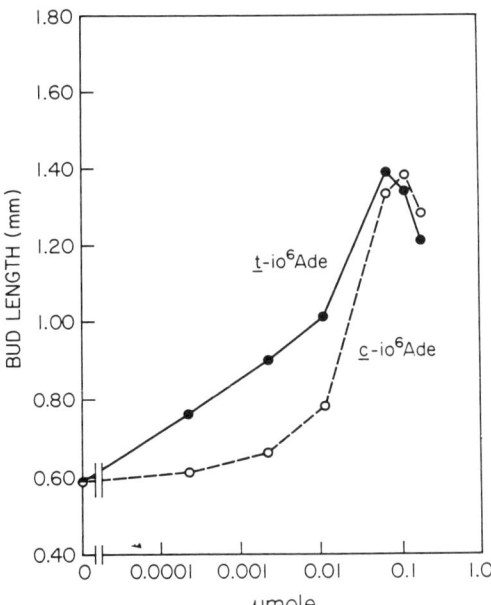

Fig. 5. Differential response of lateral bud 4 to the geometric isomers trans-zeatin (t-io⁶ Ade) and cis-zeatin (c-io⁶ Ade)

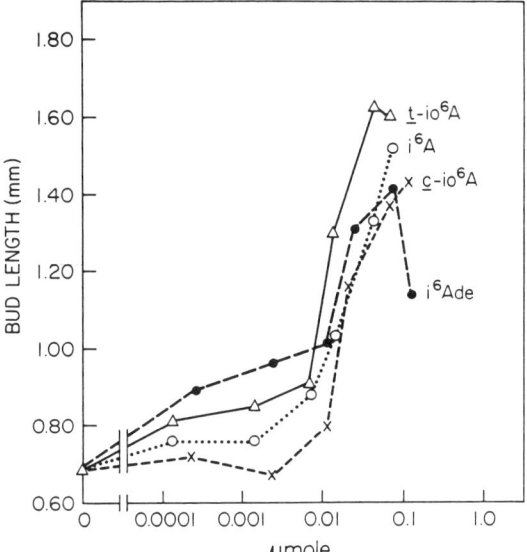

Fig. 6. Effectiveness of ribonucleosides i⁶A, t-io⁶A and c-io⁶A compared with i⁶Ade in releasing bud 4 from apical dominance

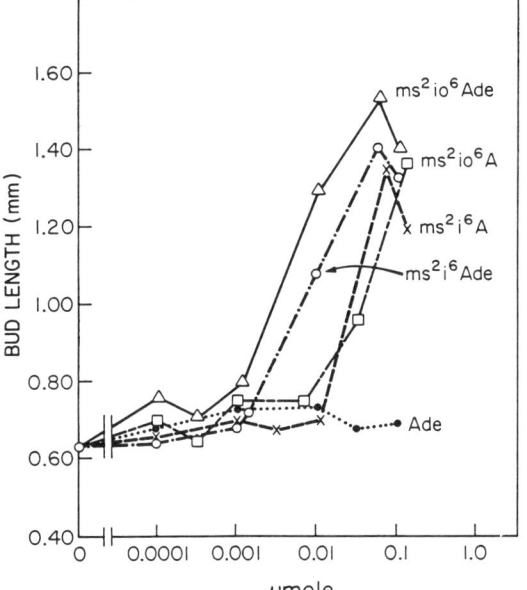

Fig. 7. Effects of 2-methylthio derivatives of the free bases and ribonucleosides of i⁶Ade and io⁶Ade on the release of lateral bud 4 from apical dominance

Of the anticytokinins, pyrrolo[2,3-d]-, pyrazolo[3,4-d]- and pyrazolo [4,3-d]pyrimidines were active in releasing buds. A summary of activities of 4-substituted pyrrolo-[2,3-d]pyrimidines is included in Table 2. Their relative activities are roughly parallel to their relative activities as anticytokinins in the tobacco bioassay.

A substituent in the 2-position appears to be essential for activity in releasing buds, as it is for anticytokinin activity of pyrrolo[2,3-d]-pyrimidine, but the 2-methyl-

Table 2. Comparison of anticytokinin- and bud-releasing activities of 4-aminopyrrolo[2,3-d]pyrimidines with different combinations of substituents in the N^4-, 2- and 7-positions

Compound		Substituent		Minimum dosage for	
				Anticytokinin activity[a] (µMolar)	Releasing buds (µmol/bud)
Number	Chemical Name	R_1	R_2		
21	4-(3-Methyl-2-butenylamino)-2-methylthiopyrrolo[2,3-d]-pyrimidine	(isoprenyl)	CH_3S	0.24	0.009 (10 µM)[b]
22	4-(n-Pentylamino)-2-methylthiopyrrolo[2,3-d]pyrimidine	(n-pentyl)	CH_3S	0.40	0.04
23	4-(n-Hexylamino)-2-methylthiopyrrolo[2,3-d]pyrimidine	(n-hexyl)	CH_3S	0.10	0.11
24	4-Cyclopentylamino-2-methylthiopyrrolo[2,3-d]pyrimidine	(cyclopentyl)	CH_3S	0.009	0.0002
25	4-Cyclohexylamino-2-methylthiopyrrolo[2,3-d]pyrimidine	(cyclohexyl)	CH_3S	0.05	0.001
26	4-Benzylamino-2-methylthiopyrrolo[2,3-d]pyrimidine	(o-methylbenzyl)	CH_3S	10	0.004
27	4-(1,1-Dimethylpropylamino)-2-methylthiopyrrolo[2,3-d]-pyrimidine	(1,1-dimethylpropyl)	CH_3S	?	N.A.
28	4-Amino-2-methylthiopyrrolo[2,3-d]pyrimidine	H	CH_3S	N.A.	N.A.

#	Name	Structure	R	Anticytokinin activity[a]	Bud release
29	4-(p-Methyl-cyclohexylamino)-2-methylthiopyrrolo[2,3-d]pyrimidine	cyclohexyl-CH₃	CH₃S	?	0.09
30	4-(3-Methyl-2-butenylamino)-pyrrolo[2,3-d]pyrimidine	isopentenyl	H	N.A. 0.6(CK)	(10 μM)[b]
31	4-n-Hexylaminopyrrolo[2,3-d]pyrimidine	n-hexyl	H	N.A. 5.8(CK)	W.A.[b]
32	4-(n-Pentylamino)-2-methyl-pyrrolo[2,3-d]pyrimidine	n-pentyl	CH₃	N.A.[a] (CK)	0.02
33	4-(2-Ethylhexylamino)-2-methyl-pyrrolo[2,3-d]pyrimidine	2-ethylhexyl	CH₃	2.6[a] (CK)	0.02
34	4-(2-Methylpropylamino)-2-methyl-pyrrolo[2,3-d]pyrimidine	isobutyl	CH₃	25.9[a]	N.A.
35	4-Cyclobutylamino-2-methylthio-pyrrolo[2,3-d]pyrimidine	cyclobutyl	CH₃S	ca. 0.01	?
36	4-Cyclopentylamino-2-methyl-pyrrolo[2,3-d]pyrimidine	cyclopentyl	CH₃	0.8[a]	0.002
37	4-Cyclohexylamino-2-methyl-pyrrolo[2,3-d]pyrimidine	cyclohexyl	CH₃	0.9[a]	0.02
38	4-(n-Pentylamino)-7-methyl-pyrrolo[2,3-d]pyrimidine	n-pentyl	H	N.A.	N.A.

$R_3 = CH_3$ for compound 38. For all other compounds $R_3 = H$
N.A. = No activity. CK = cytokinin-active. W.A. Weakly Active

[a] Anticytokinin activity of compounds with a 2-CH₃ substituent (32-34, 36 and 37) is based on 50% inhibition of growth in presence of 0.05 μM kinetin in the tobacco bioassay (Iwamura et al., 1979); that of all others is based on detectable inhibition of growth in presence of 0.003 μM i⁶Ade in the tobacco bioassay (Skoog et al., 1973)

[b] Tested by uptake through bases of stems

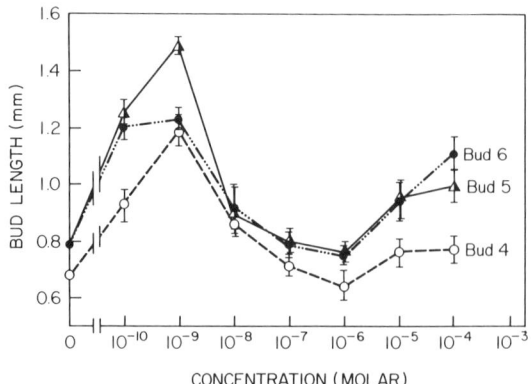

Fig. 8. Release of inhibition of lateral buds 4, 5 and 6 as a function of the i^6 Ade concentration in the nutrient medium in which the excised shoots stood. Bud lengths measured after 3 days of treatment

thio group can be effectively replaced by a 2-methyl group, as shown by tests of compounds 32-34, and 36-38 (Table 3).

Anticytokinins, like cytokinins, are also effective in releasing buds from inhibition when supplied in the nutrient solution at the base of the excised shoots.

Tests of six pyrazolo[2,3-d] and [3,4-d]pyrimidine derivatives confirmed the results obtained with the pyrrolo[2,3-d]pyrimidines. Only those with anticytokinin- or cytokinin activity were active in releasing buds from apical dominance.

The above anticytokinins also counteracted the development of buds which had been released from apical dominance by decapitation, and in general when used to release buds, they brought about shorter buds than did the analogous cytokinin-active N^6-substituted adenines.

4-Cyclopentylamino- and 4-cyclohexylamino-2-methylthiopyrrolo[2,3-d]-pyrimidines and also 4-n-pentylamino- and 4-(2-ethylhexylamino)-2-methylpyrrolo[2,3-d]-pyrimidines, which inhibit callus growth but enhance bud formation in the tobacco bioassay, also promoted growth of lateral buds to greater final lengths than did anticytokinins with other 4-substituents but not as much as did the cytokinin-active compounds.

The capacity to release lateral buds from apical dominance, as contrasted with the promotion of development of the lateral buds once they have been released, appears to be specific for cytokinins and antagonists. Treatments with abscisic-, gibberellic- or indoleacetic acid did not release lateral buds from apical dominance, although the latter two hormones will promote development of lateral buds after they have been released by cytokinins.

4 Conclusion

The parallel behavior of the cytokinins and anticytokinins in their ability to release lateral buds from apical dominance indicates that their actions in this function and in promoting or inhibiting growth, respectively, in callus may be exerted at separate metabolic sites.

Table 3. Comparisons of anticytokinin- or cytokinin activities with bud-releasing activities of 4-substituted-2-methylpyrrolo[2,3-d]-pyrimidines

Compound Number	Substituent in position N⁴-	2-	Cytokinin activity (μM)[a]	Anti-cytokinin activity ($I_{50}, \mu M$)[b]	Relative anti-cytokinin activity[c]	Relative bud-releasing activity[d]
36	4-Cyclopentyl	CH_3	–	0.07	1	0.05
33	4-(2-Ethylhexyl)	CH_3	10	2.00	0.035	0.013
37	4-Cyclohexyl	CH_3	–	0.90	0.078	0.012
32	4-n-Pentyl	CH_3	10	NA	NA	0.010
34	4-(2-Methylpropyl)	CH_3	–	20.90	0.003	NA
38[e]	4-n-Pentyl	H	NK	NK	NK	NA

NA = Not Active; NK = Not Known
[a] Cytokinin activity is the concentration required for maximum callus growth
[b] Anticytokinin activity is expressed as the concentration (μM) which gave 50% inhibition (I_{50}) of tobacco callus growth in presence of 0.05 μM kinetin. Both cytokinin and anticytokinin activities are as reported by Iwamura et al. (1979)
[c] Relative anticytokinin activity is based on that of compound 36 as reported by Iwamura et al. (1979)
[d] Relative bud-releasing activity is based on that of i^6 Ade
[e] Compound 38 had a methyl group in the 7-position instead of the 2-position

Acknowledgments. The authors are grateful to Professor N.J. Leonard, University of Illinois-Champaign-Urbana, for the cytokinins tested and to Professor S.M. Hecht, University of Virginia for cytokinin antagonists. Professor H. Iwamura, University of Kyoto kindly provided the 2-methylpyrrolo[2,3-d]-pyrimidines. This work was supported in part by National Science Foundation Research Grant BMS 72-02226 and by the Research Committee of the Graduate School with Funds from the Wisconsin Alumni Research Foundation.

References

Abdul Ghani AKB (1980) Effects of cytokinins, antagonists and related analogs on the ability to release lateral buds of *Pisum sativum* L. cv Alaska from apical dominance. Ph D Thesis, Univ Wisconsin, Madison

Davidson DR (1971) The shift of dominance in *Pisum sativum* from the terminal bud to the lateral bud through release of apical dominance with 6-(3-methyl-2-butenylamino)purine followed by indole-3-acetic acid and gibberellic acid treatments. Ph D Thesis, Univ Wisconsin, Madison

Hecht SM, Leonard NJ, Schmitz RY, Skoog F (1970) Cytokinins: Synthesis and growth-promoting activity of 2-substituted compounds in the N^6-isopentenyladenine and zeatin series. Phytochemistry 9: 1173-1180

Iwamura H, Masuda N, Koshimizu K, Matsubara S (1979) Cytokinin-agonistic and antagonistic activities of 4-substituted-2-methylpyrrolo-[2,3-d]pyrimidines, 7-deaza analogs of cytokinin-active adenine derivatives. Phytochemistry 18: 217-222

Leonard NJ, Hecht SM, Skoog F, Schmitz RY (1969) Cytokinins: Synthesis, mass spectra, and biological activity of compounds related to zeatin. Proc Nat Acad Sci USA 63: 175-182

Schmitz RY, Skoog F, Hecht SM, Bock RM, Leonard NJ (1972a) Comparison of cytokinin activities of naturally occurring ribonucleosides and corresponding bases. Phytochemistry 11: 1603-1610

Schmitz RY, Skoog F, Playtis AJ, Leonard NJ (1972b) Cytokinins: Synthesis and biological activity of geometric and position isomers of zeatin. Plant Physiol 50: 702-705

Skoog F (1973) A survey of cytokinins and cytokinin antagonists with reference to nucleic acid and protein metabolism. Biochem Soc London Symp 38: 195-215

Skoog F, Leonard NJ (1968) Sources and structure/activity relationships of cytokinins. In: Wightman F (ed) Biochemistry and physiology of plant growth regulators. Runge Press, Ontario, Canada, pp 1-18

Skoog F, Schmitz RY (1979) Biochemistry and physiology of cytokinins. In: Litwack G (ed) Biochemical actions of hormones (vol VI. Academic Press, London New York, pp 335-413

Skoog F, Hamzi Q, Szweykowska AM, Leonard NJ, Carraway KL, Fujii T, Helgeson P, Loeppky RN (1967) Cytokinins: Structure/activity relationships. Phytochemistry 6: 1169-1192

Skoog F, Schmitz RY, Bock RM, Hecht SM (1973) Cytokinin antagonists: Synthesis and physiological effects of 7-substituted 3-methylpyrazolo[3,4-d]pyrimidines. Phytochemistry 12: 25-37

Skoog F, Schmitz RY, Hecht SM, Frye RB (1975) Anticytokinin activity of substituted pyrrolo-[2,3-d]pyrimidines. Proc Nat Acad Sci USA 72: 3508-3512

Section 3
Cytokinin Hormone Receptors: Methods and Results

Photoaffinity Labeling Reagents to Detect Binding of Cytokinins: Chemical Limitations

R. Mornet[1], J.B. Theiler[2], and N.J. Leonard[2]

1 Introduction

Photoaffinity labeling has been a powerful technique for studying ligand-receptor interactions. The versatility of this technique has been illustrated in several reviews (Chowdhry and Westheimer, 1979); (Bayley and Knowles, 1977); (Katzenellenbogen, 1976); (Cooperman, 1975); (Creed, 1974); (Knowles, 1972).
   Photoaffinity labeling depends upon the design of a ligand L', analog of the natural ligand L, that may similarly associate to the receptor R and possesses a photochemically activable substituent. Photolysis generates a reactive intermediate *L', which may bind covalently to the receptor R, hopefully specifically and before diffusion away from the site of association.

$$R + L' \underset{k_{-1}}{\overset{k_1}{\rightleftharpoons}} R\text{---}L' \xrightarrow{h\nu} R\text{---}*L' \underset{k_{-3}}{\overset{k_2}{\underset{k_3}{\rightleftharpoons}}} \begin{array}{c} R\text{--}L' \\ R + *L' \end{array}$$

Applications of such photogenerated reagents include the location of particular macromolecules within a biological assembly or organelle, the identification of a receptor molecule, the location of a ligand-binding site within a receptor, investigation of structure/function relationships, etc.
   The possibility of choice of the photolabile substituent and of its position in the ligand L', may allow the selection of conditions that maximize effective labeling. These choices have to be made according to the following criteria: the precursor has to be readily synthesized, chemically stable, and photolyzable at long wavelength for avoiding as much as possible modification of the receptor during the irradiation; the photogenerated intermediate has to be highly reactive, to have an extremely short half-life, and not to be susceptible to rearrangement; moreover it is highly desirable that it be nonselective in its covalent attachment to the receptor, i.e., the labeling pattern should be independent of the relative chemical reactivities of different sites toward the intermediate.

---

[1] Laboratoire de Chimie Organique, I.R.S.T., Université d'Angers, 49045-Angers Cedex, France
[2] Roger Adams Laboratory, School of Chemical Sciences, University of Illinois, Urbana, IL 61801, USA

2 Azido-Substituted Cytokinins as Potential Photoaffinity Reagents

The azido substituent has been one of the most used photolabile groups in photoaffinity labeling experiments. Upon photolysis, it generates a nitrene which, in its singlet state, may bind covalently to a neighboring molecule by insertion in bonds of any atom with hydrogen.

$$R-N_3 \xrightarrow{h\nu} R-\overline{N} + N_2 \xrightarrow{\overset{(I)}{-A-H}} R-NH-A-$$

$$(A = C, O, N, S...)$$

Potential photoaffinity labeling reagents, azido derivatives of the natural cytokinins $N^6(\Delta^2$-isopentenyl) adenine $1a$, $2a$, and zeatin $1b$, $2b$, or N^6-benzyladenine $1c$, $2c$, 3, 4, have been synthesized and tested for their biological activity.

Except N^6-(4-azidobenzyl)adenine 3, all these compounds were found quite active in the tobacco callus bioassay (Theiler et al., 1976); (Mornet et al., 1979), and furthermore, 8-azido-N^6-benzyladenine $2c$ was reported to have "full cytokinin activity" in the moss protonema test and the tobacco cell suspension bioassay (Sussman and Kende, 1977).

Thus, almost all these compounds are expected to satisfy the requirement of photoaffinity labeling reagents by behaving as natural ligands.

3 Chemical Properties of Azido-Substituted Cytokinins

When protected from light, the azido derivatives *1-4* are quite stable at room temperature, either in solution or in solid form. They can be worked under red light without any damage.

In order to delimit their possible use as photoaffinity labeling reagents, we have studied the photolytic behavior of these azides (Mornet et al., 1979). This work should be considered as an initial approach to the problem, and its conclusions only as directions to interpret, rather than to predict, the results of further photoaffinity labeling experiments.

The azido compounds *1-4* were photolyzed with a high pressure mercury lamp in a Pyrex immersion apparatus cooled internally by running water. They were dissolved in water-ethanol (50:50) (20 mg/600 ml of solution), and in addition the more soluble *1a, 1b, 2a, 2b* were photolyzed in saturated aqueous solution. The photolysis reactions were monitored by the change in the UV spectra of the solutions.

The mixtures of photoproducts, most of them being new products, were analyzed by electron impact (EI) or field desorption (FD) mass spectrometry (MS), after solvent evaporation. High resolution mass spectrometry (HRMS) allowed us to establish their molecular formulas, from the studies of the molecular ions. Elucidation of the nature of these photoproducts was based on a comparison of the observed results with those expected according to predictable reactions of nitrenes with the solvents (Lwowski, 1970); (Reiser and Wagner, 1971), and on mass spectral data.

3.1 The Azidoazomethine-Tetrazoles Equilibria in 2-Azidopurines

The times for complete photolysis, in the conditions described above, were usually about 25 min for 8-azido derivatives *2*, and 50 min for 2- and 4-azidobenzyl derivatives *3* and *4*. 2-Azido compounds *1* needed very long times, and generally some starting material was still detected in the photoreaction mixture after 3 h.

This slower rate of photolysis with 2-azido derivatives *1* may be attributed to the occurrence of the azido-azomethine-tetrazoles equilibria, commonly observed with 2-azidopurines (Temple et al., 1966) (Scheme 1) and detected by spectroscopic methods.

The reaction of photolysis is expected to involve only the azido isomer, and formation of this isomer from the tetrazolo forms is probably rate-controlling. Thus, some complications may occur in the mechanism of photoaffinity labeling reactions with 2-azido-substituted cytokinins *1*, because of the equilibria. Moreover, the long times of photolysis needed in these cases may increase the extent of secondary photoreactions.

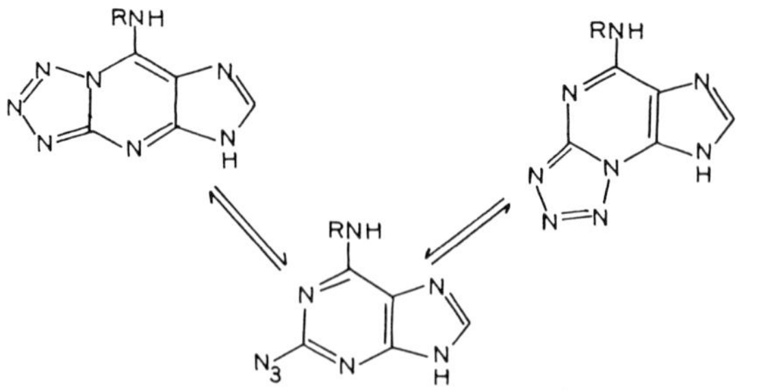

Scheme 1. Azidoazomethine-tetrazoles equilibria in 2-azidopurines *1*

3.2 Photoproducts from Azido-Substituted Cytokinins and Solvents

$$R-\bar{\underline{N}}_3 \xrightarrow[-N_2]{h\nu} R-N \text{ (singlet)} + R-\bar{\underline{\ddot{N}}} \cdot \text{(triplet)} \xrightarrow{\substack{C_2H_5OH \\ H_2O}} R-NX$$

1-4 (R = unsubstituted cytokinin)

At least two reactions were expected to occur with nitrenes photogenerated from azidopurines. The first was abstraction of two hydrogen atoms from the solvents by triplet nitrenes, leading to the formation of amino derivatives ($NX = NH_2$). The second, essential for efficient photoaffinity labeling, was insertion of singlet nitrenes into O-H and C-H bonds.

The common reaction pathways were elucidated from the photolysis of the readily available 2-azido-N^6-benzyladenine *1c*. The results observed with this reagent served as references for the analysis of the photoproducts from 2-azidopurines *1b* and *1c*, and then from the other azido derivatives *2*, *3*, and *4*.

3.2.1 Photolysis of 2-Azido Derivatives 1

Besides the expected amino derivatives ($NX = NH_2$), products resulting from insertion reaction in the OH bonds of water ($NX = NHOH$) and of ethanol ($NX = NH-OC_2H_5$) were detected in the mixture of photoproducts from the 2-azidopurines derivatives *1*. No mass spectral evidence was found of other insertion products concerning the ethanol molecule. Thus, like other nitrenes (Reiser and Wagner, 1971), those derived from 2-azidopurines might be highly selective in their insertion reactions, at least between O-H and C-H bonds.

However, insertion in the Cα-H bond of ethanol cannot be ruled out as a possible pathway for explaining the formation of a third type of photoproduct ($NX = NH-C_2H_5$) (Scheme 2). The second pathway possibly giving the same product was shown to be possible by irradiating the 2-amino product from 2-azido-N^6 benzyladenine *1c*, in the presence of acetaldehyde in ethanol.

We can observe that in photoaffinity labeling, such a kind of product ($NX = NH-C_2H_5$), although obtained by a complex mechanism, would result from the covalent attachment of the photoaffinity reagent to the substrate, as well as a true insertion product.

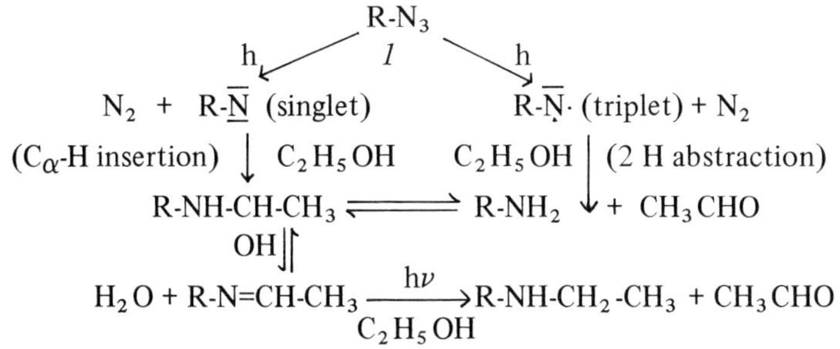

$$\text{R-N}_3 \; \mathit{1}$$
$$h\nu \swarrow \searrow h\nu$$

$$\text{N}_2 + \text{R-}\overline{\underline{\text{N}}}\text{ (singlet)} \qquad \text{R-}\overline{\text{N}}\cdot \text{ (triplet)} + \text{N}_2$$

$$(\text{C}_\alpha\text{-H insertion}) \downarrow \text{C}_2\text{H}_5\text{OH} \qquad \text{C}_2\text{H}_5\text{OH} \downarrow (\text{2 H abstraction})$$

$$\text{R-NH-CH-CH}_3 \rightleftharpoons \text{R-NH}_2 + \text{CH}_3\text{CHO}$$
$$\text{OH} \updownarrow$$
$$\text{H}_2\text{O} + \text{R-N=CH-CH}_3 \xrightarrow[\text{C}_2\text{H}_5\text{OH}]{h\nu} \text{R-NH-CH}_2\text{-CH}_3 + \text{CH}_3\text{CHO}$$

Scheme 2. Possible pathways from 2-azidopurines *1* to 2-ethylaminopurines (R = 2-purinyl)

3.2.2 Photolysis of 8-Azido Derivatives 2

Comparable qualitative results were observed with 8-azidopurines derivatives *2*, as with their 2-azido analogs. The only remarkable difference occurred in the photolysis of 8-azidozeatin *2b*, where an additional photoproduct was detected, with the molecular formula of the nitrene. Because no corresponding product had been observed with the other 8-azidopurines *2a, 2c*, we made the hypothesis that we obtained the product *5*, resulting from the intramolecular insertion of the nitrene into the O-H bond located in the N^6-substituent chain.

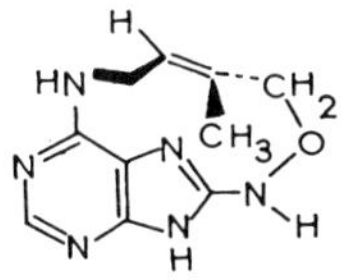

5

3.2.3 Photolysis of N^6-(4-Azidobenzyl)Adenine 3

The EI mass spectrum of the mixture of photoproducts from N^6-(4-azidobenzyl)-adenine *3* was found almost similar to the spectrum of the 2 H abstraction product (NX = NH_2). The other expected products were characterized by barely detectable peaks. FDMS confirmed the formation of insertion products analog to those observed in the two above series.

However, phenylnitrene is known to react with nucleophiles like amines (Doering and Odum, 1966) or alcohols (Sundberg and Smith, 1971) to produce 2-substituted 3-H azepines. Thus, photoproducts from the azidophenyl derivative *3* that could be considered as insertion products of the intermediate nitrene in O-H bonds were probably 3-H azepines with the same molecular formulas (Scheme 3).

3.2.4 Photolysis of N^6-(2-Azidobenzyl)Adenine 4

Five-membered ring closure by the trapping of a singlet aromatic nitrene with a well-positioned amino-N has been shown to be a very effective process (Mc Robbie et al., 1976). In the photolysis of N^6-(2-azidobenzyl)adenine *4*, this reaction probably oc-

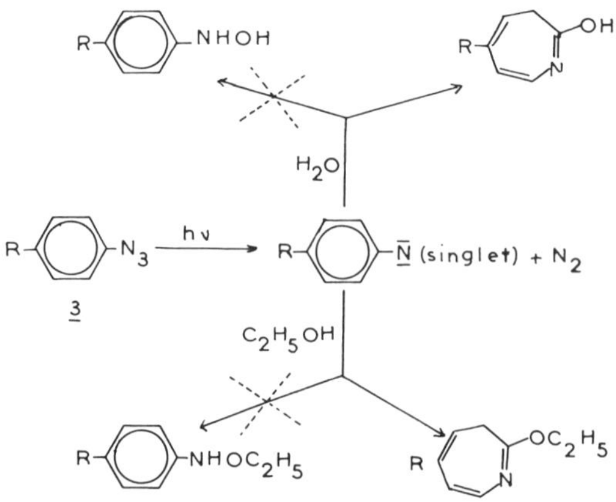

Scheme 3. Photolysis of N^6-(4-azidobenzyl)adenine. Probable reactions of singlet nitrene (R = 6-purinylmethyl)

curred to a large extent (Scheme 4). Apart from this observation, the photoproducts were found to be qualitatively very similar to those observed with the 4-azidobenzyl isomer *3*.

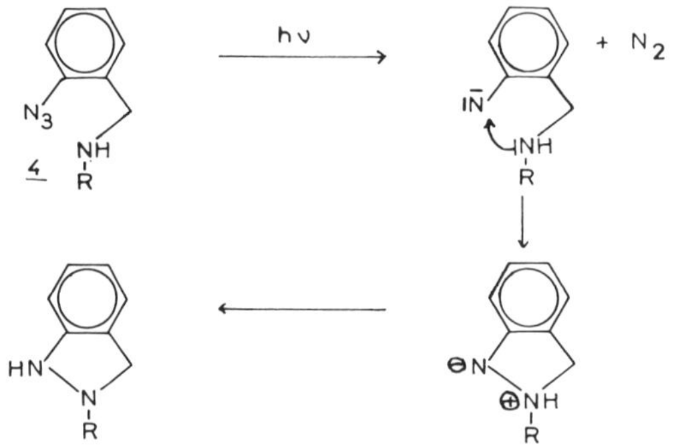

Scheme 4. Probable intramolecular photoreaction of N^6-(2-azidobenzyl)adenine *4* (R = 6-purinyl)

4 The Future of Azido-Substituted Cytokinins as Photoaffinity Labeling Reagents

4.1 Potential Uses of Azido-Substituted Compounds *1-4*

Considering their ability to make covalent bonds with solvent molecules upon photolysis, the biologically active cytokinins *1, 2, 4*, are expected to bind covalently to cytokinin association sites in photoaffinity labeling conditions. Probably, proximity effects in the receptors will improve the rate of covalent reactions which can be suspected to be low in the photolysis of some of these reagents in the only presence of solvents.

Preliminary tests made on a cytokinin-binding protein isolated recently (Moore, 1979), using 2-azido-N^6-(Δ^2-isopentenyl)adenine *1a*, gave significant results (Mornet et al., 1979). When photolyzed in the presence of the protein, this reagent caused an important decrease of the protein-binding capacity. This result is explained by an efficient binding of the reagent at or near the active site, and demonstrates the potential utility of the reagent.

Considering their biological activity, and their short-time photolysis, the 8-azido derivatives *2* appears to be the best candidates for photoaffinity labeling. But all the reagents *1-4* should be considered as potentially useful. It may be valuable to have on hand this series of azido-substituted cytokinins that differ only in the position of the photolabile group, for defining the orientation of binding of the cytokinin to its receptor. To accomplish this goal, the experiments would include the analysis of the position of covalent attachment to the receptor, on photolysis of each azido-substituted cytokinin. Further, it might be possible to compare specific and nonspecific binding, by determining the position of covalent attachment of a biologically active reagent (e.g., *4*) to the position of attachment of a biologically inactive analog (*3*).

However, we have to keep in mind the intrinsic limitation of these reagents resulting from nitrene selectivity in insertion reactions.

4.2 Possibilities of Double-Labeling

In some photoaffinity labeling experiments, like identification of a receptor or mapping studies, the photosensitive reagent has to be radioactively labeled. With this aim, directions have been given for the synthesis of 8-azido-2-[^{3}H]-N^6-benzyladenine [^{3}H]*2c* (Sussman and Kende, 1977).

An alternative approach of double labeling is fluorescent photoaffinity labeling (FPAL) (Dreyfuss et al., 1978). No fluorescent species were characterized in the photolysis products from 2-azido-N^6-(Δ^2-isopentenyl)adenine *1a* (Theiler, 1977) and to our knowledge no fluorescence studies have been made with any other azido-substituted cytokinin. Cytokinin analogs with "stretched-put" adenine moiety *6* (Sprecker et al., 1976), which are fluorescent compounds, should provide, through azido substitution, an access to fluorescent photoaffinity labeling.

For radioactive labeling, it is necessary to have on hand simple and efficient synthetic methods, and desirable to introduce the radioactive label as late as possible. Tritium labeling is generally requested to attain a sufficient specific activity for the detection of cytokinin binding sites.

Apart from 8-azido-2-[^{3}H]-N^6-benzyladenine[^{3}H^6]2c whose synthesis, if possible, cannot be extended to further analogs 2b, 2c, the simplest solution for synthesis of radiolabeled azido-substituted cytokinins seems to have the azido substituent and the radioactive label in each part of the molecule, i.e., the adenine moiety and the N^6-chain. Highly radioactive N^6-benzyl substituents are available (Sussman and Firn, 1976) (Fox et al., 1979). Such a radioactive substituent may be easily introduced in a 2-azido-purine derivative, by reacting the benzylamine with 2,6-diazidopurine 7, which is readily obtained from 2,6-dichloropurine 8, a commercial product (Mornet et al., 1979) (Scheme 5).

Such an efficient procedure, directed to the synthesis of 8-azido derivatives 2 is yet to be discovered.

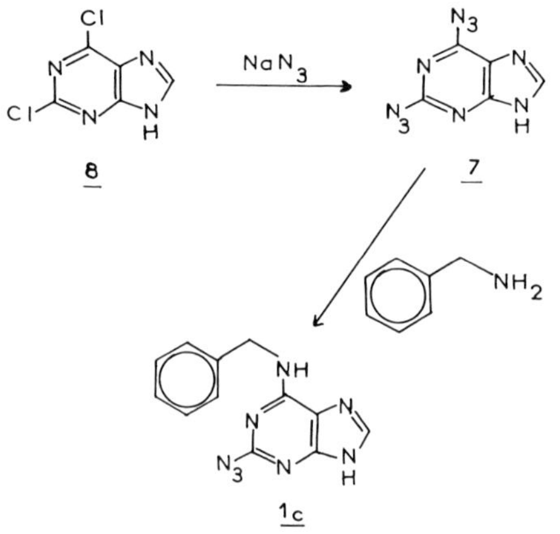

Scheme 5. An easy access to 8-azido-N^6-benzyladenine 1c suitable for radioactive labeling

5 Conclusions

The existence of azido-substituted cytokinins makes the technique of photoaffinity labeling accessible to biologists working in the cytokinin field. Improvements in the synthetic methods of these reagents, a better knowledge of their properties, and synthesis of new reagents are working areas through which chemists may help this technique to reach its full efficiency.

References

Bayley H, Knowles JR (1977) Photoaffinity labeling. Methods Enzymol 46: 69-114

Chowdhry V, Westheimer FG (1979) Photoaffinity labeling of biological systems. Annu Rev Biochem 48: 293-325

Cooperman BS (1975) Photoaffinity labeling of proteins and more complex receptors. In: Smith KC (ed) Aging, carcinogenesis and radiation biology. Plenum Press Corp, New York, p 315

Creed D (1974) Photochemical probes for biological interactions. Photochem Photobiol 19: 459-462

Doering WE von, Odum RA (1966) Ring enlargement in the photolysis of phenyl azide. Tetrahedron 22: 81-93

Dreyfuss G, Schwartz K, Blout ER, Barrio JR, Liu FT, Leonard NJ (1978) Fluorescent photoaffinity labeling: Adenosine 3',5'-cyclic monophosphate receptor sites. Proc Natl Acad Sci USA 75: 1199-1203

Fox JE, Erion JL, Mc Chesney JD (1979) A new intermediate in the synthesis of a tritiated cytokinin with high specific actiivity. Phytochemistry 18: 1055-1056

Katzenellenbogen JA (1976) Affinity labeling as a technique in determining hormone mechanisms. In; Litwack G (ed) Biochemical action of hormones, vol IV. Academic Press, London, New York, p 1

Knowles JR (1972) Photogenerated reagents for biological receptor-site labeling. Acc Chem Res 5: 155-160

Lwowski W (1970) Nitrenes. Wiley-Interscience, New York, pp 185-224

Mc Robbie IM, Meth-Cohn O, Suschitzky H (1976) Competitive cyclizations of singlet and triplet nitrenes Part 1. Cyclization of 1-(2-nitrophenyl) pyrazoles. Tetrahadron Lett 925-928; part 2. Cyclization of 2-nitrophenyl-thiophens, -benzothiazoles and -benzimidazoles. Tetrahedron Lett 929-932

Moore FH (1979) A cytokinin-binding protein from Wheat-Germ. Isolation by affinity chromatography and properties. Plant Physiol 64: 595-599

Mornet R, Theiler JB, Leonard NJ, Schmitz RY, Moore FH, Skoog F (1979) Active cytokinins. Photoaffinity labeling agents to detect binding. Plant Physiol 64: 600-610

Reiser A, Wagner HM (1971) Photochemistry of the azido group. In; Patai S (ed) The chemistry of the azido group. Wiley-Interscience, New York pp 441-501

Sprecker MA, Morrice AG, Gruber BA, Leonard NJ, Schmitz RY, Skoog F (1976) Fluorescent cytokinins: stretched-out analogs of N^6-benzyladenine and N^6-(Δ^2-isopentenyl)adenine. Phytochemistry 15: 609-613

Sundberg RJ, Smith RH (1971) Nucleophilic aromatic substitution during deoxygenation. Deoxygenation of nitrosobenzene by triethyl phosphite in alcohols. J Org Chem 36: 295-300

Sussman MR, Firn R (1976) The synthesis of radioactive cytokinin with high specific activity. Phytochemistry 15: 153-155

Sussman MR, Kende H (1977) The synthesis and biological properties of 8-azido-N^6-benzyladenine, a potential photoaffinity reagent for cytokinin. Planta 137: 91-96

Temple C, Thorpe MC, Coburn WC, Montgomery JA (1966) Studies on the azidoazomethine-tetrazole equilibrium. IV-Azidopurines. J Org Chem 31: 935-938. Temple C, Kussner CL, Montgomery JA (1966) V-2- and 6-azidopurines. J Org Chem 31: 2210-2215

Theiler JB (1977) Synthesis of purines derivatives as probes of hormones receptors. Ph D Thesis, Univ Illionis-Urbana-Champaign

Theiler JB, Leonard NJ, Schmitz RY, Skoog F (1976) Photoaffinity-labeled cytokinins. Synthesis and biological activity. Plant Physiol 58: 803-805

Uptake of [^{14}C]-8-Azido-N^6-Benzyladenine, a Radioactive Photosensitive Cytokinin, by the Cells of the Moss *Funaria hygrometrica* L.

R. Miassod[1]

1 Introduction

High affinity sites for cytokinins that appeared to have some of the properties expected for hormone receptors have been shown to occur in tobacco cells (Sussman and Kende, 1978) wheat germ (Fox and Erion, 1975; Polya and Davies, 1978; Moore, 1979) and pumpkin cotyledons (Rousselt and Fox, 1979). Low affinity sites for cytokinins were also observed in much higher frequency. From their characteristics it was proposed that they were unrelated to the biological action of the hormone. All these cytokinin-binding sites have been studied with subcellular fractions from plant homogenates. Since it is now known that cytokinins have a strong tendency to bind to any substrate, even unrelated to biological materials, such as talc (Sussman and Kende, 1978; Gardner et al., 1978), the question then arose of whether the observed ratio of low affinity sites to high affinity sites was true or had been favored during the preparation of the subcellular fractions, thus hampering any attempt to characterize cellular receptors for cytokinins.

One possible experimental approach that could overcome this difficulty would be to covalently label the cytokinin-binding proteins in situ, prior to any cell fractionation procedure, in a biological material made of both cytokinin target and non target cells. Covalently labeled cytokinin-binding proteins would be expected only in the target cells. Indeed, procedures to prepare photosensitive derivatives of cytokinins were available (Theiler, 1977; Sussman and Kende, 1978). It was also known for a long time (Bopp, 1968) that some cells of the protonemata of the moss *Funaria hygrometrica* specifically form gametophores in response to cytokinin treatment, whereas other cells of the same organ do not. Several reports aimed at the characterization of cytokinin-binding proteins in this material have appeared (Brandes and Kende, 1968; Erichsen et al., 1977; Gardner et al., 1978). Therefore it seemed of interest to prepare a radioactive photosensitive cytokinin and to investigate its uptake by the moss cells by performing autoradiographies.

Abbreviations: BA: N^6-benzyladenine; 8AzBA: 8-azido-N^6-benzyladenine; 8Az9RBA: 8-azido-N^6-benzyl-9-(1-ethoxyethyl)-adenine; BrBA: para-bromo-N^6-benzyladenine; i^{6}A: N^6-(Δ^2-isopentenyl)adenine.

[1] Laboratoire de Biochimie Végétale, UER de Luminy, Université d'Aix-Marseille II, 70 route Léon Lachamp, 13288 Marseille Cédex 2, France

2 Material and Methods

2.1 Chemical Syntheses

BrBA was prepared from 6-methylthiopurine and *p*-bromobenzylamine as described by Okumura et al. (1959). The crude product was recrystallized from methylene chloride and further purified on Sephadex LH 20 (Table 1). Unlabeled 8AzBA and 8Az-9RBA were synthesized according to Theiler (1977) and purified on Sephadex LH 20 (Table 1). The azido-derivatives displayed UV characteristics and a mass spectrometry fragmentation pattern identical to the records. [^{14}C]-8AzBA was synthesized according to the four step-route, as reported by Theiler (1977), with modifications. The radioactive precursor was [methylene-^{14}C]-benzylamine, HCl from Amersham (53 mCi/mmol).

Table 1. *Relative R_F of some purine derivatives on Sephadex LH 20 resin*. The derivatives were chromatographed on Sephadex columns developed with ethanol/water (1/2 v/v). Their relative R_F (ratio of eluting volume to the resin volume) were measured

Analog	Relative R_F value
6-methylthiopurine	1.66[a]
N^6-benzyladenine	2.66[a]
p-bromo-N^6-benzyladenine	4.97[a]
6,8-dichloropurine	1.26[b]
8-azido-N^6-benzyl-9-(1-ethoxyethyl)adenine	2.69[b]
8-azido-N^6-benzyladenine	3.30[b]

[a] Chromatography at room temperature, in a 27 x 2 cm column
[b] Chromatography at 4°C, in a 100 x 1.3 cm column

2.2 Moss Culture and Incubation with the Cytokinin Analogs

Spores of *Funaria hygrometrica* L. were germinated on an agar-solidified culture medium covered with a cellophane film, according to Brandes and Kende (1968). Eight days later, the young protonemata were transferred onto a fresh medium, still covered with the cellophane film, and grown for 10-15 additional days. The 18-23-day-old protonemata were then incubated with the cytokinin analog in the following way. A microscope slide covered with a metallic frame on which a 2-cm-wide circular aperture had been fitted was set in a glass Petri dish. The filter-sterilized analog was added to cooling autoclaved medium and the mixture poured onto the frame. The protonema was transported, over the cellophane film, on this approximately 1-ml disc of cytokinin-containing culture medium. To prevent an excessive evaporation of water from the culture medium, during the incubation period, a water-saturated paper disc was set under the microscope slide and the Petri dish was sealed with parafilm.

2.3 Estimation of the Radioactivity Accumulated in the Moss Protonemata

The protonemata were immersed in 3 ml of water and sonicated for 30 s with a MSE Ultrasonifier. 0.5 ml of hydrogen peroxide was added and digestion allowed to proceed overnight at 40°C. Ten ml of the liquid scintillation mixture Instagel (Packard Co) were then added and the radioactivity was estimated with an Intertechnique scintillation spectrometer.

2.4 Autoradiography of the Moss Protonemata

After incubation with the cytokinin analogs the moss protonemata were irradiated for 10 min, at a distance of 1 cm, with a common UV hand-held lamp (UVS-15 Mineralight, Ultra-violet Products, Inc. San Gabriel, CAL USA). Then they were fixed either with the "liquid technique" or the "non liquid technique". The "liquid technique" consisted of three successive rinses of the protonemata by immersing the organ in 20 ml of distilled water for 5 min, then loading the moss protonemata on microscope slides and dipping the slides in ethanol-glacial acetic acid-formaldehyde 4% (6/1/3, v/v) for 30 s, according to Hahn (1968). The microscope slides were then brought back to water through successively lowered concentrations of the fixing solution.

The "non liquid technique" consisted in transferring the moss protonemata, resting on the cellophane film, onto a new agar-solidified medium (three times, 1 h each time), then loading the protonemata on a microscope slide and heating the loaded slide for a few seconds. The microscope slides were then cooled on ice and the protonemata covered with the liquified sensitive emulsion NTB 3 from Kodak, with the loop technique of Caro et al. (1962). A rapid gelification of the emulsion was thus ensured. The protonemata were stored at 4°C in light-proof boxes for a few days. The emulsion was developed with Microdol X, from Kodak, at 24°C and for 8 min under constant magnetic stirring, and fixed.

3 Results and Discussion

3.1 Synthesis of a Radioactive Photosensitive Derivative of Cytokinin

3.1.1 Choice of a Radioactive Precursor and of a Route to Synthesize a Radioactive Photosensitive Derivative of BA

The radioactive compound needed for this study had to display a broad chemical reactivity and to be a very active cytokinin in order to bind covalently to any cytokinin-binding protein(s) with a high specificity and efficiency. This was precisely the case for two derivatives of BA, 2-azido-N^6-benzyladenine and 8-azido-N^6-benzyladenine, the synthesis of which had been reported (Theiler et al., 1976; Theiler, 1977; Sussman and Kende, 1978). These photosensitive derivatives were known to be active cytokinins and it is notorious that the nitrene group generated upon UV-irradiation of the azido group is highly reactive. 8AzBA was thought to be a better choice than the 2-azido-derivative for the synthesis of a radioactive probe because of the higher biological activity of 8AzBA and of the additional complication of the azido-methine-tetrazole equilibrium in the case of the 2-azido-derivative. The rationale to set a procedure for the synthesis of the radioactive 8AzBA was the following. (1) The high lability of the azido group precluded the use of tritiated precursors or the realization of a catalytic exchange reaction between tritium and an adequate substituent introduced in the molecule. (2) Since it was intended to feed the moss protonemata with the radioactive cytokinin for a long time, a labeling of the purine ring was not allowed; any labeling on the purine moiety would have led to the build-up of a radioactive pool of nucleoside triphosphates, via the catabolism of the radioactive analog by the cells, and therefore to the covalent labeling of the ribonucleoproteins. (3) Lastly, the availability of already published routes for synthesis, as well as of commercial radioactive precursors, had to be taken into account. It was thus decided to prepare [^{14}C]-labeled 8AzBA according to the route reported by Theiler (1977), the radioactive precursor being [^{14}C]-benzylamine.

3.1.2 Synthesis and Purification of [^{14}C]-8AzBA

[^{14}C]-8AzBA was synthesized from 6,8-dichloropurine and [^{14}C]-benzylamine, HCl, as starting reagents. The purine precursor had been discontinued by chemical companies and was therefore synthesized from 4,5-diamino-6-hydroxypyrimidine, via 6,8-dihydroxypurine, and then purified. [^{14}C]-8AzBA was then obtained via [^{14}C]-8-chloro-N^6-benzyladenine, [^{14}C]-8-chloro-N^6-benzyl-9-(1-ethoxyethyl)adenine and [^{14}C]-8-azido-N^6-benzyl-9-(1-ethoxyethyl)adenine according to the route reported by Theiler (1977) for unlabeled 8AzBA, but with modifications. The detailed procedure for synthesis will be reported elsewhere (manuscript in preparation).

Figure 1a shows the purification of [^{14}C]-8AzBA. The largest peak of radioactive material corresponded to [^{14}C]-8AzBA, as demonstrated by its cochromatography with standard 8AzBA on Sephadex LH 20 columns developed with ethanol/water (1/2 v/v) or on TLC plates developed with various solvents. Also, its UV-spectra before and after UV-irradiation (Fig. 2b) were identical to those reported for unlabeled 8AzBA. The radioactive derivative was as active as the unlabeled compound in the moss bioassay.

3.1.3 Other Potential Choices to Probe Cytokinin-Binding Protein(s) in Situ

Methods of synthesizing various unlabeled azido-derivatives of BA, zeatin, and i^{6}A were published while the present work was in progress (Mornet et al., 1979). The reactive group was introduced in the molecules as a substituent on C^2, C^8 or on the lateral chain. The biological activity of these derivatives was tested and the compounds generated by UV-irradiation of the molecules in alcoholic-aqueous solutions were identified (Mornet et al., 1979). From these data the question of which of the photosensitive derivatives would be the most appropriate choice to probe cytokinin-binding protein(s) in situ could be discussed. The BA derivatives substituted on the lateral chain appear to be of limited value in the tobacco callus bioassay because of their low biological activity, as compared to the parental product. There are chemical limitations for the synthesis of the C^2-substituted derivatives. In the case of the C^8-derivative of zeatin, it has been shown that the nitrene group inserts in the molecule

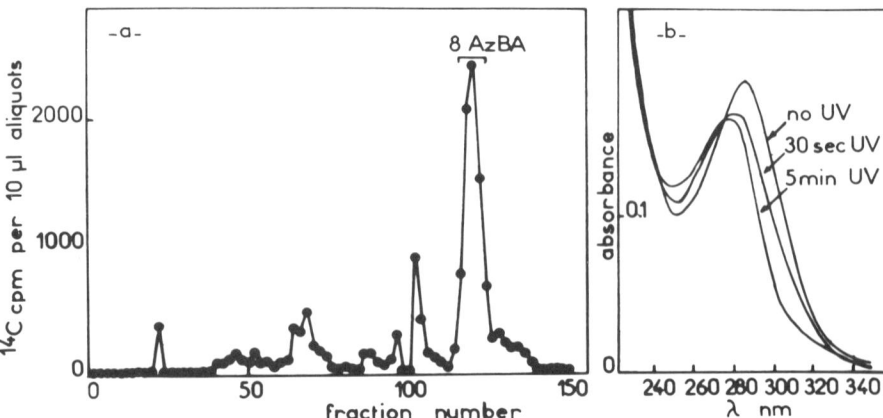

Fig. 1a, b. Purification of crude [^{14}C]-8-azido-N^6-benzyladenine on Sephadex LH 20 (a) and UV spectra of the purified compound (b). 0.12 μmol of [^{14}C]-8AzBA, obtained from the initial condensation of 1.5 μmol [^{14}C]-8AzBA with 4.5 μmol 6,8-dichloropurine, was chromatographed, at 4°C and in the absolute dark, on a 100 × 1.3 cm glass column of Sephadex LH 20. The solvent used was ethanol/water (1/2 v/v). Fractions of 4.3 ml were collected. Three nmol of the purified [^{14}C]-8AzBA were UV-irradiated for various times in 1 ml ethanol/water (1/2 v/v) and the corresponding UV spectra of the solution recorded

itself, therefore at the expense of the surrounding molecules. 8AzBA and the 8-azido-derivative of i^6A do not suffer any of these disadvantages and therefore can be presently taken as the best compounds to prepare a radioactive photosensitive derivative of cytokinin. It appeared convenient to synthesize [^{14}C]-8AzBA, because of the availability of a route for the synthesis, starting from a commercial radioactive precursor, although a procedure to obtain [^{14}C]-labeled 8-azido-N^6-(Δ^2-isopentenyl) adenine could probably be achieved.

3.2 Response of the Moss Protonemata to a Simultaneous Treatment of the Cells with the Radioactive Probe and BrBA or 8Az9RBA

3.2.1 The Experimental Approach

It was intended to vizualize covalent complexes between the probe and cytokinin-binding macromolecules on autoradiograms. Therefore the moss protonemata had to be fed with the radioactive probe, to be UV-irradiated to allow the covalent linkage of the photosensitive derivative to any surrounding molecule, and then to be rinsed prior to autoradiography to allow the exit of the unreacted molecules. This experimental approach would give a picture of cytokinin-binding sites in situ. However, it would not discriminate between high affinity sites, expected to be observed only in the target cells, and low affinity sites expected in all the moss cells. Therefore it was modified as follows: the moss protonemata would be given together the radioactive probe and an unlabeled analog displaying the following properties. This analog would be stereochemically related to BA, would be an inactive cytokinin analog and would not disturb the response of the moss cells to the radioactive probe. The expected result would be the selection of high affinity sites on the autoradiograms, the unlabeled analog having competed with the labeled probe only for low affinity sites.

3.2.2 Effect of BrBA or 8Az9RBA on the Uptake of [^{14}C]-8AzBA by Moss Cells

[^{14}C]-8AzBA was fed to moss protonemata at a concentration ranging from 10^{-8} M to 10^{-6} M, either alone or together with BrBA or 8Az9RBA. The level of radioactivity of the organs was estimated after various times. The results of a typical experiment are reported in Fig. 2. [^{14}C]-8AzBA was continuously taken up from the culture medium over a 40-h incubation period, whatever had been the hormonal treatment. The slowing down of the uptake by 15-20 h of incubation was probably due to the alternate periods of light and dark of the culture room, since it disappeared when the experiment was repeated under continuous light or in complete darkness. BrBA and 8Az9RBA did not inhibit the uptake of the radioactive probe but, on the contrary, stimulated it. The stimulation was consistently observed for short incubations and concentrations of [^{14}C]-8AzBA from 10^{-8} M to 10^{-7} M.

3.2.3 Effect of a Treatment of the Moss Protonemata with BrBA, or 8Az9RBA, on Their Ability to Respond to a Cytokinin Treatment

The biological activity of BrBA and 8Az9RBA was tested in the cytokinin moss assay and compared to that of BA. The results are reported in Fig. 3 (inserted graph in Fig. 3a). 8Az9RBA was completely inactive. BrBA displayed a weak activity and only at a high concentration. The moss protonemata still formed gametophyte buds when they were incubated with BrBA for 24 h, then with BrBA and various amounts of 8AzBA for an additional 36 h (Fig. 3a). A similar result was obtained when the first incubation of the organs with the bromo-derivative alone was omitted (Fig. 3b). Gametophyte buds also formed in response to a treatment with various amounts of 8AzBA, when 8Az9RBA was added to the culture medium (Fig. 3c). A very similar

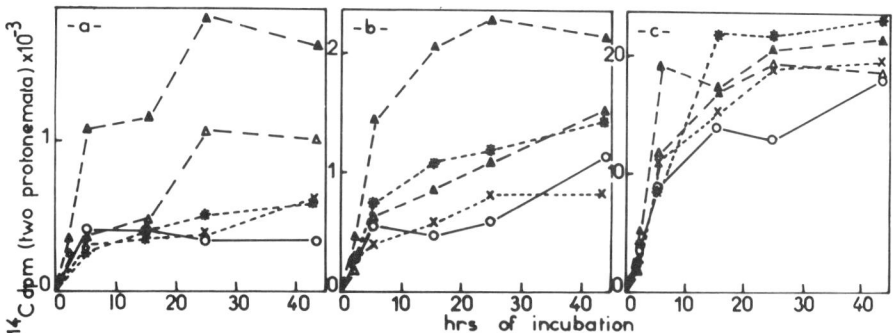

Fig. 2a-c. Effect of p-bromo-N^6-benzyladenine and 8-azido-N^6-benzyl-9-(1-ethoxyethyl)adenine on the uptake of [^{14}C]-8-azido-N^6-benzyladenine by moss protonemata. The protonemata were incubated for various periods with a given amount of [^{14}C]-8AzBA (o—o). 10^{-6} M BrBA(x- -x), 10^{-5} M BrBA (*- -*), 10^{-5} M 8Az9RBA (△- -△) or 10^{-4} M 8Az9RBA (▲- -▲) were added to the culture medium in parallel experiments. [^{14}C]-8AzBA concentration was 10^{-8} M (a), 10^{-7} M (b) and 10^{-6} M (c)

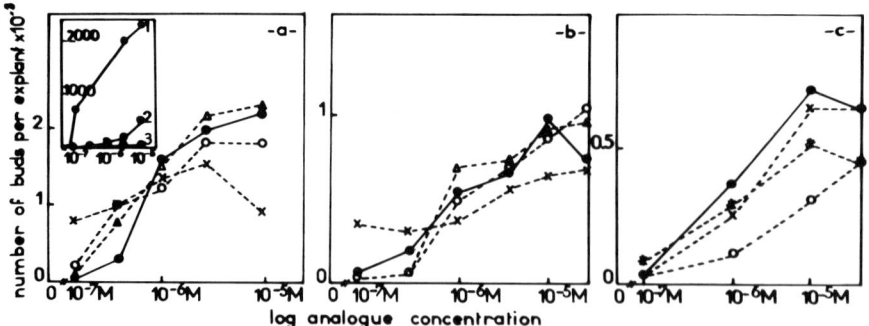

Fig. 3 a-c. Effects of BrBA and 8Az9RBA on the response of moss protonemata to a treatment with 8AzBA. Moss protonemata were incubated with increasing concentrations of 8AzBA (●—●), for 36 h. In parallel experiments a given amount of BrBA (**b**) and 8Az9RBA (**c**) was added together with various concentrations of 8AzBA. The fixed concentration of the bromo or the 9-substituted analog was 10^{-7} M (△--△), 10^{-6} M (○--○), 10^{-5} M (x--x) or 10^{-4} M (*--*). **a** was as in **b**, except that the protonemata were submitted to a 24-h pretreatment with the BrBA, at the same concentration as the one used for the final incubation. Gametophyte buds were then counted. The *inserted figure* refers to the total number of buds formed in response to a treatment of the protonemata with BA (*curve 1*), BrBA (*curve 2*) or 8Az9RBA (*curve 3*) alone

result was obtained when these experiments were repeated with BA instead of 8AzBA as the active cytokinin (manuscript in preparation). However, the response of the moss to the active cytokinin was lowered for the highest tested concentration of BrBA (10^{-5} M) and for 10^{-6} M 8Az9RBA.

BrBA, which had been shown to fully compete with [^{3}H]-BA for nonbiological binding to a particulate fraction from tobacco cells (Sussman and Kende, 1978), and 8Az9RBA, the immediate precursor of 8AzBA in the route of synthesis, thus fulfilled the required properties for the unlabeled cytokinin analog to be used in that study.

3.3 Autoradiographic Study of the Distribution of [^{14}C]-8AzBA Among the Cells of Moss Protonemata in the Presence, or Absence, of BrBA or 8Az9RBA

3.3.1 Autoradiograms of Moss Protonemata Incubated with [^{14}C]-8AzBA Alone

The protonemata were incubated with [^{14}C]-8AzBA under various conditions and the distribution of the probe within the plants was investigated by autoradiography, using the "liquid technique". The background of silver grains over the microscope slides virtually equalled zero, bearing evidence of the efficiency of the rinse operation. Quantitation of the data was therefore possible and the results are reported in Fig. 4. Most of the silver grains were located at the periphery of the round-shaped organs, above caulonema cells, although some were also present on the central part of the protonemata above chloronema cells. Counting of the silver grains showed that the analog was concentrated in the three terminal cells of caulonema filaments when the protonemata were incubated with [^{14}C]-8AzBA for a short period (16 h). [^{14}C]-8AzBA was found to be present in a larger number of caulonema cells (6-7 terminal

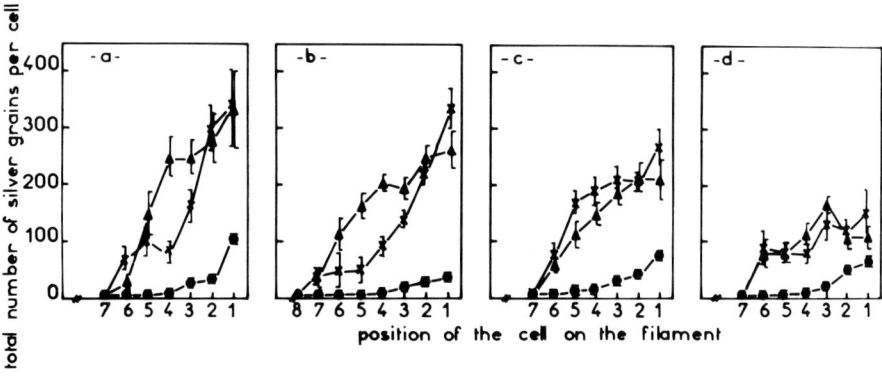

Fig. 4 a-d. Distribution of silver grains among the caulonema cells of autoradiograms of moss protonemata incubated with [^{14}C]-8AzBA. Moss protonemata were incubated with 10^{-7}M (a, b) or 10^{-6}M (c, d) [^{14}C]-8AzBA for 16 h (●—●), 26 h (▲—▲) or 36 h (x--x). Incubations were performed either under alternate periods of light (16 h) and dark (8 h) (a, c), or in complete darkness (b, d). The protonemata were rinsed and fixed with the "liquid technique". Silver grains were counted on the cells of 10-15 filaments and averaged (± standard error)

cells of caulonema filaments) when the incubation was prolonged to 26-36 h. Similar results were obtained whatever the analog concentration (10^{-7}M or 10^{-6}M) and the experimental procedure for growth (alternate light and darkness or continuous darkness): in all cases there appeared to be a decreasing gradient of the hormone concentration from the tip to the central part of the caulonema filaments.

This was an unexpected result since it had been previously shown with the same material that another radioactive cytokinin [^{3}H]-BA) was preferentially accumulated in the caulonema cells known to form buds (cells number 5-7 from the tip of the primary caulonema filaments) as well as in the buds themselves. What was seen on the autoradiograms obtained in the present study was probably complexes between [^{14}C]-8AzBA and high affinity sites, in the cytokinin-target cells, and complexes between [^{14}C]-8AzBA and other sites not related to the morphogenetic effect of the hormone and located in all cells, especially the actively dividing ones, namely the tip cells of the caulonema filaments. The ratio of the former to the latter increased with the duration of the hormone treatment. The discrepancy between the two studies could be due to a difference in the affinity of the two cytokinins used for these biologically meaningless binding sites. Alternatively it could be related to the use of a "liquid technique" for fixing the protonemata, in the present study, instead of the "non liquid technique" of Brandes and Kende (1968).

3.3.2 Autoradiograms of Moss Protonemata Incubated with [^{14}C]-8AzBA, in the Presence of BrBA

Moss protonemata were preincubated with BrBA, then incubated with [^{14}C]-8AzBA, still in the presence of the bromo-derivative. An autoradiographic experiment was carried out. Most of the silver grains were located over the caulonema cells, as had been the case when BrBA was absent from the incubation medium. Quantitative ana-

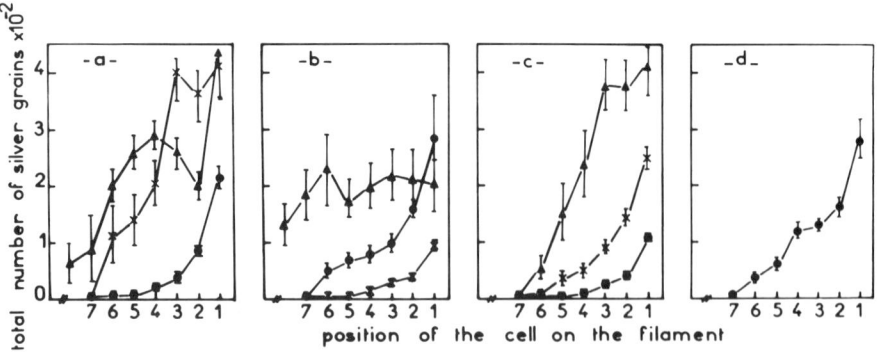

Fig. 5a-d. Distribution of silver grains among the caulonema cells of autoradiograms of moss protonemata incubated with [^{14}C]-8AzBA and BrBA. Moss protonemata were preincubated with 3.10^{-7}M BrBA for 48 h, then incubated with 10^{-7}M (a, b) or 10^{-6}M (c, d) [^{14}C]-8AzBA for 16 h (●—●), 26 h (▲—▲) or 36 h (x—x), still in the presence of BrBA. Incubations with the radioactive analog were conducted under alternate periods of light and dark (a, c) or in darkness (b, d). The protonemata were rinsed and fixed with the "liquid technique". Silver grains were counted over the cells of 10-15 filaments and the counts averaged (± standard error)

lysis pointed to a decreasing gradient of silver grains along the caulonema filaments, from the tip to the basis, whatever were the BrBa and [^{14}C]-8AzBA concentrations, the duration of the incubation and the experimental procedure for growth. Results are illustrated in Fig. 5 for 3.10^{-7}M BrBA, 10^{-7} or 10^{-6}M [^{14}C]-8AzBA and 16 to 36 h.

BrBA thus seemed to be inefficient in removing [^{14}C]-8AzBA from "nonbiological" sites in the moss cells in vivo, while being able to compete with [^{3}H]-BA for low affinity binding sites in a particulate fraction of tobacco cells, in vitro. This could be due to a lower affinity of BrBA for those sites than 8AzBA itself or, as suggested above, to the inadequacy of the fixing procedure used in that study.

3.3.3 Autoradiograms of Moss Protonemata Incubated with [^{14}C]-8AzBA, in the Presence of 8Az9RBA

An autoradiographic study of moss protonemata that had been incubated with 10^{-7}M [^{14}C]-8AzBA and 10^{-6}M 8Az9RBA was performed. The "nonliquid technique" was used to rinse and fix the protonemata. Preliminary results showed that the silver grains were virtually absent from chloronema cells and accumulated only in caulonema cells number 5-7. Quantitative analysis of these experiments is now in progress.

This result suggests that 8Az9RBA efficiently competed for biologically meaningless binding of the radioactive probe to cellular components, so that only a binding to sites involved in the response of the target cells to the hormone was observed. Although confirmation of this result is needed, it would well agree with the prefential binding of [^{3}H]-BA to cytokinin target cells of the caulonema filaments (Brandes and Kende, (1968) and with the evidence brought for specific proteins binding BA in those cells (Erichsen et al., 1977). The photosensitive derivative thus appeared to behave quite similarly to the parental cytokinin, BA, both with regard to the early events involved in the hormonal action (interaction with receptor-structures) and the late ones (the bud formation). Therefore, the use of $[^{14}C]$-8AzBA as a radioactive photosensitive probe to look for cytokinin-receptor(s) seems to be valuable, provided it is kept in mind that the probe, as also the parental cytokinin, may interact with cellular components other than hormone receptor(s). Application of the present experimental approach to cultured tobacco cells is presently in progress.

References

Bopp M (1968) Control of differentiation in fern-allies and bryophytes. Ann Rev Plant Physiol 19: 361-380
Brandes H, Kende H (1968) Studies on cytokinin-controlled bud formation in moss protonemata. Plant Physiol 43: 827-837
Caro LG, Van Tubergen RP, Kolb JA (1962) High-resolution autoradiography. I Methods. J Cell Biol 15: 173-188
Erichsen J, Knoop B, Bopp M (1977) On the action mechanism of cytokinins in mosses:caulonema-specific proteins. Planta 135: 161-168
Fox JE, Erion JL (1975) A cytokinin-binding protein from higher plant ribosomes. Biochim Biophys Res Commun 64: 694-700
Gardner G, Sussman MR, Kende H (1978) In vitro cytokinin binding to a particulate fraction from protonemata of *Funaria hygrometrica*. Planta 143: 67-73
Hahn H (1968) Die Wirkung von Cytokinine auf die Knospenbildung isolierter Caulonemen von *Funaria hygrometrica*. Dissertation, Hannover
Moore III FH (1979) A cytokinin-binding protein from wheat germ. Isolation by affinity chromatography and properties. Plant Physiol 64: 594-599
Mornet R, Theiler JB, Leonard NJ, Schmitz RY, Moore FH, Skoog F (1979) Active cytokinins. Photoaffinity labeling agents to detect binding. Plant Physiol 64: 594-599
Okumura FS, Kotani Y, Ariga T, Masumura M, Kuraishi S (1959) Syntheses of kinetin analogues. Bull Chem Soc Japan 32: 883-887
Polya GM, Davis AW (1978) Properties of a high-affinity cytokinin-binding protein from wheat germ. Planta 139: 139-147
Roussell DL, Fox JE (1979) Properties of a cytokinin binding protein isolated from pumpkin cotyledons. In: 10th Int Conf Plant Growth Substances, Madison WIS USA, p 46
Sussman MR, Kende H (1978) The synthesis and biological properties of 8-azido-N^6-benzyladenine, a potential photoaffinity reagent for cytokinin. Planta 137: 91-96
Theiler JB (1977) Synthesis of purine derivatives as probes of hormone receptors. PhD Thesis, Univ Wisconsin, Madison
Theiler JB, Leonard NJ, Schmitz RY, Skoog F (1976) Photoaffinity-labeled cytokinins. Synthesis and biological activity. Plant Physiol 58: 803-805

Cytokinins and the Hormone Receptor Concept

D. Klämbt[1]

1 Introduction

Cytokinins are hypermodified adenines. They are synthesized by alkylation of A-nucleotides within RNA molecules. The cytokinin action is correlated with the monomer nucleotide, nucleoside or the most active base status. There appear to be various routes for incorporation of intact cytokinin-nucleotides into macromolecular RNA (Murai et al., 1977; Teyssendier de la Serve and Jouanneau, 1979), but this incorporation seems to be unrelated to the cytokinin action.

Each signal, which enforces a reply, has to be recognized. This is a very general widely acknowledged cellular feature. Recognition may vary in specificity but it has to occur. Since the work on structural relationships of substituted adenines to cytokinin action by Skoog et al. (1967) we have a better understanding of structural limitations for cytokinins with high activity. These results predict a precise discrimination between related adenine-derivatives by every cytokinin-sensitive cell.

In this context Berridge et al. (1970) reported cytokinin binding onto ribosomes. The characteristics for this cytokinin affinity were described in more detail by Fox and Erion since 1975, Polya and coworkers since 1978, and Moore III (1979) for wheat germ proteins. The specificity of these proteins with cytokinin affinity has, however, become questionable since the data of Polya and Bowman (1979) has been known. They showed no correlation of cytokinin binding and cytokinin activity. Chung et al. (1979) reported cytokinin binding onto ribosomes in *Mercurialis* and Takegami and Yoshida (1975) described a protein with cytokinin affinity from tobacco leaves. One big gap in our knowledge has been the inability to demonstrate any functional relationship between these proteins and any presumed hormone action.

My own research reported hitherto only at meetings in Lausanne (1975), Edinburgh (1978), and Madison (1979) has not been any more conclusive. We purified very specific proteins from wheat germ by affinity chromatography. We were then able to describe a function of these proteins in a a-tRNA transfer reaction onto ribosomes. However, although the experiment was attempted many times, we were unable to determine the cytokinin binding kinetics of the proteins. The difference in my own results with the published data is dependent on the varying fractions of the S-100 extract from wheat germ. The published data are related to proteins with cytokinin binding activity

---

Abbreviations: IPA: N^6 (Δ^2 isopentenyl)adenosine, IPAde: N^6 (Δ^2 isopentenyl)adenine, SDs: sodium dodecylsulfate.

[1] Botanisches Institut, Universität Bonn, Meckenheimer Allee 170, 5300 Bonn 1, FRG

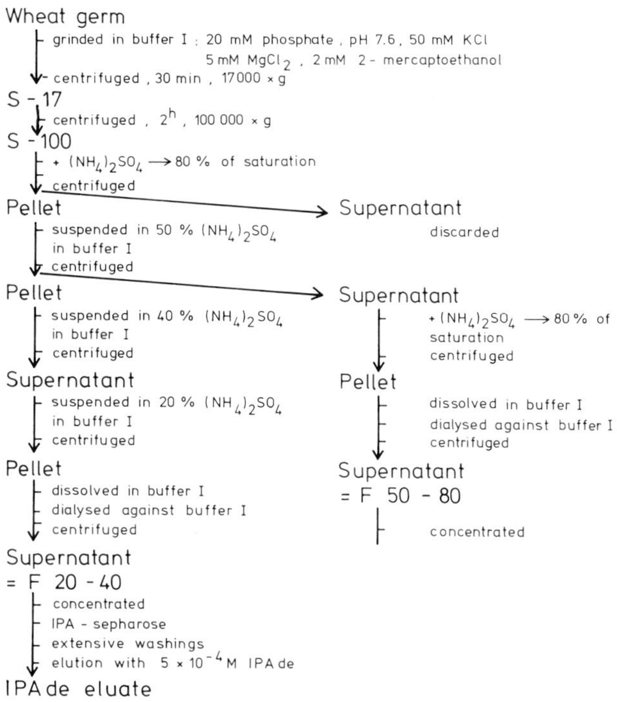

Fig. 1. Preparation procedures

```
Wheat germ
  ├ grinded in buffer I : 20 mM phosphate, pH 7.6, 50 mM KCl
  │                      5 mM MgCl_2, 2 mM 2-mercaptoethanol
  ↓ centrifuged, 30 min, 17000 × g
S - 17
  ↓ centrifuged, 2^h, 100 000 × g
S - 100
  ├ + (NH_4)_2SO_4 → 80 % of saturation
  ↓ centrifuged
Pellet ─────────────────────────→ Supernatant
  ├ suspended in 50 % (NH_4)_2SO_4         discarded
  │ in buffer I
  ↓ centrifuged
Pellet ─────────────────────────→ Supernatant
  ├ suspended in 40 % (NH_4)_2SO_4    ├ + (NH_4)_2SO_4 → 80 % of
  │ in buffer I                       │   saturation
  ↓ centrifuged                       ↓ centrifuged
Supernatant                         Pellet
  ├ suspended in 20 % (NH_4)_2SO_4    ├ dissolved in buffer I
  │ in buffer I                       ├ dialysed against buffer I
  ↓ centrifuged                       ↓ centrifuged
Pellet                              Supernatant
  ├ dissolved in buffer I             = F 50 - 80
  ├ dialysed against buffer I         ├ concentrated
  ↓ centrifuged
Supernatant
  = F 20 - 40
  ├ concentrated
  ├ IPA - sepharose
  ├ extensive washings
  ↓ elution with 5 × 10^{-4} M IPAde
IPAde eluate
```

precipitated by $(NH_4)_2SO_4$ above 50% of saturation. We used proteins precipitated by $(NH_4)_2SO_4$ between 20% and 40% of saturation, as shown in Fig. 1.

All these data make the work on cytokinin receptors frustrating and doubtful (Trewavas, 1979). But as mentioned above, cells have to recognize all signals to which they are required to react. For high specific recognition of these signals, cells seem to have no other molecules apart from proteins. Therefore we have to insist on looking for such receptors.

2 Material and Methods

Wheat germ was obtained from Ellmühle-Kampffmeyer[2], Köln. The main preparation scheme is shown in Fig. 1. For all experiments we used the F20-40 directly for the IPAde eluate or an 1.0 M NaCl eluate of the affinity column = IPA-sepharose (Hermwille et al., 1981). The various samples were separated by gel filtration on ACA 54 and by polyacrylamide gel electrophoresis under natural or denaturating conditions according to Maizel (1971). Gel slabs were stained with Coomassie blue R 250. Cytokinin binding was assayed by gel filtration on Sephadex G25, according to Moore III (1979).

---

[2] We thank Mr. Scharloh for the generous gift

3 Results and Discussion

The IPAde eluates consisted of 60% protein of MW 50,000. Gel electrophoresis under denaturing conditions results in 2-3 polypeptide chains of MW 16,000 to 18,000 (Fig. 2). Under natural conditions, the main proteins of the IPAde eluates are separated in several bands. They were highly regular (Fig. 3). Treating these protein bands, stained or unstained, with SDS and 2-mercaptoethanol, a second electrophoresis under denaturing conditions results in the same protein monomers. We therefore conclude that the main proteins of the IPAde eluates from the affinity column are trimeric isoproteins.

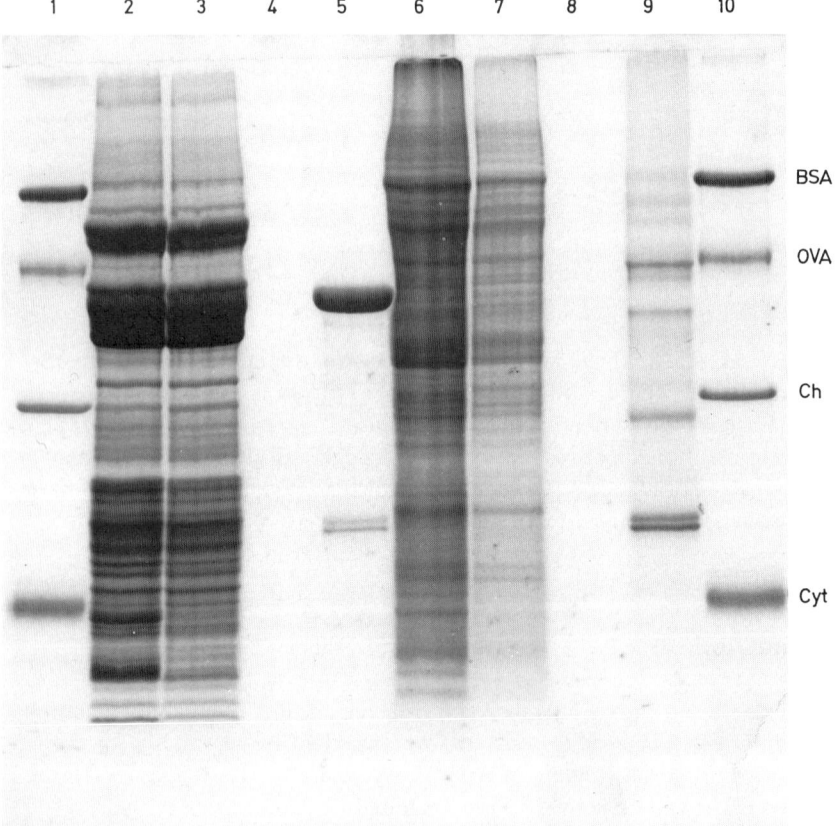

Fig. 2. Polyacrylamide gel electrophoresis under denaturing conditions of F50-80 and F20-40 and their fractions by the use of affinity chromatography. Tracks *1* and *10* Marker proteins: *BSA* bovine serum albumin; *OVA* ovalbumin; *Ch* chymotrypsinogen A; *Cyt* cytochrome c, Tracks 2-5 F50-80: Track 2 5µl F50-80 out of 13 ml ≙ 250g wheat germ, Track *3* 5 µl flow through, Track *4* washings 1/8 of the last 20 ml, Track *5* IPAde eluate 1/12 of the total (30 ml); Tracks 6-9: F20-40; Track 6 5 µl F20-40 out of 20 ml ≙ 60g wheat germ, Track *7* 5 µl flow through, Track *8* washings 1/3 of the last 20 ml, Track *9* IPAde eluate 1/3 of the total (30 ml)

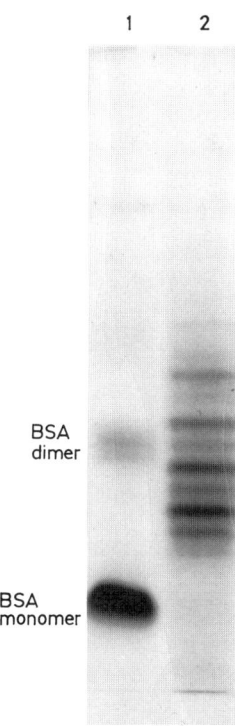

Fig. 3. Polyacrylamide gel electrophoresis under natural conditions of IPAde eluate after affinity chromatography of F20-40. Track *1* bovine serum albumin, Track *2* IPAde eluate

Further analysis has shown that our "tentative" receptors do exist also as trimeric molecules, when they are free of cytokinins. We applied the F20-40 onto an ACA 54 column and analyzed the resulting fractions for cytokinin binding, as well as for the polypeptide chains with the molecular weight of 16,000 to 18,000. Cytokinin binding ($K_d = 2 \times 10^{-6}$ M) and the typical protein bands of the monomeric molecules clearly interrelated. Peak fractions run off the column slightly before ovalbumin (MW 45,000) (Fig. 4). What then is the binding situation of the trimeric proteins within the affinity adsorbent? To help resolve this question we applied F20-40 onto IPA-sepharose. After extensive washing, the column was eluted with 1.0 M NaCl instead of 5×10^{-4} M IPAde. The resulting eluate was concentrated by membrane filtration and applied to an ACA 54 column previously equilibrated with 1.0 M NaCl in buffer. The same solvent was used to develop the column. The resulting fractions were analyzed as mentioned above. Actually, there appeared to be no binding activity, but the polypeptides with molecular weights of 16,000 to 18,000 were found in fractions eluting between chymotrypsogen A (MW 25,000) and cytochrome c (MW 12,400). These results indicate that the cytokinin affinity proteins probably bind onto IPA-sepharose as each single subunit.

To verify this statement we had to prove whether 1.0 M NaCl causes dissociation of the trimeric proteins. Therefore a concentrated IPAde eluate of the affinity column was dialyzed against 1.0 M NaCl in buffer and applied to an ACA 54 column equilibrated and finally developed with 1.0 M NaCl in buffer. The resulting fractions pro-

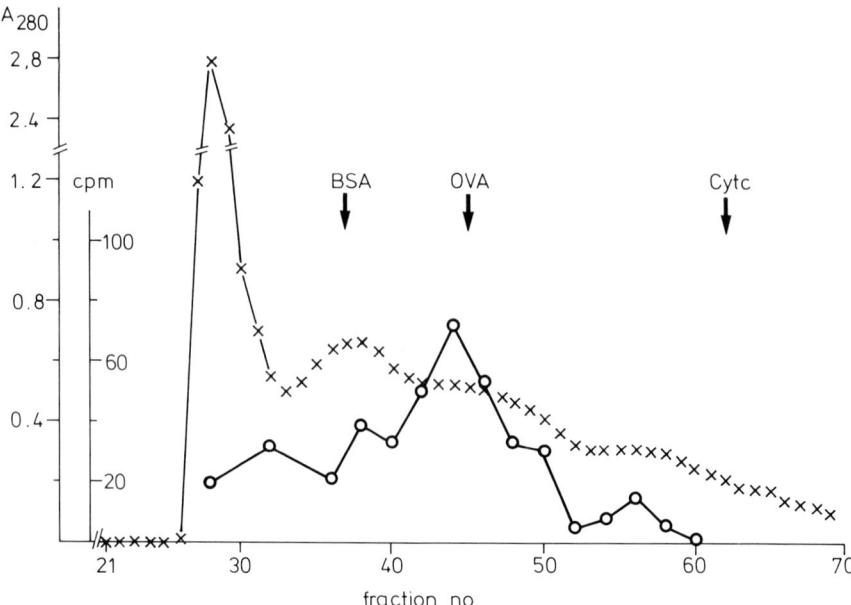

Fig. 4. Gel filtration on ACA 54 of F20-40. xx—x A_{280} o—o IPAde binding assayed according to Moore III (1979). ↓ Positions of marker proteins: *BSA* bovine serum albumin; *OVA* ovalbumin; *Cytc* cytochrome c

duced no binding activity. However, protein bands on SDS gel electrophoresis at molecular weights of 16,000 to 18,000 were eluted in fractions earlier than ovalbumin.

We conclude therefore that the presence of 1.0 M NaCl does not dissociate the trimeric cytokinin-binding proteins, but inhibits the association of the subunits. These results unequivocally demonstrate that the molecular status of proteins with cytokinin affinity is monomeric with respect to their binding onto the affinity adsorbent. In addition we may interpret from these data that each subunit possesses at least one binding site for cytokinins.

The IPAde eluate from the affinity column is taken as a specific preparation of proteins with cytokinin affinity. Unfortunately these proteins, when prepared in such a fashion, no longer specifically bind to cytokinins. Nevertheless the IPAde eluates significantly increase the activity of the elongation factor EF1 from wheat germ in the transfer reaction of phenylalanyl-tRNA onto poly U-ribosomes and lysyl-tRNA onto poly A-ribosomes (Klämbt and Schwabe, 1981). These results correlate well with data already published on the effect of cytokinin on in vitro protein synthesis (Klämbt, 1976, 1977).

4 Conclusion

Cytokinins, as other plant hormones, are micromolecular signals. They require specific macromolecular recognition structures. Recognizing one signal is the first step in hor-

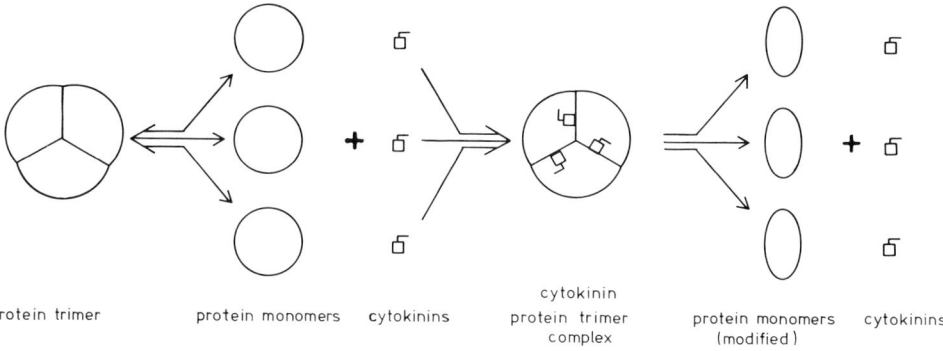

Fig. 5. Scheme of the "one-way receptor" hypothesis

mone action. There is no point in repeated recognition by the same receptor molecule. If we assume a "one-way receptor" system this would give cells the opportunity to keep their ability to react to hormones under genetic control. A "one-way receptor" bound to its hormone, necessarily has to be modified to an inactive protein, which no longer binds hormone. This may be possible by a conformational change, accompanied eventually by a biochemical modification. Fig. 5 depicts the described hypothesis.

All the reported data are in agreement with the hypothesis. However, the crucial evidence to prove that the trimeric isoproteins, separated by affinity chromatography, are actually the required receptor remains elusive.

It is hoped that by the used of immunological methods attempts will be made to separate the trimeric isoproteins to yield high purity proteins without any contact with the cytokinins. Hopefully, these samples will conclusively support the data and the proposed interpretation.

Acknowledgments. This work was granted by the Deutsche Forschungsgemeinschaft. I thank Miss Nicola Vogel for reading the English manuscript.

References

Berridge MV, Ralph RK, Letham DS (1970) The binding of kinetin to plant ribosomes. Biochem J 119: 75-84
Chung SR, Durand R, Durand B (1979) Differential cytokinin binding to dioecious plant ribosomes. FEBS Lett 102: 211-215
Fox, JE, Erion JL (1975) A cytokinin binding protein from higher plant ribosomes. Biochem Biophys Res Commun 64: 694-700
Fox JE, Erion JL (1977) Cytokinin-binding proteins in higher plants. In: Pilet PE (ed) Plant growth regulation. Springer, Berlin Heidelberg New York, pp 139-146
Hermwille E, Schwabe U, Klämbt D (1981) Cytokinin affinity Proteins. I. Preparation of a cytokinin-sepharose for affinity chromatography. In preparation
Klämbt D (1976) Cytokinin effects on protein synthesis of in vitro systems of higher plants. Plant Cell Physiol 17: 73-76
Klämbt D (1977) Cytokinin and cell metabolism. In: Pilet PE (ed) Plant growth regulation. Springer Berlin Heidelberg New York, pp 154-160

Klämbt D, Schwabe U (1981) Cytokinin affinity proteins. II. Isolation and characterization of cytokinin affinity proteins from wheat germ. In preparation

Maizel JV (1971) Polyacrylamide gel electrophoresis of viral proteins. In: Maramorosch K, Koprowski H (ed) Methods in virology. Academic Press, London New York, pp 179-246

Moore III FH (1979) A cytokinin-binding protein from wheat germ. Plant Physiol 64: 594-599

Murai N, Taller BJ, Armstrong DJ, Skoog F, Micke MA, Schnoes HK (1977) Kinetin incorporated into tobacco callus ribosomal RNA and transfer RNA preparations. Plant Physiol 60: 197-202

Polya GM, Davis AW (1978) Properties of a high-affinity cytokinin-binding protein from wheat germ. Planta 139: 139-147

Polya GM, Bowman JA (1979) Ligand specificity of a high affinity cytokinin-binding protein. Plant Physiol 64: 387-392

Skoog F, Hamzi HQ, Szweykowska AM, Leonard NJ, Carraway KL, Fujji T, Helgeson JP, Loeppky RN (1967) Cytokinins: Structure/activity relationships. Phytochemistry 6: 1169-1192

Takegami T, Yoshida K (1975) Isolation and purification of cytokinin binding protein from tobacco leaves by affinity column chromatography. Biochem Biophys Res Commun 67: 782-789

Teyssendier de la Serve B, Jouanneau JP (1979) Preferential incorporation of an exogenous cytokinin N^6-benzyladenine into 18s and 25s ribosomal RNA of tobacco cells in suspension culture. Biochimie 62: 913-922

Trewavas A (1979) What is the molecular basis of plant hormone action? TIBS N 199-201

Yoshida K, Takegami T (1977) Isolation of cytokinin binding protein from tobacco leaves by bioaffinity chromatography and its partial characterization. J Biochem 81: 791-799

The Current Status of Cytokinin-Binding Moieties

P. Keim, J. Erion, and J.E. Fox[1]

1 Introduction

Several years ago we described (Fox and Erion, 1975, 1977) the properties of a soluble protein (CBF-1) isolated from wheat germ which binds cytokinins with relatively high affinity and specificity. This study was stimulated by an earlier report from investigators in Letham's laboratory (Berridge et al., 1970) who described nonsaturable, multisite, reversible binding of kinetin to ribosomes isolated from *Brassica pekinensis* leaves. Although Scatchard plots of binding were not presented and the investigators were unable to solubilize a binding substance, their data indicate a low affinity, nonspecific binding of the kind seen in wheat germ ribosomes (Fox and Erion, 1977) at the high cytokinin concentrations used in their study. However a wheat embryo protein having a relatively high affinity for cytokinins has recently been studied in two other laboratories (Polya and Davis, 1978; Polya and Bowman, 1979; Moore, 1979) and is apparently identical with CBF-1.

A low molecular weight polypeptide (4000 daltons) from tobacco leaves which reportedly binds cytokinins has been described (Takegami and Yoshida, 1975; Yoshida and Takegami, 1977) but its affinity for cytokinins ($Kd = 4 \times 10^{-5}$ M) is so weak as to cast doubts on its biological role as a cytokinin receptor. A slightly larger cytokinin binding moiety (mw = 8,500) but with much higher affinity for cytokinins ($Kd = 9 \times 10^{-7}$ M) was recently isolated from tobacco tissue cultures (Chen et al., 1980). High affinity cytokinin binding moieties have also been solubilized from pumpkin cotyledons (Roussel and Fox, unpubl.) and bean embryos (Starr and Fox, 1978). A particulate fraction from tobacco cells which binds bzl^6 Ade ($Kd = 1 \times 10^{-7}$ M) has been reported (Sussman and Kende, 1978). Thus cytokinin-binding moieties are widespread in nature, but in no instance has their physiological function been delineated. Recently striking effects of cytokinins on the respiration of isolated mitochondria have been described (Miller, 1979, 1980); we report here isolation of a particulate fraction which sediments with the mitochondria and which binds cytokinins with higher affinity than any so far reported. Curious and as yet unexplained is the fact that all cytokinin-binding entities so far reported have much lower affinity for zeatin

---

Abbreviations: CBF-1, wheat germ cytokinin binding protein; bzl^6 Ade, 6-benzylaminopurine; i^6 Ade, 6(Δ^2-isopentenylamino) purine; io^6 Ade, Zeatin; OH-PA, 6 (5'-hydroxy-n-pentylamino) purine; i^6 Ado, N^6 (Δ^2-isopentenyl)adenosine; io^6 Ado, Zeatin Riboside

[1] Departments of Biochemistry and Botany, University of Kansas, Lawrence, KS 66045, USA

(io^6 Ade) than for other cytokinins. We report here studies on the binding of hydroxylated cytokinins in an effort to obtain further information on the role of the hydroxyl group in cytokinin activity. In addition further studies on the nature of the binding site of CBF-1 with the aid of a radiolabeled cytokinin photoaffinity probe are described and the distribution of CBF-1 as judged by reactivity with anti-CBF-l antibody is presented.

2 Methods and Materials

2.1 Isolation of a Wheat Germ Cytokinin-Binding Protein (CBF-1)

Wheat germ ribosomes were isolated as previously described (Fox and Erion, 1975) and a 50mM potassium acetate extract was prepared (Erion and Fox, 1980). The cytokinin binding protein CBF-1 was isolated from this preparation either by affinity chromatography (Erion et al., 1978; Erion and Fox, 1980) or by a conventional procedure involving chromatography on phosphocellulose, DEAE Biogel A and Sephadex G-200 (Figs. 1, 2, 3). The purified protein had an apparent mol. wt. of 155,000 as judged by Sephadex G-200 chromatography (Fig. 3), but SDS polyacrylamide gel electrophoresis revealed the presence of four subunits whose aggregate mol. wt. was judged to be 183,000 (Erion and Fox, 1980).

2.2 Cytokinin-Binding Assay

The specificity of ligand binding to cytokinin binding moieties was determined by equilibrium dialysis with the aid of a very high specific activity radiolabeled cytokinin (Fox et al., 1979). In most experiments the concentration of the radiolabeled probe, ^{3}H-bzl^6 Ade sp. act. 27 Ci/mmol, was 1×10^{-9} M. The binding of unlabeled cytokinins or analogs was determined by competition studies with ^{3}H-bzl^6 Ade. Assays were performed in plastic vials with 0.5 ml of a buffer (25 mM Tris-HCl pH 7.5, 50 mM KCl, 0.1 mM EDTA and 4 mM 2-mercaptoethanol) containing 5-25 µg of highly purified CBF-l inside the dialysis bag and 4.5 ml of buffer containing 1×10^{-9} M of the radiolabeled ligand together with varying concentrations of the unlabeled competitor outside. After an 18 h dialysis period at 4°C, 0.4 ml aliquots were removed from both inside and outside the dialysis bag and assayed for radioactivity by liquid scintillation spectrometry. Binding constants were estimated by assuming that the concentration of competitor required to yield 50% competition of the radiolabeled ligand binding to CBF-l is equal to the dissociation constant (K_d). This assumption is valid provided the labeled ligand concentration is much less than the K_d of the binding protein for the ligand, and also providing that the competitor interacts with the same site as the ligand (i.e., competitive interaction) (Cheng and Prusoff, 1973).

A very useful equation for studies of this type can be derived from classical treatments of enzyme inhibitors which show first order kinetics. If L is radioative ligand, I is nonlabeled competitor and P is protein, then it can be shown that the ratio between bound competitor [PI] and total protein, or total binding sites (Pt) can be described by the following equation:

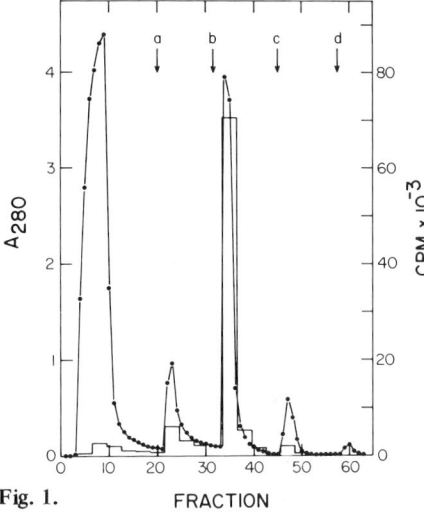
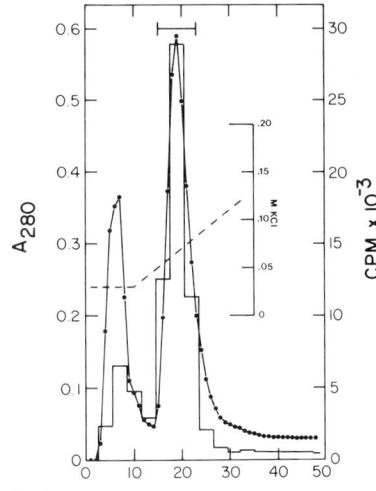

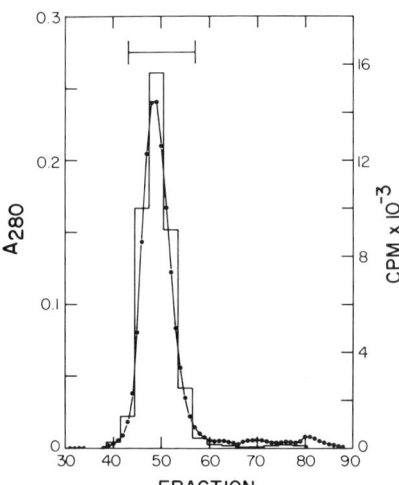

Fig. 1. Phosphocellulose chromatography of ribosomal CBF-1. 250 mg of protein was loaded on a Whatman P-11 1.6 x 20 cm phosphocellulose column. The protein was eluted with a step gradient in a buffer made (a) 0.15 M KCl (b) 0.5 M KCl (c) 0.75 M KCl and (d) 1.0 M KCl. CPM (*bar graph*) represents bound bzl^6 Ade (tritium labeled sp. Act. 27 ci/mmol) per 0.4 ml aliquot from three combined 3.5 ml fractions subjected to equilibrium dialysis (conc. of free bzl^6 Ade was 1×10^{-9} M)

Fig. 2. DEAE Biogel chromatography of CBF-1. Sixty-nine mg of protein which eluted from the 0.5 M KCl step from a phosphocellulose column (see Fig. 1) was applied to the top of a 1.6 x 15 cm column of DEAE Biogel A (BioRad). The column was developed with 25 ml of buffer followed by a linear gradient of KCl in buffer. Binding of bzl^6 Ade (*bar graph*) was determined as described in the legend to Fig. 1

Fig. 3. Sephadex G-200 chromatography of CBF-1. Twenty-five mg of protein representing the major cytokinin binding fraction eluting from a DEAE-Biogel A Column (see Fig. 2) was applied to the top of a column of Sephadex G-200, 2.5 x 95 cm, and eluted in buffer. Binding of bzl^6 Ade was determined (*bar graph*) as described in the legend to Fig. 1

$$\frac{[PI]}{[Pt]} = \frac{\{I\}}{K_I + \{I\} + \frac{K_I\{L\}}{K_d}}$$

We are interested in the special case where the concentration of the competitor is equal to its binding constant, $[I] = K_I$. In the sigmoid curve which is generated by plotting concentration of a nonlabeled competitor against binding of the radio labeled ligand, it is desirable to have the point where $[I] = K_I$ fall on the straightest part of the curve, i.e., at the 50% competition level. An examination of the equation above demonstrates that this situation will occur when the concentration of labeled ligand is considerably smaller than its Kd so that this term essentially drops out of the equation. Thus for $[I] = Ki$, the equation reduces to:

$$\frac{PI}{Pt} = \frac{1}{2}$$

In some instances, limitations in the specific activity of the radiolabel available prevent the experimenter from utilizing as low a concentration as is desirable. In a recent study on an auxin binding protein (Cross and Briggs, 1978) for example, the labeled ligand was used at a concentration near its Kd in a competition study and a correction factor was applied. The above equation, however, suggests that when $[L] = Kd$ then

$$\frac{PI}{Pt} = \frac{1}{3} \quad \text{when } [I] = K_I.$$

i.e., that the K_i of the competitor is equivalent to that concentration of I which reduces the binding by 1/3. This equation can be used to rapidly calculate the K_I of potential competitors over a wide range of [L] where Kd is known.

2.3 Source of Reagents

The meta, ortho, and para isomers of N^6-hydroxybenzyladenine and the respective 9-ribonucleosides were kindly suuplied by Dr. R.W. Morris (Oregon State University, Corvallis). A cytokinin not previously described, 6(5′-hydroxy-n-pentylamino) purine (OH-PA) was synthesized by the condensation of 5-amino-pentanol with 6-chloropurine by the general method we have previously used (Fox, 1966). The cytokinin photoaffinity probe, {methylene-^{14}C} 2-azido-6-benzylaminopurine was prepared as previously described (Keim and Fox, 1979, 1980). An antibody to CBF-l was generated and purified as previously described (Dobbins, 1978; Dobbins and Fox, 1978).

3 Results

3.1 The Effects of Hydroxylation on Cytokinin Binding to CBF-1

As we have previously reported (Fox and Erion, 1977) the hydroxylated cytokinin trans-zeatin binds significantly to CBF-1 only at concentrations 100 times higher than the corresponding nonhydroxylated analog i^6Ade (Table 1); ribosyl zeatin has a similar relationship with i^6Ado. In order to investigate the effects of the hydroxyl group on binding in more detail, a number of hydroxylated analogs of active cytokinins were subjected to competition binding studies and bioassays for cytokinin activity. The newly synthesized analog, 6 (5'-hydroxy-n-pentylamino) purine (OH-PA) (see Sect. 2) has about the same cytokinin activity in several standard tissue culture and chlorophyll retention tests as pentylaminopurine. The same is true of dihydrozeatin, an isomer of OH-PA. However, the two hydroxylated cytokinins interact with CBF-1 only very weakly (Table 1). That the effect of the hydroxyl group on cytokinin binding to CBF-1 is not simple and straightforward, however, can be seen in the results of competition binding studies with the *ortho, meta* and *para* isomers of hydroxybenzaladenine (Table 1). The *ortho* hydroxybenzyladenine binds at least as well as bzl^6 Ade and binding of the *meta* isomer is only slightly reduced. However, the *para* isomer binds nearly 100 times less tightly to CBF-1 than bzl^6 Ade. These isomers of hydroxybenzyladenine have also been tested for cytokinin activity in standard bioassays. The *meta* isomer has proved to be the most active compound in support of the growth of cytokinin requiring soybean and tobacco tissues as well as in the cucumber cotyledon bioassay for cytokinins (Fletcher and McCullagh, 1971). The *ortho* hydroxy isomer, while inferior to the *meta* in all the bioassays, exhibited considerably more cytokinin activity than the *para* except in the tobacco tissue bioassay where they were about the same. The 9-ribonucleosides of these hydroxybenzyladenines show the same relative order of binding affinity and biological activity as the free bases.

Table 1. Dissociation constants of the complex formed between CBF-1 and various cytokinins

Compound	$K_d \times 10^7$
6-o-hydroxybenzylaminopurine	6.0
bzl^6 Ade	6.5
6-m-hydroxybenzylaminopurine	7.5
6-n-pentylaminopurine	11.0
bzl^6 Ado	15.8
I^6 Ade	17.0
I^6 Ado	63.0
OH-PA	500
6-p-hydroxybenzylaminopurine	500
trans-io^6 Ade	1580
trans-io^6 Ado	> 2000
dihydrozeatin	> 2000

Dissociation constants were estimated from plots of competition binding studies performed as described in Sect. 2

3.2 Photoaffinity Labeling Studies

The radiolabeled cytokinin photoaffinity probe synthesized in 1978 in our laboratory (Erion et al., 1978; Keim and Fox, 1979; Keim and Fox, 1980) was used at a concentration of 6×10^{-7}M to label a highly purified sample of CBF-l (Fig. 4). This preparation of the binding protein bound one molecule of bzl^6Ade per molecule of protein at saturation (when mol. wt. of CBF-l was assumed to be 180,000) and exhibited four subunits on SDS-polyacrylamide gel electrophonesis (Fig. 4). The results of the photoaffinity study (Fig. 1) indicate covalent binding of the cytokinin probe to each of the four subunits. A detailed description of the kinetics of this interaction has been published elsewhere (Keim and Fox, 1980).

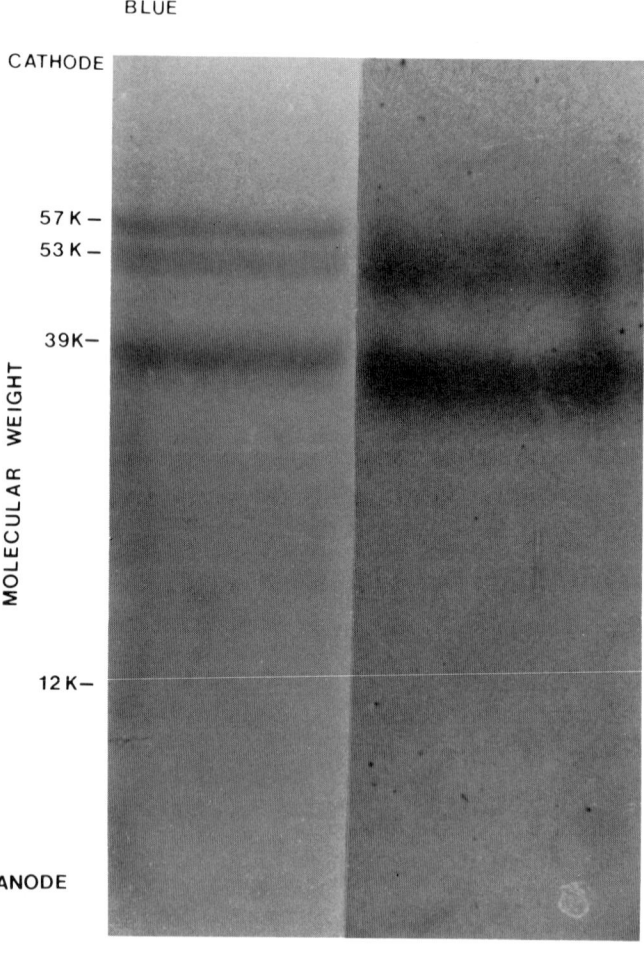

Fig. 4. A highly purified sample of CBF-l (0.9 mg in 1.0 ml buffer) was incubated with 6×10^{-7}M [^{14}C] 2-azido-6-benzylaminopurine in a light tight room at 4°C under a photographic safe red lamp for 1 h. The sample was then irradiated in a quartz cuvette placed 10 cm from a G.E. 15 W germicidal lamp (G15T8) for 5 min, the sample then precipitated and washed five times with cold acetone and resolubilized in 1% sodium dodecyl sulfate. SDS-polyacrylamide gel electrophoresis was carried out by conventional procedures. Gels were either stained with coomasie blue or subjected to fluorography (Bonner and Laskey, 1974)

3.3 The Biological Distribution of CBF-1

With the aid of an anti CBF-1 rabbit antibody previously described (Dobbins and Fox, 1978) and double immunodiffusion techniques, we have tested a number of plant preparations for moieties which are antigenically related to CBF-1 from wheat germ. Extracts from dried seeds of plants closely related to wheat such as barley and oats show antigenically similar proteins (Fig. 5) while others taxonomically more distant, e.g., corn and coconut milk do not. Using similar techniques we have been able to show that within the wheat seed CBF-1 is present both in the embryo and the scutellum. However, the binding protein either disappears or falls to undetectable levels during germination of the seed. We have never been able to detect CBF-1 in any of the vegetative parts of the wheat plant. However, a protein which is at least immunologically identical with CBF-1 appears in the earliest floral stages before fertilization associated with the ovule and the surrounding tissues of the gynoecium. The protein is not present in the surrounding glumes and bracts or other parts of the flower. After fertilization as the seed develops, CBF-1 clearly remains associated with the embryo and is never detected in the endosperm.

3.4 Isolation of a New High Affinity, Particulate, Cytokinin-Binding Moiety from Mung Bean Seedlings

Seedlings of mung beans (*Phaeolus aureus* Roxb.) were grown at 26°C for 6 days in the dark and the hypocotyls crushed in a mortar with 2 ml/g of a solution containing 10 mM Tris HCl pH = 7.5, 10 mM KCl 10 mM KH_2PO_4, 0.1 mM EGTA and 0.1 mM DTT. The resulting homogenate was centrifuged at 2000 g for 30 min, the supernatant decanted and centrifuged a second time for 30 min at 20,000 g. The pellets were resuspended in the same buffer along with 1 mg/ml bovine serum albumin either in the presence or absence of 0.1% Triton X-100 and subjected to equilibrium dialysis against ^{3}H-bzl^{6}Ade over a wide range of concentrations. No significant binding of the cytokinin occurred in the absence of Triton in the experiment, but with the inclusion of 0.05% to 0.5% Triton X-100 (present both inside and outside the dialysis bag) a large excess of radioactivity was retained in the dialysis bag after an 18 h incubation at 4°C. Above 1% Triton X-100 cytokinin binding was greatly reduced. High affinity cytokinin binding occurred both in the low speed pellet as well as in the 2,000-20,000 g pellet. Each pellet may be presumed to contain etioplasts recognized by their β-carotene absorption spectrum, mitochondria detected by cytochrome c oxidase activity and contaminating membrane fragments. High affinity cytokinin binding correlated with cytochrome c oxidase activity but not with β-carotene absorption. The results from a series of binding studies with bzl^{6}Ade over concentrations ranging from 1×10^{-10}M to 1×10^{-4}M were analyzed by the method of Scatchard (1949) (Fig. 6) yielding a biphasic plot characteristic of a high affinity binding site in the presence of multiple low affinity sites. The low affinity portion of the curve has been carried out another two orders of magnitude without showing saturation.

Fig. 5. Double immunodiffusion of various plant extracts with rabbit serum containing anti-CBF-1 antibody. Rabbits were inoculated with 2 mg each of highly purified CBF-1 in Freunds adjuvant by injections into the foot pads, intramuscularly at several locations, intraperitoneally and subcutaneously at two week intervals for two months followed by a booster directly into the blood stream. Serum was collected after 3 months and contained a component which precipitated CBF-1 but none of several control proteins in double immunodiffusion tests. Control serum from the same rabbits collected before inoculation with CBF-1 contained no detectable anti-CBF-1. Plant extracts were prepared by grinding seeds in the buffer used to extract CBF-1 (Fox and Erion, 1975). *eIF-2* is protein synthesis initiation factor 2 purified from wheat germ ribosomes

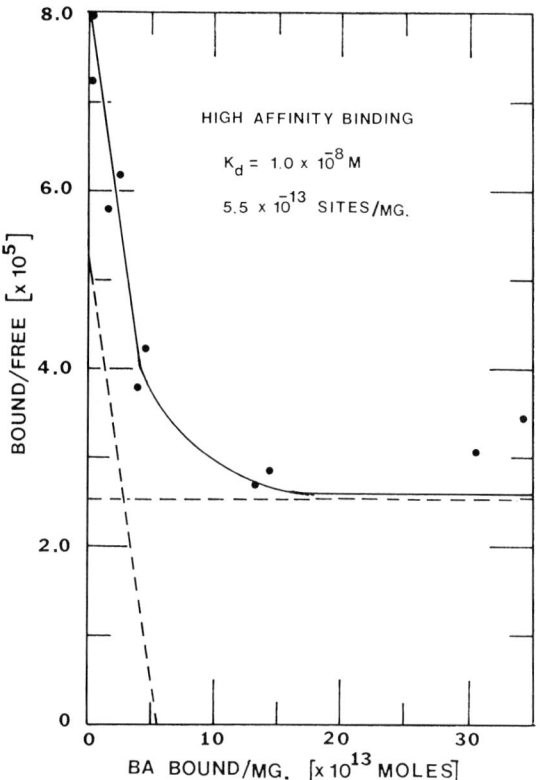

Fig. 6. Scatchard plot of concentration-dependent binding of bzl^6 Ade to a particulate fraction from mung bean (2,000-20,000 g pellet resuspended in buffer). See text for details of preparation. Experimental points were obtained and calculations made as described previously (Fox and Erion, 1975)

4 Discussion

4.1 The Binding Assay

In this study, as in all of our previous work on a cytokinin-binding protein, we have used equilibrium dialysis to determine binding. Although this technique is somewhat more cumbersome than other widely used procedures, it has been shown that in the presence of nonspecific binding all the available separation procedures with the exception of equilibrium dialysis include a part of the nonspecifically bound ligand in the estimate of unbound ligand concentration (Blondeau and Robel, 1975). In the most widely used method for studying plant hormone binding, which was pioneered by Hertel et al. (1972) the biological extract is incubated twice with radioactive ligand under identical conditions except that in the second a very high level of unlabeled ligand is added. It has been widely accepted that substracting counts bound in the second incubation from the first gives a reliable measure of specific binding. However Blondeau and Robel (1975) report that this assumption is theoretically invalid and can lead to large errors in the calculation of the number of specific binding sites. We have therefore avoided this approach.

4.2 Why Does Zeatin Bind so Poorly to Cytokinin "Receptors"?

As noted in the Introduction, all of the high affinity cytokinin binding moieties so far reported, including CBF-l, interact only weakly with zeatin. To be sure, zeatin has long been somewhat of an anomaly; its activity is an exception to the general rule that polar groups in the side chain reduce activity of a cytokinin (Strong, 1958). Our findings suggest that the hydroxyl group itself substantially reduces the affinity of the cytokinin for CBF-l since we see about 100-fold less binding both with zeatin and with OH-PA compared with their nonhydroxylated counterparts. The relationship between binding and hydroxylation is complex, however; our results with positional isomers of hydroxybenzyladenine indicate that the geometry of the hydroxyl group with respect to the rest of the molecule greatly influences binding (as well as biological activity).

In several cytokinin-requiring or -influenced systems zeatin is considerably less active than other cytokinins, (e.g., Kuhnle et al., 1977) and it has been suggested that some exogenously supplied cytokinins may need to be metabolically modified in order to achieve growth promotion (Dyson et al., 1972; Mok et al., 1978). It is conceivable that the hydroxyl group of zeatin serves a function distinct from its cytokinin activity at the site of action, perhaps preserving it from enzymatic destruction or coding those cytokinin active molecules which are to be discriminated from other 6-substituted purines in the cell. This hypothesis leads to the consequence that zeatin would be dehydroxylated before interacting with a receptor at a growth active site. However, none of the cytokinin receptors yet isolated have a known biological function; it may be that a receptor for zeatin has simply escaped detection up to now. The isolation of a new cytokinin-binding particulate from mitochondria-enriched fractions reported in this paper may allow us to relate cytokinin binding to function in view of the profound cytokinin effects on mitochondria reported by Miller (1979, 1980). This material has the highest affinity for cytokinins yet reported ($K_d = 1 \times 10^{-8}$ M for bzl^6-Ade). The fact that a low level of detergent is required for interaction suggests a binding moiety buried in a membrane or present on the inner surface of a membrane bilayer. Supporting this view is the fact that although zeatin itself is only weakly active in the system described by Miller, lipid soluble esters of zeatin are as active as any other cytokinin (Miller personal communication).

4.3 How Many Cytokinin Binding Proteins Exist and What is Their Biological Function?

As noted in the Introduction several relatively high affinity cytokinin binding moieties have now been described in various laboratories, and there are at least three and possibly more distinct entities. The wheat embryo protein described by Polya and Davis (1978) and Moore (1979) is almost certainly identical with CBF-l (Fox and Erion, 1975, 1977) although some controversy still exists concerning the mw. and number of subunits. However, as we have shown in this paper, CBF-l is restricted to wheat and related plants although similar cytokinin binding proteins may exist in the seeds of other species. Furthermore, CBF-l either does not occur in the vegetative parts of the wheat plant or else is present at very low levels even though cytokinin-modulated activities must certainly take place. How is one to account for the multiple roles of cytokinins which affect nearly every phase of plant growth and development from seed germination to seed formation? It is possible that a unique receptor exists which accounts for the biological role of cytokinins at a specific growth active site. However the current evidence suggests that multiple cytokinin receptors may occur in higher plants acting as transducers for the numerous and diverse activities of this group of plant hormones.

5 Summary

Because zeatin has a very low affinity for any of the cytokinin binding moieties so far described, several hydroxylated cytokinins were tested for biological activity and their affinity for a wheat germ cytokinin binding protein (CBF-l). Although in general the OH group greatly impedes binding, the effect is complex and depends upon the position of the hydroxyl group in the cytokinin molecule. With the aid of a rabbit-derived anti-CBF-l antibody, it was shown that CBF-l exists only in wheat and closely related species such as oats and barley but not in other plants. Within the wheat plant the protein is found in the embryo and unfertilized ovule and surrounding tissues but not the endosperm or vegetative parts of the plant. Experiments with a radiolabeled cytokinin photoaffinity probe indicate all four subunits of CBF-1 take part in binding a molecule of cytokinin.

Finally we report here a particulate moiety associated with a mitochondrial fraction which binds cytokinins with a higher affinity (Kd = 1 x 10^{-8} M for bzl^{6}Ade) than any yet reported.

References

Berridge MV, Ralph RK, Letham DS (1970) The binding of kinetin to ribosomes. Biochem J 119: 75-84
Blondeau JP, Robel P (1975) Determination of protein-ligand binding constants at equilibrium in biological samples. Eur J Biochem 55: 375-381
Bonner WM, Laskey RA (1974) A film detection method for tritium labeled proteins and nucleic acids in polyacrylamide gels. Eur J Biochem 46: 83-88
Chen CM, Melitz DK, Petschow B, Eckert RL (1980) Isolation of cytokinin binding protein from plant tissues by affinity chromatography. Eur J Biochem 108: 379-387
Cheng Y, Prusoff WH (1973) Relation between the inhibition constant (K_1) and the concentration of inhibitor which causes 50 per cent inhibition (I_{50}) of an enzymatic reaction. Biochem Pharmacol 22: 3099-3108
Cross JW, Briggs WR (1978) Properties of a solubilized microsomal auxin-binding protein from coleoptiles and primary leaves of *Zea mays*. Plant Physiol 62: 152-157
Dobbins JM (1978) Studies on Cytokinin binding factor 1 and protein synthesis *in vitro*. Ph D Thesis, Univ Kansas, Lawrence
Dobbins JM, Fox JE (1978) Purification of an antibody to a cytokinin binding protein. Plant Physiol Suppl 61: 10
Dyson W, Fox JE, McChesney JD (1972) Short term metabolism of urea and purine cytokinins. Plant Physiol 49: 406-413

Erion JL, Fox JE (1981) Purification and properties of a protein which binds cytokinin-active 6-substituted purines. Plant Physiol 156-162

Erion JL, Keim P, Roussell D, Fox JE (1978) Cytokinin binding proteins: isolation by affinity chromatography and active site studies. Plant Physiol Suppl 61: 10

Fletcher RA, McCullogh D (1971) Cytokinin-induced chlorophyll formation in cucumber cotyledons. Planta 101: 88-90

Fox JE (1966) Incorporation of a kinin N^6-benzyladenine into soluble RNA. Plant Physiol 41: 75-82

Fox JE, Erion JL (1975) A cytokinin binding protein from higher plant ribosomes. Biochem Biophys Res Commun 64: 695-700

Fox JE, Erion JL (1977) Cytokinin binding proteins in higher plants. In: Pilet P (ed) Plant growth regulation. Proc 9th Int Conf Plant Growth Substances. Springer, Berlin Heidelberg New York, pp 139-146

Fox JE, Erion JL, McChesney JD (1979) A new intermediate in the synthesis of a tritiated cytokinin with high specific activity. Phytochemistry 18: 1055-1056

Hertel R St, Thomson K, Russo VEA (1972) In vitro auxin binding to particulate cell fractions from corn coleoptiles. Planta 107: 325-340

Keim P, Fox JE (1979) Modification of the active site of a cytokinin binding protein. Plant Physiol Suppl 63: 32

Keim P, Fox JE (1980) Interaction of a radiolabeled cytokinin photoaffinity probe with a receptor protein. Biochem Biophys Res Commun 96: 1325-1334

Kuhnle JA, Fuller G, Corse J, Mackey B (1977) Antisenescent activity of natural cytokinins. Physiol Plant 41: 14-21

Miller CO (1979) Cytokinin inhibition of respiration by cells and mitochondria of soybean, *Glycine max* (L.) Merrill. Planta 146: 503-511

Miller CO (1980) Cytokinin inhibition of respiration in mitochondria from six plant species. Proc Natl Acad Sci USA 77: 4731-4735

Mok MC, Mok DWS, Armstrong DJ (1978) Differential cytokinin structure-activity relationships in *Phaseolus*. Plant Physiol 61: 72-75

Moore FH III (1979) A cytokinin-binding protein from wheat germ. Isolation by affinity chromatography and properties. Plant Physiol 64: 594-599

Polya GM, Bowman JA (1979) Ligand specificity of a high affinity cytokinin-binding protein. Plant Physiol 64: 387-392

Polya GM, Davis AW (1978) Properties of a high affinity cytokinin-binding protein from wheat germ. Planta 139: 139-147

Scatchard G (1949) The attraction of proteins for small molecules and ions. Ann NY Acad Sci 51: 660-672

Starr AM, Fox JE (1978) Cytokinin binding of a subcellular fraction from imbibed bush bean seed. Plant Physiol Suppl 61: 10

Strong FM (1958) Topics in Microbial Chemistry. Wiley and Sons, New York, pp 98-157

Sussman MR, Kende H (1978) In vitro cytokinin binding to a particulate fraction of tobacco cells. Planta 140: 251-259

Takegami T, Yoshida K (1975) Isolation and purification of cytokinin binding protein from tobacco leaves by affinity column chromatography. Biochem Biophys Res Commun 67: 782-789

Yoshida K, Takegami T (1977) Isolation of cytokinin binding protein from tobacco leaves by bioaffinity chromatography and its partial characterization. J Biochem 81: 791-799

Section 4
Cytokinin in Protein Synthesis and RNA Metabolism

Control of Growth by Cytokinin: An Examination of Tubulin Synthesis During Cytokinin – Induced Growth in Cultured Cells of Paul's Scarlet Rose

D.E. Fosket, L.C. Morejohn, and K.E. Westerling[1]

1 Introduction

Cytokinins are characterized as hormones which control cell proliferation in cultured plant tissues (Skoog et al., 1965). The best currently available evidence suggests that cytokinins regulate the passage of cells through their division cycle by regulating events which are necessary for the transition from G2 to mitosis (Jouanneau, 1971; Fosket, 1977). Cytokinins also have been shown to bring about both qualitative and quantitative changes in protein synthesis in various plant systems (Jouanneau, 1971; Short et al., 1974; Tepfer and Fosket, 1978), and, in the case of cultured soybean cells, these changes were shown to precede cytokinin-induced cell division (Fosket and Tepfer, 1978). If these quantitative and qualitative effects on protein synthesis are necessary for cytokinin-induced cell division, then it should be possible to identify and determine the function of those proteins which are required for cell divisions and whose synthesis is dependent upon cytokinin. One of the proteins whose synthesis was strongly promoted by cytokinins in cultured soybean cells had a molecular weight near 55,000 daltons, approximately the molecular weight of the subunits of tubulin (Fosket et al., 1977). Since tubulin is the principle structural component of the microtubules which comprise the mitotic spindle (Hepler and Palevitz, 1974), we have begun an investigation to determine whether or not tubulin synthesis is regulated by cytokinin.

2 Materials and Methods

2.1 Cell Culture

We obtained cell cultures derived from stem tissue of Paul's scarlet rose from Dr. A. Marcus, Institute for Cancer Research, Philadelphia. The cells were cultured both as a suspension and on agar-solidified medium (Nesius et al. 1972) in which zeatin at 0.5 μM was the only cytokinin. Batch suspension cultures were prepared by diluting 10 ml of a stationary phase culture with 140 ml of fresh medium. The cultures were grown in the dark at 20°C on a Labline orbital shaker which completed 130 rpm. The cells also were maintained on agar-solidified medium at 20°C in the dark. In either case the

---

[1] Department of Developmental and Cell Biology, University of California, Irvine, CA 92717, USA

cultures were transferred routinely every 14 to 21 days. Over the course of the stationary phase (day 12-21) cell viability, as measured by phenosafranine dye exclusion (Widholm, 1972) declined from 93% to 80%.

Changes in cell number and fresh weight were determined by aseptically removing aliquots of the culture at intervals after inoculation. The cells were collected on Miracloth with aspiration and then weighed. The cells were placed in a mixture of 10% nitric acid + 10% HCl and cell number was determined with a hemocytometer. The growth constants $k = \ln(W/W_0)$ and $k = \ln(N/N_0)$ were calculated, where N = cell number after time t, N_0 = initial cell number, W = aggregate cell fresh weight at time t, and W_0 = initial aggregate cell fresh weight.

2.2 Polysome Isolation

Cells were collected on Miracloth, washed with a cold cell washing buffer (100 mM Tris-HCl pH 8.5 + 80 mM KCl + 10 mM $MgCl_2$) and resuspended in cold ribosome extraction buffer (cell wash buffer + 2.5 mM dithiothreitol and 250 mM RNase-free sucrose) at a ratio of 2 ml of buffer per gm fresh weight of cells. The cell suspension was added dropwise to a pool of liquid N_2 in a large mortar where it was ground to a fine powder. The frozen mass was allowed to thaw partially before homogenization with a motor-driven Teflon-glass tissue grinder. The homogenate was filtered through Miracloth, centrifuged at 12,000 g for 15 min at 4°C and the supernatant was layered over 1.5 ml of 1.8 M RNase-free sucrose in ribosome resuspension buffer (40 mM Hepes pH 7.6 + 50 mM KCl + 5 mM $MgCl_2$) and centrifuged for 90 min at 58,000 rpm in a Beckman fixed angle 65 rotor. The ribosomal pellet was resuspended in a minimum volume of resuspension buffer and stored frozen in liquid N_2 until needed.

2.3 Wheat Germ Cell-Free Protein Synthesizing System

A wheat germ cell-free protein synthesizing system was employed to translate the extracted polysomes as described by Roberts and Paterson (1973) and Roman et al. (1976). The reaction mixture contained 1.2 mM ATP, 0.1 μM GTP, 5.5 mM creatin phosphate, 0.2 mg/ml creatine phosphokinase, 80 μM spermidine phosphate, 2.5 mM magnesium acetate, 33 mM KCl, 110 mM potassium acetate and 19 amino acids, minus methionine, each at 50 μM, all in a 20 mM Hepes buffer pH 7.5. The assays also contained a variable amount of wheat germ S23, [^{35}S]-methionine and between 25 and 50 μg of polysomal RNA. The reaction mixture was inoculated for 1 h at 25°C whereupon the reaction was stopped by placing the tubes in an ice bath. A 10 μl aliquot was removed from each assay mixture for determination of incorporation by means of the standard hot TCA digest procedure. The remaining assay mixture was made 1 mM in EDTA, 50 μg of pancreatic RNase was added and the mixture was incubated at 30°C for 15 min.

2.4 Tubulin Isolation

Tubulin was isolated from bovine brain tissue by three different methods. These include both the Shelanski et al. (1973) and the Sloboda et al. (1976) modifications of the Borisy and Olmsted (1972) procedure in which tissue was homogenized in cold 0.1 M Mes or 0.05 M Pipes buffer, pH 6.9 containing 1 mM EGTA, 1 mM or 0.5 mM GTP and 0.5 mM Mg^{2+}. The homogenate was centrifuged at 40,000 rpm in a Beckman 60 Ti rotor at 4°C for 75 mm. The supernatant was diluted 1:1 with buffer containing 8 M glycerol, incubated for 30 min at 30°C, and centrifuged at 36,000 rpm in the 60 Ti rotor at 25°C for 60 min. The pellet was resuspended in cold buffer without glycerol with the aid of a Dounce homogenizer, incubated in ice for 30 min, and centrifuged as above at 4°C for 30 min. The pellet was discarded while the supernatant was made 4M in glycerol and stored in aliquots at −20°C. Before use the tubulin was allowed to assemble again by incubating this mixture at 30°C for 30 min. The solution was centrifuged at 36,000 rpm for 60 min at 25°C, the supernatant discarded, and the microtubules were depolymerized by resuspending in cold buffer and incubating the suspension at 4°C for 30 min.

Tubulin also was isolated by the procedure of Asnes and Wilson (1979). The procedure is similar to that of Shelanski et al. (1973) and Sloboda et al. (1976) in that it relies on repeated rounds of warm-assembly and cold-disassembly to purify the tubulin. The chief differences are in the composition of the extraction buffer which contained 100 mM glutamate and 20 mM phosphate at pH 6.75. After centrifuging, the cold homogenate is made 0.5 mM in $MgCl_2$, 1 mM in EGTA and 2.5 mM in GTP by adding a 10X "polymix" and incubated at 30°C for 30 min to allow tubulin assembly (in the absence of glycerol).

2.5 Electron Microscopy

Samples of material pelleted by centrifugation were prepared for electron microscopy by negative staining. A 5 μl sample was placed on a 0.5% formvar-coated, carbon-overcoated, copper grid (200 mesh) followed by a 5 μl drop of 8% gluteraldehyde. After 10 min of fixation, the grid was drained with filter paper and stained with 10 μl of 1% phosphotungstic acid (pH 7.0) for 5-10 s. The staining solution was removed with filter paper and the grid was allowed to air dry. Specimens were examined with a Zeiss 9S-2 electron microscope with an acceleration voltage of 60KV, a 300 μm condenser aperture and a 30 μm objective aperture. Measurements of the structures observed were made directly from photographic enlargements.

2.6 In Vivo Labeling and Extracts of PSR Protein

Cells were exposed to [^{35}S]-methionine for 20 min, collected on Miracloth, washed with cold 0.05 M Tris-HCl buffer, pH 7.6 containing 1 mM methionine and weighed. The cells were homogenized with a Dounce tissue homogenizer in SDS extraction buffer [5% SDS + 1% β-mercaptoethanol + 1 mM PMSF + 0.1% lactoalbumin hydrolysate + 5% glycerol in 0.05 M Tris-HCl buffer, pH 7.6] at a ratio of 1 ml of buffer/gm of cells. The samples were centrifuged at 12,000 g for 15 min at room temperature and the pellet was discarded. The supernatant was made 55% saturated in ammonium sulfate and the samples were centrifuged at 12,000 g for 15 min. The precipitated protein-SDS conjugate formed a waxy pellicle on the surface of the tube which was removed. Most of the cellular proteins precipitated with the addition of the ammonium sulfate and they were not dealt with further in the work reported in this paper. Proteins which remained in solution after adding ammonium sulfate were precipitated with 2 volumes of acetone at $-20°C$. The precipitated proteins were collected by centrifugation and redissolved in SDS-polyacrylamide gel electrophoresis sample buffer (2% SDS + 0.1 M dithiothreitol + 10% glycerol in 0.08 M Tris-HCl buffer pH 6.9).

2.7 Polyacrylamide Gel Electrophoresis and Autoradiography

Proteins were separated by SDS-polyacrylamide gel electrophoresis as described by Studier (1973). Samples in SDS-gel electrophoresis sample buffer were loaded into wells in a 5% acrylamide stacking gel and electrophoresed at 80V until the front entered the 10% acrylamide separating gel, whereupon the voltage was increased to 150V. Electrophoresis was continued until the tracking dye reached the bottom of the gel. The gel was fixed and stained in 0.025% Coomassie Brilliant Blue R in methanol: acetic acid: water (5:1:5). After destaining in methanol: acetic acid: water (5:1:5) the gel was incubated in 3 gel volumes of Enhance (New England Nuclear Fluorographic scintillant NEF-966), the gel was dried, and an autoradiogram of the gel was made by exposing it to Kodak RP Royal X-Omat medical X-ray film at $-76°C$.

3 Results

3.1 Regulation of PSR Cell Growth by Cytokinin

3.1.1 The Cytokinin Dose Response

The role of cytokinin in regulating the growth of cultured cells of Paul's scarlet rose (PSR) was examined. Stationary phase PSR cells growing on agar-solidified medium were used to inoculate plates of fresh media which either lacked a cytokinin or contained zeatin at concentrations ranging from 1 nM to 1 μM. Twenty-one days later the plates were harvested for determination of fresh weights. The results of this experiment demonstrated that the extent of growth of PSR cells which occurred over the culture period was a function of the cytokinin concentration. In the absence of zeatin the cell mass increased by a factor of approximately 5, while the optimal zeatin concentration (0.5 μM) brought about a 30-fold increase in fresh weight. The lowest zeatin concentration tested (1 nM) brought about a 13-fold increase in fresh weight, roughly a 2.6 increment over the growth observed in the absence of zeatin (Fig. 1). Although considerable growth occurred through one transfer on medium lacking zeatin, this growth in the absence of exogenous cytokinin could not be sustained through subsequent transfer to media lacking cytokinin (Fosket, 1980).

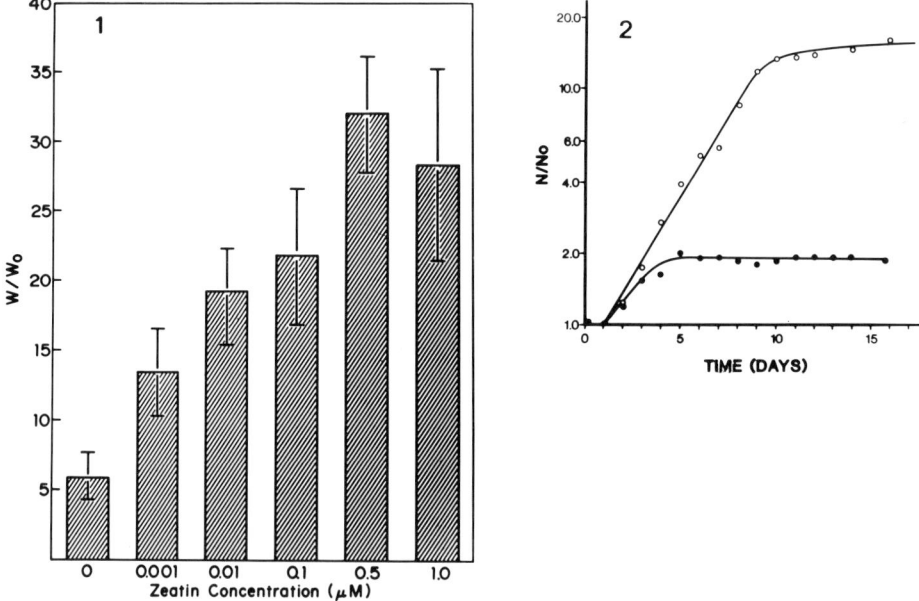

Fig. 1. The growth of PSR cells in response to cytokinin. Agar-solidified media in Petri plates, containing various concentrations of zeatin were inoculated with stationary phase cells. Three weeks later the fresh weight of the cells was determined (Fosket, 1980)

Fig. 2. The growth of PSR cells in suspension culture as a function of cytokinin. A stationary phase culture of PSR cells was diluted 15-fold with fresh medium which either contained 0.5 µM zeatin (o—o) or lacked any cytokinin (•—•). At intervals thereafter 10 ml aliquots were removed from the cultures for determinations of cell number (Fosket, 1980)

3.1.2 Growth Kinetics

The growth kinetics of the PSR cell suspension were influenced by exogenous cytokinin. A pronounced lag phase was observed when stationary phase cells were diluted with fresh medium. Cytokinin had no effect on the duration of the lag phase. Logarithmic growth began sometime between the 1st and 2nd day after dilution both in the presence and absence of cytokinin. However, in the absence of cytokinin the log growth phase was brief, and the cells entered stationary phase after a minimum of 4-5 days in culture during which a 2-5-fold increase in cell number occurred. In the presence of cytokinin the log growth phase persisted until the 8th to 12th day of culture during which a 10- to 15-fold increase in cell number took place (Fig. 2) (Fosket, 1980).

3.1.3 Growth of PSR Subclones

In an effort to obtain a more cytokinin-responsive strain of PSR cells, the parent clone was subcloned and the response of these various subclones to cytokinin was determined. At the time it was subcloned, the parental cell line exhibited growth constants ($k = \ln W/W_o$) of 3.5 in medium containing 0.5 µM zeatin and of 1.6 in medium lacking a cytokinin. The average growth constants of the 30 subclones tested was 3.1 on media containing zeatin, with a range of 1.1 to 4.0. When cultured on medium

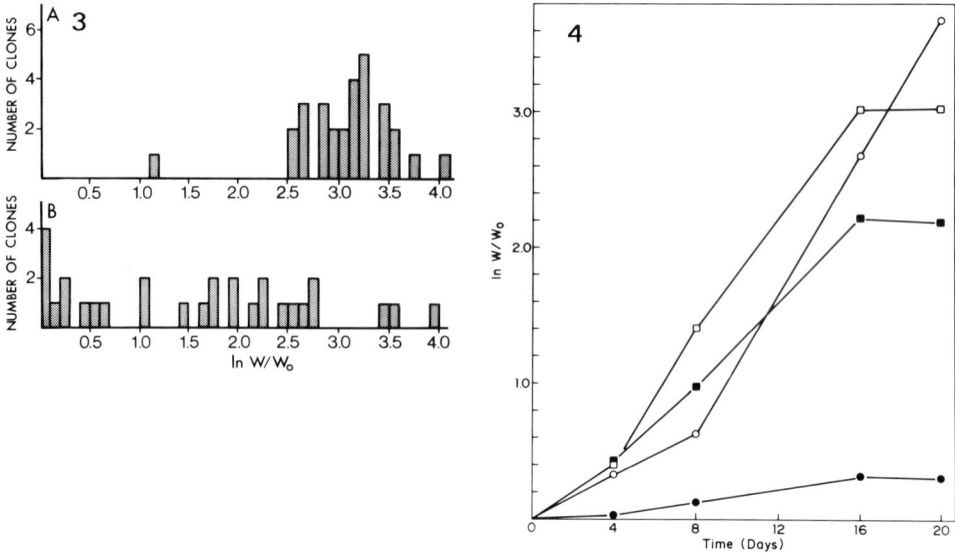

Fig. 3A, B. Determination of the growth constants of subclones derived from parental PSR strain. The growth constants (k = ln W/Wo) of 30 PSR subclones were determined after 3 weeks of culture on medium containing or lacking a cytokinin. **A** The growth constants of clones when cultured on medium containing 0.5 μM zeatin. **B** The growth constants of the same clones when cultured on medium lacking a cytokinin (Fosket, 1980)

Fig. 4. The growth kinetics of subclone I-1A in suspension culture as influenced by cytokinin. Stationary phase cell suspensions of subclone I-1A which had been cultured for 21 days in medium containing 0.5 μM zeatin or lacking a cytokinin were diluted 15-fold with fresh medium which either contained 0.5 μM zeatin or lacked a cytokinin. Samples were taken at intervals thereafter for determination of fresh weights. Each *point* represents an average of two determinations: □—□ = zeatin-cultured cells diluted with zeatin-containing medium; ■—■ = zeatin-cultured cells diluted with medium lacking a cytokinin; ○—○ = zeatin-deprived cells diluted with medium containing zeatin; ●—● = zeatin-deprived cells diluted with medium lacking a cytokinin (Fosket, 1980)

lacking a cytokinin the same subclones exhibited an average growth constant of 1.5, with a range of 0 to 3.9 (Fig. 3). Although some of these subclones appeared to be much more cytokinin-dependent for growth than others, generally the subclones exhibiting little or no growth on medium lacking cytokinin also grew less vigorously on medium containing a cytokinin. The more vigorously growing subclones in the presence of a cytokinin also tended to give substantial growth in the absence of a cytokinin. The data provided in Fig. 4 demonstrate that even these apparent cytokinin autotrophs required cytokinin for sustained growth through repeated subcultures (Fosket, 1980).

3.1.4 Effect of Previous Exposure to Cytokinin Upon the Growth Response of PSR Cells

Washing the cells before transferring them to fresh medium lacking a cytokinin did not reduce their growth response in the 1st transfer to medium lacking a cytokinin. In the data shown in Table 1, cells of the vigorously growing clone I-1A were collected on

Miracloth filters, washed with fresh medium lacking a cytokinin and transferred to media either containing or lacking a cytokinin. When the cells of the inoculum had been cultured in medium lacking a cytokinin, there was virtually no growth.

However, when the cells of the inoculum had been cultured in medium containing 0.5 µM zeatin they grew nearly as well whether the fresh medium contained or lacked zeatin and washing them with fresh medium lacking a cytokinin seemed to have no effect on their growth response (Fosket, 1980).

Table 1. The effect of pretreatment on the response of subclone I-1A to cytokinin. After 8 days in culture in medium which either contained 0.5 µM zeatin or lacked a cytokinin, cells were collected on Miracloth filters and washed with three 50 ml aliquots of fresh medium lacking a cytokinin. Petri plates of agar-solidified medium which either contained 0.5 µM zeatin or lacked a cytokinin were inoculated with approximately equal amounts of the washed cells. Some plates were harvested immediately to determine the fresh weight of the inoculum, while other plates were grown for 3 weeks before the tissue fresh weight was determined

Inoculum cultured	ln W/W_0 after culture on medium:		
In medium:	Lacking zeatin	+0.5 µM zeatin	% of +Z growth in absence of cytokinin
Lacking zeatin	0.017	2.608	0.65
+0.5 µM zeatin	2.541	2.660	95.5

3.1.5 Logarithmic Growth Rates as a Function of the Cytokinin Concentration

Although cytokinin seemed to affect the duration of the log phase of PSR cell growth, the zeatin concentration did not initially affect the logarithmic growth rate. In Fig. 5 we see the growth kinetics of PSR cells when a stationary phase culture was diluted with fresh medium containing one of three different zeatin concentrations (1.0, 0.1, and 0.01 µM). The log rate of growth was the same at these three concentrations during the first 7 to 8 days of the culture period. Only after the 8th day was a new growth rate established in the cultures with a growth-limiting zeatin concentration (0.01 µM). All three cultures continued to grow logarithmically beyond the 8th day, but the logarithmic growth rate was reduced in the presence of 0.01 µM zeatin (Fig. 5) (Fosket, 1980).

3.1.6 Mitotic Activity in Response to Cytokinin

These studies demonstrated that, although PSR cells require cytokinin for growth, the role of cytokinin in regulating their growth may not be demonstrable for many days after the cells are transferred to fresh medium. In order to see if we could obtain a more rapid growth response to cytokinin, stationary phase PSR cells were diluted with fresh medium lacking a cytokinin. Five days later when the culture had entered stationary phase, cytokinin was injected into the flask to a final concentration of 0.5 µM. As shown by the data presented in Fig. 6, cytokinin induced a wave of mitotic activity which appeared approximately 24 h after the hormone was added to the culture. Fur-

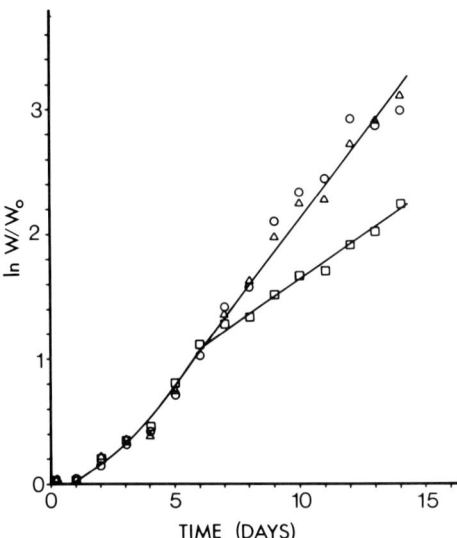

Fig. 5. The effect of cytokinin concentration on the growth kinetics of PSR cell suspension cultures. Stationary phase cultures of PSR cells were diluted 15-fold with fresh medium which contained one of three different zeatin concentrations, 1.0 μM (o—o), 0.1 μM (△—△), or 0.01 μM (□—□). At daily intervals thereafter the cultures were sampled aseptically and the fresh weight of the sampled cells was determined. Each *point* represents the average of two determinations. The data are expressed as the natural log of the ratio of the average observed fresh weight (W) to the fresh weight of cells at the time of dilution (W_o), plotted as a function of time (Fosket, 1980)

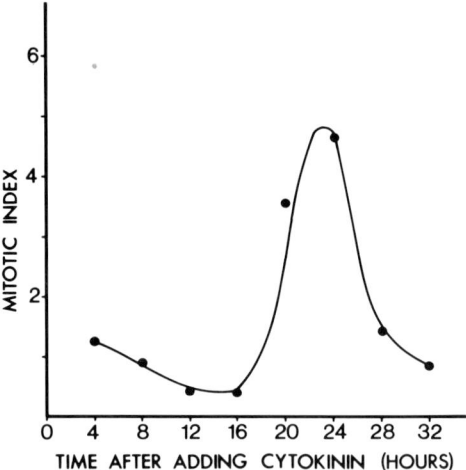

Fig. 6. Mitotic activity of a PSR cell culture after cytokinin treatment. PSR cells were cultured for 5 days in medium lacking a cytokinin, whereupon sufficient zeatin stock solution was added to make the culture 0.5 μM in zeatin. At intervals thereafter 10 ml aliquots of the culture were removed aseptically, the cells were collected on Miracloth and Feulgen-stained squashes were prepared from which the mitotic indices of the samples were determined. The mitotic index, M.I. = # mitotic cells/total cells counted x 100

thermore, the culture resumed growth with kinetics similar to that of control cultures which were continuously exposed to medium containing 0.5 μM zeatin.

3.2 Effects of Cytokinin on Protein Synthesis in PSR Cells

The data presented in Fig. 6 demonstrate that PSR cells respond fairly rapidly to exogenous cytokinin when the hormone has clearly become growth-limiting. We examined the possibility that this cytokinin-induced growth response was preceded or accompanied by quantitative and/or qualitative changes in protein synthesis. Preliminary work had demonstrated that [^{35}S]-methionine was rapidly taken up by PSR cells and incorporated into protein. The data presented in Table 2 show that cells cultured in medium containing cytokinin for 6 days incorporated nearly three times as much [^{35}S]-methionine into protein during a 20-min exposure to the isotope than did cells which were cultured in medium lacking a cytokinin. Furthermore, when the cytokinin-deprived cells were given zeatin at 0.5 μM during the 24 h prior to their exposure to [^{35}S]-methionine at day 6, nearly as much radioactivity was incorporated into protein as in the control cells which were cultured continuously in medium containing zeatin.

Table 2. [^{35}S]-Methionine incorporation into PSR cell protein. Stationary phase cells were diluted with fresh medium which either lacked a cytokinin or contained 0.5 μM zeatin. Five days later the cells were collected on Miracloth, the cytokinin-deprived cells were washed with fresh medium which either contained or lacked 0.5 μM zeatin. The cultures originally diluted with fresh medium containing zeatin also were collected on Miracloth, washed with fresh medium containing zeatin and returned to fresh medium containing 0.5 μM zeatin. Twenty-four hours later [^{35}S]-methionine (New England Nuclear NEG-009T, specific activity = 1219.8 Ci/m mol) was injected into the flasks to give 0.47 μ Ci/ml of medium. Twenty minutes later the cells were harvested, weighed and protein was extracted as described in Section 2. Incorporation of [^{35}S]-methionine into the extracted protein was determined by standard hot TCA digests and scintillation counting

Cultural conditions	cpm incorporation/ gm cell fr. wt.	% of Control
Cultured continuously in medium containing zeatin	57,036	–
Cultured continuously in medium lacking zeatin	21,602	37.9%
Cultured without zeatin for 5 days + 24 h in zeatin-containing medium	54,452	95.5%

The [^{35}S]-methionine-labeled proteins were ammonium sulfate fractionated and separated by SDS-gel electrophoresis. An autoradiogram of the gel (Fig. 7) revealed that the spectra of proteins synthesized by the zeatin-treated cells were the same whether the cells were exposed to the hormone only during the last 24 h of the 6-day culture period or whether they were cultured continuously in medium containing 0.5 μM zeatin. In contrast, cells cultured in medium lacking a cytokinin synthesized a distinctively different spectrum of proteins. Among these differences was a protein with a molecular weight near 55,000 daltons which was synthesized by cytokinin-treated cells but not by cells cultured in medium lacking zeatin.

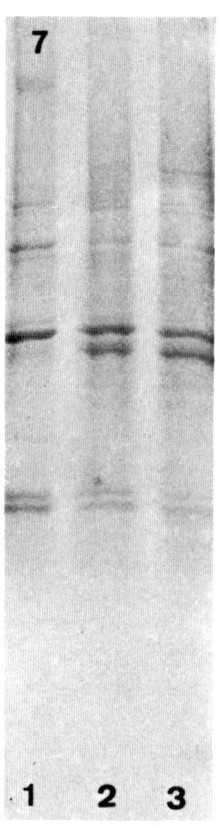

Fig. 7. Autoradiograph of in vivo [^{35}S]-methionine-labeled PSR cell proteins. Cells were cultured and exposed to [^{35}S]-methionine as described in Table 2. The labeled proteins were extracted, electrophoresed, and the gel was autoradiographed as described in Section 2. Lane *1* proteins from zeatin-deprived cells; lane *2* protein from cells grown continuously in medium containing zeatin; lane *3* proteins from cells which were grown 5 days in medium lacking zeatin followed by 1 day in medium containing 0.5 µM zeatin before labeling

3.3 PSR Tubulin Synthesis in Relation to Cytokinin

3.3.1 In Vitro Assembly of PSR Microtubule-like Structures

We attempted to isolate PSR cell tubulin by various procedures that have been developed to isolate tubulin from bovine brain. Direct in vitro assembly of PSR tubulin by the procedures of Shelanski et al. (1973), or Asnes and Wilson (1979) or Sloboda et al. (1976) were largely unsuccessful. In each case the high-speed supernatant became turbid when incubated at 30°C for 30 min, and the product assembled under these conditions frequently resembled microtubules when examined by electron microscopy (Fig. 8), but the assembled structures were not depolymerized by cold or colchicine treatments. Furthermore, SDS slab gel electrophoresis of the products showed no enrichment for bands comigrating with bovine brain tubulin (Fig. 9). The PSR crude supernatant exhibited two bands corresponding roughly in electrophoretic mobility to the α and β subunits of bovine brain tubulin, but the material assembled at 30°C showed no enrichment for these polypeptides.

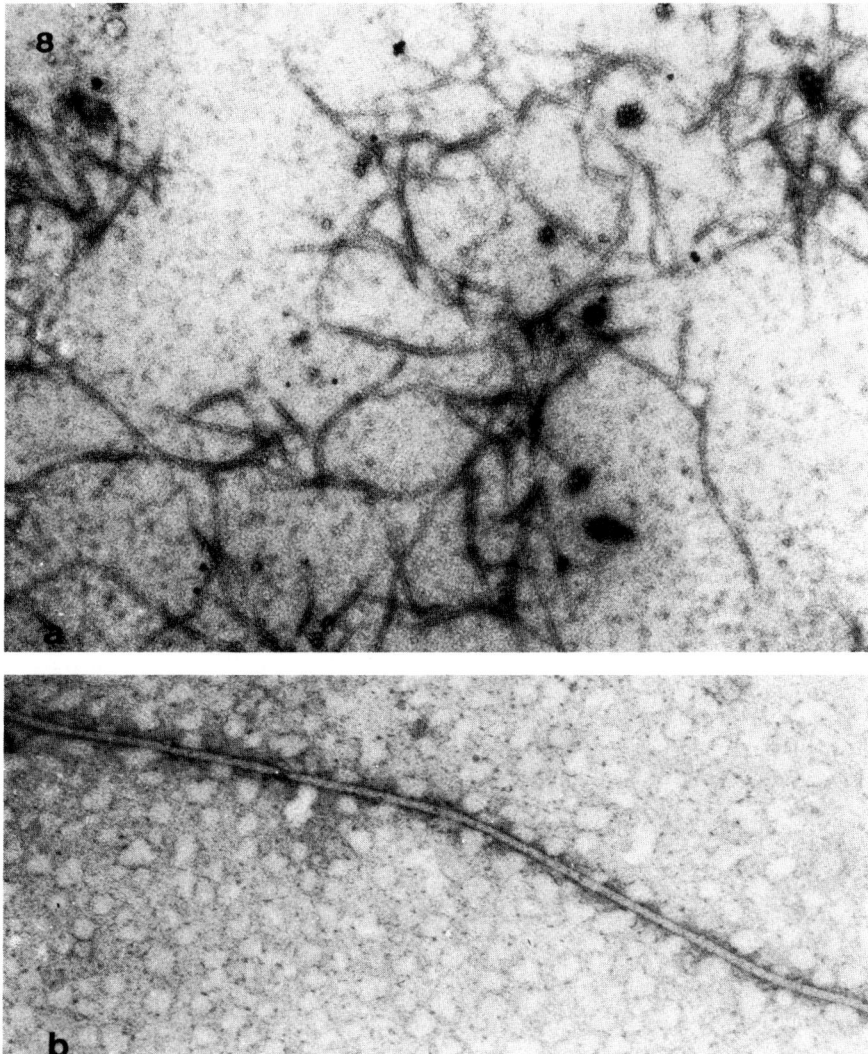

Fig. 8 a, b. Electron micrographs of the products assembled in vitro from PSR cell homogenates. The material obtained by centrifugation after the incubation of PSR cell homogenates at 30°C for 30 min was negatively stained and examined by electron microscopy as described in Section 2: **a.** Branched linear aggregates observed in pelleted material obtained by the procedure of Sloboda et al. (1976). **b.** Tubular structures averaging 190 ± 10 Å observed in pelleted material obtained by the in vitro assembly procedure of Asnes and Wilson (1979), both 84,000 x

3.3.2 Vinblastine Sulfate Precipitation of PSR Cell Proteins

Vinblastine sulfate has been used to precipitate and isolate tubulin from animals as diverse as sea urchins and mammals (Bryan, 1972; Berry and Shelanski, 1972). Therefore an attempt was made to precipitate tubulin from the high speed supernatant of a PSR cell homogenate with vinblastine sulfate. As is shown by Fig. 10, a wide spectrum of proteins was precipitated from PSR cell high-speed supernatant by this alkaloid. However, none of these polypeptides had a mobility on SDS gels similar to that of the subunits of bovine brain tubulin.

3.3.3 Assembly of Bovine Brain Tubulin in PSR Cell Extracts

Early log phase PSR cells were exposed to 0.5 μCi/ml of [^{3}H]-leucine for 18 h. These cells were washed with L-GNP (20 mM sodium phosphate + 100 mM sodium glutamate, pH 6.75) at 4°C and then homogenized in the same buffer. The homogenate was centrifuged at 36,000 rpm in the Beckman type 65 rotor and the supernatant was mixed with a solution of bovine brain tubulin prepared by three rounds of assembly and disassembly by the procedure of Asnes and Wilson (1979). This mixture was allowed to undergo two additional rounds of assembly and disassembly and the radioactivity in the various fractions was determined by scintillation counting. The results, shown in Table 3 indicated that PSR tubulin did not co-assemble with bovine brain microtubules under these conditions.

Further work on the effects of PSR cell homogenates on bovine brain tubulin assembly in vitro demonstrated that the plant cell homogenate contained factor(s) which inhibited the assembly of the bovine brain tubulin. In the experiment shown in Fig. 11, one of three dilutions of PSR cell high-speed supernatant or dilutions of a deproteinized PSR high speed supernatant were added to bovine brain tubulin which had gone through three rounds of assembly-disassembly and the tubulin was assembled a fourth time. The experiment demonstrated that increasing amounts of either the high-speed supernatant or the deproteinized PSR cell extract inhibited bovine brain tubulin assembly. In addition, the high-speed PSR cell supernatant contained a proteolytic activity which brought about the degradation of bovine brain microtubule-associated proteins (MAP's) and effect which could be overcome only partially by adding the proteolytic inhibitor PMSF to the extracts at a concentration of 1 mM (data not shown). Examination of the bovine brain microtubules assembled in the presence of PSR cell supernatant by electron microscopy showed that the microtubule morphologies were abnormal (Fig. 12).

3.3.4 Coassembly of Bovine Brain Tubulin with PSR Cell Translation Products

We have succeeded in obtaining coassembly of in vitro synthesized PSR polypeptides with bovine brain tubulin. Polysomes were isolated from log phase cells and translated in a wheat germ cell free protein synthesizing system as described in Section 2. The translation products were treated with RNase and mixed with twice assembled and disassembled bovine brain tubulin. As shown by the data presented in Table 4, substantial radioactivity from the PSR polysome translation products was conserved through two additional rounds of assembly and disassembly. The experiment was repeated with cells which were cultured for 5 days in medium containing 0.5 μM zeatin as well as cells cultured for the same time in medium lacking a cytokinin. Polysomes were isolated from the cells, translated in the wheat germ cell-free protein synthesizing system, and the [^{35}S]-methionine-labeled translation products were combined with bovine brain tubulin and allowed to reassemble under the conditions described by Sloboda et al. (1976). The microtubules formed were collected by centrifugation, depolymerized with cold and allowed to assemble again. After centrifugation

Table 3. Attempted coassembly of PSR cell and bovine brain tubulins. Early log phase PSR cells were exposed to 0.5 μ Ci/ml of [^{3}H]-leucine in culture medium for 18 h. The cells were washed with L-GNP extraction buffer (20 mM Na phosphate + 100 mM Na glutamate, pH 6.75) at 4°C, weighed and homogenized in the same buffer at 4°C. After centrifugation at 40,000 rpm in a Beckman 60 Ti rotor for 1 h at 4°C the supernatant was diluted with 3 volumes of twice-assembled and disassembled bovine brain tubulin in L-GNP prepared according to Asnes and Wilson (1979) or with 3 volumes of bovine serum albumin in L-GNP so that the final solution contained 6 mg/ml of protein as determined by Hartree's (1972) method. An aliquot of 10x polymix (see Sect. 2) was added to each solution and they were incubated at 30°C for 30 min. The flocculant material that formed during this period was collected by centrifugation, resuspended in L-GNP and depolymerized by cold treatment. The supernatant was again incubated at 30°C for 30 min and the assembled material was collected by centrifugation. Hot TCA digests were performed on samples of the first and second pellets as well as aliquots of the high-speed supernatant. Radioactivity was determined by scintillation counting

Sample	1st High Speed Supernatant, total cpm	cpm recovered in 1st pellet	cpm in cold-insensitive material	cpm in 2nd pellet
PSR extract + bovine brain tubulin	2.472x10^6	14,520	10,641	38
PSR extract + bovine serum albumin	2.472x10^6	20,580	10,356	45

the radioactivity retained by the pelleted microtubules was determined by scintillation counting. With the zeatin-treated cells, 0.73% of the total radioactivity incorporated in vitro pelleted with the microtubules while in the zeatin-deprived cells 0.86% of the total counts incorporated were retained by twice-assembled microtubules. Autoradiography of the proteins after separation on SDS gels revealed a polypeptides with mobilities similar to those of bovine brain tubulin subunits (data not shown).

4 Discussion

4.1 Cytokinin as a Regulator of PSR Cell Growth

The clone of PSR cells used for these investigations clearly is cytokinin-dependent for growth, as is shown by the data presented in Figs. 1-6. However, it was difficult to demonstrate this cytokinin dependence in the initial response of stationary phase cells upon transfer to fresh medium lacking a cytokinin (Fig. 2) or upon transfer to media containing zeatin at different concentrations (Fig. 5). In these cases the growth kinetics of the cells were independent of cytokinin initially. Only after many days in culture, after the cells had entered the log growth phase, did the absence of cytokinin, or a growth-limiting cytokinin concentration in the medium cause the cells to enter stationary phase prematurely.

These results suggest that the PSR cells do not enter stationary phase in the standard batch culture growth cycle in cytokinin-containing medium because they have exhausted the medium cytokinin, but rather because other medium components,

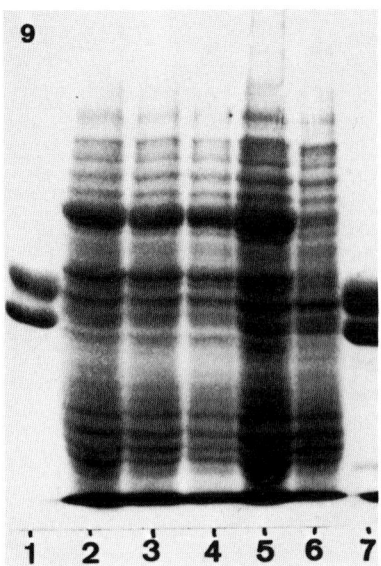
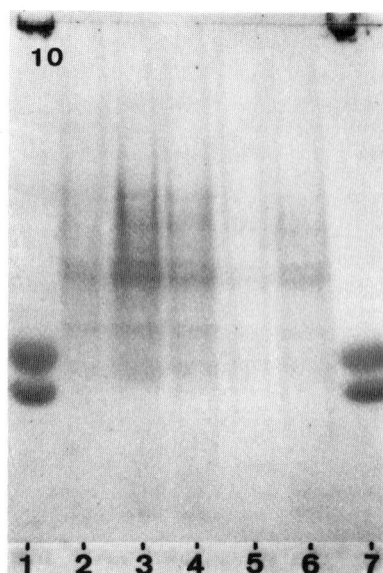

Fig. 9. SDS gel electrophoretic separation of PSR proteins. An attempt was made to isolate tubulin from log phase PSR cells by the procedure of Sloboda et al. (1976). Proteins were prepared for SDS gel electrophoresis from samples taken at several steps in the procedure. The lanes of the slab gel were loaded with comparable amounts of protein as follows: *1* 3 times assembled bovine brain tubulin; *2* crude PSR cell homogenate after centrifugation at 100,000 *g* for 1 h at 4°C; *3* PSR cell protein after concentrating the supernatant three fold by Amicon filtration; *4* PSR cell proteins after desalting the Amicon filtrate by Sephadex G-25 chromatography; *5* soluble proteins remaining after incubating the PSR cell extract at 30°C for 30 min and pelleting the assembled material by centrifugation; *6* Proteins recovered from the material pelleted by centrifugation after incubation of the extract at 30°C for 30 min; *7* three-times assembled bovine brain microtubule protein

Fig. 10. SDS gel electrophoretic separation of PSR cell proteins precipitated by vinblastine sulfate. Log phase PSR cells were homogenized in 0.05 M Pipes buffer, pH 6.9 containing 10 mM EGTA and 0.5 mM $MgCl_2$ (=PM buffer of Sloboda et al., 1976) which also contained 1 mM PMSF and 0.1 mg/ml RNase A. The homogenate was centrifuged at 45,000 rpm in the Beckman type 65 rotor for 1 h at 4°C. The supernatant was desalted on a Sephadex G-25 column which had been pre-equilibrated with PM buffer. The excluded volume was made 4M in glycerol, 2mM in GTP, 10 mM in $MgCl_2$ and 2 mM in vinblastine sulfate and incubated at 4°C for 3 h. The material that precipitated was collected by centrifugation, and dissolved in SDS gel electrophoresis sample buffer. Lanes *1* and *7* 3 times assembled bovine brain microtubule protein; lanes *2-5* vinblastine sulfate precipitated PSR cell protein at various loadings

possibly nutrients, have become growth-limiting. Since washing the stationary phase cells prior to transfer did not prevent them from growing in medium lacking a cytokinin, we conclude that the cells either took up the cytokinin from the medium during the previous culture cycle on medium containing zeatin and stored it, or that the hormone induced a "growth potential" which was not exhausted by one culture cycle. Some of this stored cytokinin, or "growth potential" would remain in the stationary phase cells where it could be utilized when they were transferred to fresh medium, overcoming the nutritional block to further growth.

In any event, the PSR cells prematurely entered stationary phase when they were cultured in medium lacking a cytokinin. At this time the cytokinin-deprived cells

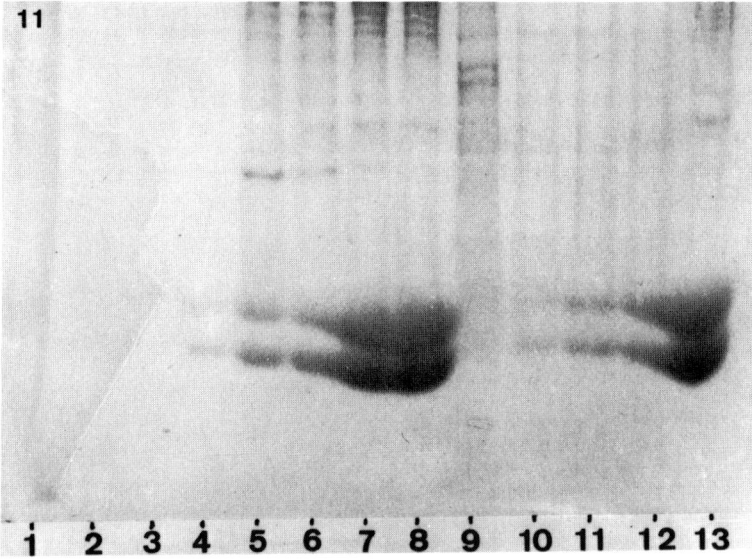

Fig. 11. SDS gel electrophoretic separation of bovine brain tubulin assembled in the presence of PSR cell extract. Log phase PSR cells were collected on Miracloth washed with cold L-GNP buffer (Asnes and Wilson, 1979) and then homogenized in the same buffer. The homogenate was centrifuged at 100,000 g for 1 h at 4°C and the supernatant was divided into two aliquots. The proteins were precipitated from one aliquot with acetone at −20°C, the acetone was removed from the supernatant by flash evaporation, the solution was reconstituted to its original volume with deionized water and aliquots of this solution were either used directly, diluted 1:1 or 1:4 with L-GNP. The part of the PSR homogenate which was not deproteinized was either used directly or diluted 1:1 or 1:4 with L-GNP. A three-times assembled bovine brain tubulin solution in L-GNP, prepared according to Asnes and Wilson (1979) was combined with one quarter of its volume of the deproteinized PSR cell extract or with the PSR cell homogenate, 10x polymix was added and the tubulin was allowed to assemble. The assembled microtubules were collected by centrifugation, each sample was dissolved in 500 μl of SDS gel electrophoresis sample buffer, and 25 μl of the resultant solution were loaded onto the gels as follows: *1* Pellet obtained from PSR cell homogenate alone; *2-4* bovine brain tubulin assembled in the presence of PSR cell homogenate either undiluted (*2*), diluted 1:1 (*3*) or diluted 1:4 (*4*) with L-GNP; *5-7*.bovine brain microtubule protein assembled in the presence of a deproteinized PSR cell extract either undiluted (*5*), diluted 1:1 (*6*) or diluted 1:4 (*7*) with L-GNP; *8* four times assembled bovine brain tubulin; *9* pellet obtained from PSR cell homogenate prepared with 1 mM PMSF; *10-12* bovine brain tubulin assembled in the presence of PSR cell homogenate prepared with 1 mM PMSF either undiluted (*10*), diluted 1:1 (*11*) or diluted 1:4 (*12*) with L-GNP; *13* four-times assembled bovine brain tubulin

became highly responsive to the hormone, as indicated by the fact that a partially synchronous wave of mitotic activity was observed approximately 24 h after the addition of zeatin to the cytokinin-deficient medium (Fig. 6). The behavior of these cells was similar to that reported for tobacco (Jouanneau, 1971) and for soybean (Fosket and Tepfer, 1978) cell cultures. The most probable explanation for this rapid response to cytokinin is that the hormone regulates some event which is necessary for G2 cells to enter mitosis (cf. Fosket, 1977).

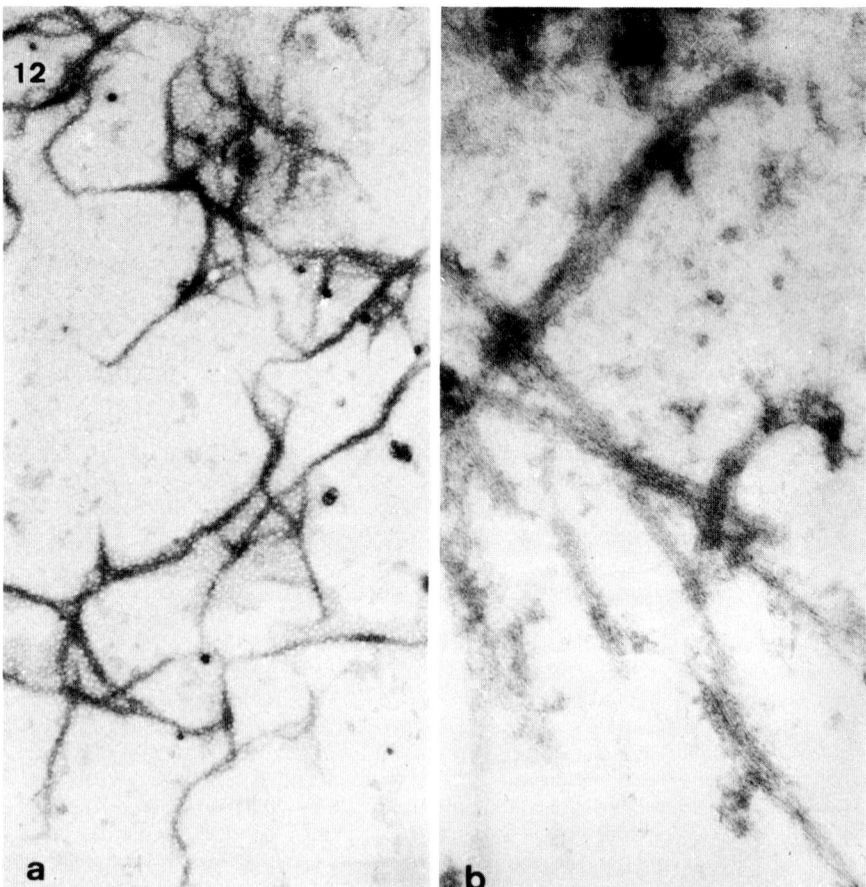

Fig. 12a, b. Electron micrographs of bovine brain microtubules assembled in the presence of PSR cell homogenates. Three-times assembled and disassembled bovine brain tubulin prepared according to the procedures of **a** Shelanski et al. (1973) or **b** Sloboda et al. (1976) was assembled in the presence of PSR cell homogenate prepared in the same buffer system, collected by centrifugation and prepared for electron microscopy as described in Section 2. **a** non-uniform diameter linear aggregates exhibiting a high degree of branching (74,000x); **b** beaded linear aggregates (120,000x)

4.2. Cytokinin as Regulator of Protein Synthesis Essential for Cell Division

The nature of the cytokinin-regulated event which is necessary for cell division is not known. However, cytokinins are well known to alter both the rate of protein synthesis (Short et al., 1974; Tepfer and Fosket, 1978) and the kinds of proteins synthesized by plant tissues (Jouanneau, 1970; Fosket et al., 1977). In the present study we show that both qualitative and quantitative changes in PSR cell protein synthesis accompany cytokinin-induced cell division (Table 2, Fig. 7). Although we have no evidence that these changes in protein synthesis and the subsequent cell division are linked in any way, Jouanneau (1975) has demonstrated that cytokinin-induced cell division in cultured tobacco cells was dependent upon protein synthesis.

Table 4. Copolymerization of bovine brain tubulin with PSR cell polysome translation products. Polyribosomes were extracted from log phase PSR cells and translated in a wheat germ cell-free protein synthesizing system as described in Section 2. Ten μl of the reaction mixture were removed for determination of incorporated radioactivity. The translation mixture was made 0.05 M in EDTA and 50 μg of pancreatic RNase was added. This mixture was incubated at 30°C for 15 min and then it was combined with 2.8 ml of three-times assembled and disassembled bovine brain tubulin prepared by the procedure of Sloboda et al. (1976). The resultant solution was made 4 M in glycerol + 2 mM in GTP and incubated at 30°C for 30 min. The assembled product was collected by centrifugation, resuspended with a Dounce homogenizer, and incubated 30 min at 4°C. Material that was not cold-sensitive was pelleted by centrifugation at 36,000 rpm in the Beckman 65 rotor at 4°C. The supernatant was made 4 M in glycerol and incubated at 30°C for 30 min. The assembled material was again collected by centrifugation and the pellet was washed. Samples were taken for determination of incorporated radioactivity by scintillation counting

Sample	Radioactivity (cpm)	% of Total
Translation mixture, total incorporation	7.54×10^6	–
Cold-insensitive material after 1st assembly	3.13×10^6	41.5%
Supernatant after 2nd assembly	9.28×10^4	1.23%
2nd pellet (microtubules) after washing	1.44×10^4	0.19%

There is good evidence that the entry of mammalian cells into the S-phase of the cell cycle is dependent upon the synthesis of a specific labile protein (Rossow et al., 1979). However, the nature of the controlling mechanism for the transition from G2 to mitosis is not known. In fact, it has been argued that G2 and mitosis are part of a determinant phase of the cell cycle which is not under external regulation (Shields and Smith, 1972). The bulk of the data supporting this hypothesis has been obtained in studies on animals cells which normally do not arrest in G2. However, in plants it is clear that some cell populations do leave the cell cycle from G2, implying that this phase of the plant cell cycle is not deterministic, but regulated (Fosket, 1977).

Our data on the qualitative changes in protein synthesis brought about by treating PSR cells with cytokinin demonstrated that at least one of the proteins synthesized in response to the hormone had a molecular weight near 55,000 daltons, the molecular weight of the tubulin monomers. Since tubulin is an essential component of the mitotic spindle (Hepler and Palevitz, 1974) and it has been shown that microtubule assembly from tubulin dimers will occur only when the tubulin concentration exceeds a critical level (Gaskin et al., 1974; Borisy et al., 1975) it seemed reasonable to investigate the possibiliy that cytokinin regulated tubulin biosynthesis. However, our data suggest that this is unlikely since translation products from both cytokinin-treated and cytokinin-deprived cells co-assembled with bovine brain tubulin with similar efficiencies.

4.3 Control of Tubulin Assembly

There are at least two ways in which the formation of the mitotic spindle microtubules might be regulated. Tubulin self-assembly has been shown to be highly sensitive to divalent cations, in particular Ca^{2+}. Hepler (1977) has proposed that membranous elements of the endoplasmic reticulum control the formation of the spindle microtubules by sequestering Ca^{2+}, thereby creating the cytoplasmic ionic conditions necessary for tubulin assembly.

Alternatively, a number of proteins have been shown to coassemble with tubulin. These have been designated variously as HMW for high molecular weight components (Murphy and Borisy, 1975), tau (Weingarten et al., 1975; Cleveland et al., 1977) and MAP's for microtubule-associated proteins (Sloboda et al., 1976). These proteins greatly accelerate the rate of tubulin self-assembly in vitro, if they are not required for microtubule formation (Murphy et al., 1977; Witman et al., 1976). Furthermore, immunofluorescence studies have demonstrated that these proteins are associated with microtubules in situ (Sherline and Schiavone, 1977; Lockwood, 1978). Since these microtubule-associated proteins are highly labile (Sloboda et al., 1976), it is possible that spindle microtubule formation in vivo could be regulated at the level of synthesis of these proteins.

Acknowledgment. Research supported by grant number 7722398 from the National Science Foundation to D.E.F.

References

Asnes CF, Wilson L (1979) Isolation of bovine brain microtubule protein without glycerol: Polymerization kinetics change during purification cycles. Anal Biochem 98: 64-73

Berry RW, Shelanski ML (1972) Interactions of tubulin with vinblastine and guanosine triphosphate. J Mol Biol 71: 71-80

Borisy GG, Olmsted J (1972) Nucleated assembly of microtubules in porcine brain extracts. Science 177: 1196-1197

Borisy GG, Marcum JM, Olmsted JB, Murphy DB, Johnson KA (1975) Purification of tubulin and associated high molecular weight proteins from porcine brain and characterization of microtubule assembly in vitro. Ann NY Acad Sci 253: 107-132

Bryan J (1972) Vinblastine and microtubules II. Characterization of two protein subunits from isolated crystals. J Mol Biol 66: 157-168

Cleveland DW, Hwo S-Y, Kirchner MW (1977) Purification of tau, a microtubule-associated protein that induces assembly of microtubules from purified tubulin. J Mol Biol 116: 207-225

Fosket DE (1977) The regulation of the plant cell cycle by cytokinin. In: Rost TL, Gifford EM Jr (eds Mechanism and control of cell division. Dowden Hutchison Ross, Stroudsburg, Pa, pp 62-91

Fosket DE (1980) The stability of the cytokinin requirement in cultured Paul's scarlet rose cells. In Vitro submitted

Fosket DE, Tepfer DA (1978) Hormonal regulation of growth in cultured plant cells. In Vitro 14: 63-75

Fosket DE, Volk MJ, Goldsmith MR (1977) Polyribosome formation in relation to cytokinin-induced cell division in suspension cultures of *Glycine max* [L.] Merr. Plant Physiol 60: 554-562

Gaskin F, Cantor CR, Shelanski ML (1974) Turbidimetric studies of the in vitro assembly and disassembly of porcine neurotubules. J Mol Biol 89: 737-758

Hartree EF (1972) Determination of protein: A modification of the Lowry method that gives a linear photometric response. Anal Biochem 48: 422-427

Hepler PK (1977) Membranes in the spindle apparatus: Their possible role in the control of microtubule assembly. In: Rost TL, Gifford EM Jr (eds) Mechanism and control of cell division. Dowden Hutchison Ross, Stroudsburg, Pa pp 212-232

Hepler PK, Palevitz BA (1974) Microtubules and microfilaments. Annu Rev Plant Physiol 25: 309-362

Jouanneau JP (1970) Renouvellement des protéines et effet spécifique de la kinétine sur des cultures de cellules de Tabac. Physiol Plant 23: 232-244

Jouanneau JP (1971) Contrôle par les cytokinines de la synchronisation des mitoses dans les cellules de Tabac. Expt Cell Res 67: 329-337

Jouanneau JP (1975) Protein synthesis requirement for the cytokinin effect upon tobacco cell division. Exp Cell Res 91: 184-190

Lockwood AH (1978) Tubulin assembly protein: Immunochemical and immunofluorescent studies on its function and distribution in microtubules and cultured cells. Cell 13: 613-627

Murphy DB, Borisy GG (1975) Association of high-molecular-weight proteins with microtubules and their role in microtubule assembly in vitro. Proc Natl Acad Sci USA 72: 2696-2700

Murphy DB, Vallee RB, Borisy GG (1977) Identity and polymerization-stimulatory activity of the nontubulin proteins associated with microtubules. Biochemistry 16: 2598-2605

Nesius KK, Uchytil LE, Fletcher JS (1972) Minimal organic medium for suspension cultures of Paul's scarlet rose. Planta 106: 173-176

Roberts BE, Paterson BM (1973) Efficient translation of tobacco mosaic virus RNA and rabbit globin 9S RNA in a cell-free system from commercial wheat germ. Proc Natl Acad Sci USA 70: 2330-2334

Roman R, Brooker JD, Seal S, Marcus A (1976) Inhibition of the translation of a 40S ribosome-Met-tRNAi^{Met} complex to an 80S ribosomal-Met-tRNAi^{Met} complex by 7-methyl guanosine-5' phosphate. Nature (London) 260: 359-360

Rossow PW, Riddle VGH, Pardee AB (1979) Synthesis of labile serum-dependent protein in early G_1 controls animal cell growth. Proc Natl Acad Sci USA 76: 4446-4450

Shelanski ML, Gaskin F, Cantor CR (1973) Microtubule assembly in the absence of added nucleotides. Proc Natl Acad Sci USA 70: 765-768

Sherline P, Schiavone K (1977) Immunofluorescence localization of proteins of high molecular weight along intracellular microtubules. Science 198: 1039-1040

Shields R, Smith JA (1972) Cells regulate their proliferation through alterations in transition probability. J Cell Physiol 91: 345

Short KC, Tepfer DA, Fosket DE (1974) Regulation of polyribosome formation and cell division in cultured soybean cells by cytokinin. J Cell Sci 15: 75-87

Skoog F, Strong FM, Miller CO (1965) Cytokinins. Science 148: 532

Sloboda RD, Dentler WL, Rosenbaum JL (1976) Microtubule-associated proteins and the stimulation of tubulin assembly in vitro. Biochemistry 15: 4497-4505

Studier WF (1973) Analysis of bacteriophage T7 early RNA's and protein on slab gels. J Mol Biol 79: 237-248

Tepfer DA, Fosket DE (1978) Hormone-mediated translational control of protein synthesis in cultured cells of *Glycine max*. Dev Biol 62: 486-497

Weingarten MD, Lockwood AH, Hwo S-Y, Kirchner MW (1975) A protein factor essential for microtubule assembly. Proc Natl Acad Sci USA 72: 1858-1862

Widholm JM (1972) The use of fluorescein diacetate and phenosafranine for determining viability of cultured plant cells. Stain Tech 47: 189-194

Witman GB, Cleveland DW, Weingarten MD, Kirschner MW (1976) Tubulin requires tau for growth onto microtubule initiating sites. Proc Natl Acad Sci USA 73: 4070-4074

The Cytokinin Control of Protein Synthesis in Plants

A. Szweykowska, E. Gwóźdź, and M. Spychała[1]

A stimulatory effect of cytokinins on protein and nucleic acid syntheses in plant tissues has been well known for a long time (Mothes et al., 1959; Wollgiehn, 1961; Roychoudhury et al., 1965; Kulaeva, 1967). A relationship of various effects of growth substances on plant development to the metabolism of nucleic acids was suggested first by Skoog (1954), and later, during the intense progress in molecular biology, a tentative idea was widely assumed that plant hormones — playing a role of effector-like factors — act primarily at a level of DNA-directed RNA synthesis. However, despite a great number of studies, this kind of action appeared rather difficult to be prove. In the case of cytokinins, the only direct evidence was provided in a paper by Matthysse and Abrams (1970) who reported that in a cell-free system with purified chromatin or DNA from pea buds, and with *Escherichia coli* RNA polymerase, in the presence of a receptor protein previously isolated from the chromatin, the addition of cytokinins increased the rate of RNA synthesis. On the other hand, numerous data exist suggesting that cytokinins can stimulate protein synthesis at a post-transcriptional or even directly at a translational level. The first assumptions about this were made after the cytokinins had been discovered in some tRNAs in the position adjacent to the anticodon (Zachau et al., 1966; Biemann et al., 1966). It was found later (Fittler and Hall, 1966; Gefter and Russell, 1969) that the presence of cytokinins in tRNA's increases the ribosome binding affinity of the aminoacylated tRNAs, facilitating the codon recognition in the polyribosomal complexes. There are also data that cytokinins can modify the function of the ribosomes themselves. They came from work by Berridge et al. (1970) who showed a binding of the cytokinins to the 80S ribosomes, with a correlation between the extent of binding and the biological activity of various cytokinins and cytokinin analogs. Fox and Erion (1975) later found two sites of cytokinin binding in the ribosomes, from which one, characterized by a high affinity, showed features of a receptor protein. Rijven (1974) reported that the ribosomes from the cytokinin-treated *Trigonella* cotyledons were more active in an initiation of the poly(U)-directed polyphenylalanine synthesis, and suggested an activation by the cytokinin of the protein initiation factors. Takegami and Yoshida (1977) reported that a cytokinin-binding protein is present in the 40S ribosomal subunit and suggested that this protein itself may be a subunit of the initiation factors for protein synthesis.

In the laboratory of Kulaeva it was found that the polyribosomal preparations isolated from benzyladenine-treated pumpkin cotyledons were twice as active in pro-

---

[1] Institute of Biology, Adam Mickiewicz University, Poznań, Poland

tein synthesis in vitro as those isolated from the untreated ones (Klyachko et al., 1973; Yakovleva et al., 1977). Klämbt (1976) showed that N^6-isopentenyl-adenine added to a cell-free system from young corn shoots and tobacco pith increased protein synthesis by about 20-30%. In Kulaeva's laboratory, however, various growth regulators, including the cytokinins, had no effect in protein synthesis when added directly to the incubation mixture of the cell-free system (Yakovleva et al., 1975). Taking into consideration that the response of the protein-synthesizing apparatus to the treatment of the cotyledons with the regulators appeared rather quickly, after 1-2 h, and that it involved a shift from mono- to polyribosomes, several possibilities were proposed as effective causes: a very quick, mass synthesis of new mRNA, a mobilization of stored templates from informosomes, or an enhanced initiation of protein synthesis leading to an intensified utilization of functioning templates (Klyachko et al., 1978). A possibility was proposed of a regulation of the shift from mono- to polyribosomes not only at a transcriptional, but also at a post-transcriptional level. Considerable evidence for this presumption was supplied by Fosket and co-workers in a series of studies on the cytokinin-induced cell divisions in suspension cultures of soybean cells (cf. Tepfer and Fosket, 1978). In these cells, the hormone-stimulated cell divisions were preceded by an enhanced polyribosome formation. The effect was not accompanied by an enhanced RNA synthesis and was not blocked by inhibitors of RNA synthesis (Muren and Fosket, 1977). No effect could be shown on the polyribosome size, and on the initiation, elongation, and termination rates of the polypeptide chains. A nonpolyribosomal poly(A)$^+$RNA synthesized in the absence of cytokinin was found to move into the polyribosomes at an enhanced rate in the presence of the hormone. A hypothesis was proposed that a fraction of the cytoplasmic mRNA is in a cryptic form in the hormone-deprived soybean cells and is made translatable upon the cytokinin stimulation leading to an increased proportion of polyribosomes in cells (Tepfer and Fosket, 1978).

The problem of the cytokinin effect on protein synthesis was studied in our laboratory using two model systems: protonema of the moss *Ceratodon purpureus* (Hedw.) Brid. and isolated cotyledons of cucumber (*Cucumis sativus* L. cv. Monastyrski). In the moss protonema, cytokinins enhance cell division (Szweykowska et al., 1971, 1972) and chloroplast development (Woźny, 1978), and they induce a mass appearance of gametophore buds (Gorton et al., 1957; Hahn and Bopp, 1968). The latter response is very specific and involves a qualitative step in the moss morphogenesis. The cucumber cotyledons specifically respond to the cytokinins with an enhanced growth (Narain and Laloraya, 1974) and − in light − with an enhanced chlorophyll synthesis and chloroplast differentiation (Fletcher and McCullagh, 1971; Woźny and Szweykowska, 1975).

In the moss protonema, an enhancement of [^{14}C]-leucine incorporation into proteins was found as soon as after 15 min of treatment with a cytokinin (N^6-isopentenyladenine, i^{6}ade). However, a more detailed analysis showed that this enhancement could be in some part a result of an intensification of the precursor uptake. Further experiments, carried out with an exclusion of the effects of i^{6}ade on the precursor influx, demonstrated a stimulation of protein synthesis after 30 min of cytokinin treatment. A pulse-chase experiment showed that the enhanced incorporation of [^{14}C]-leucine was accompanied by an enhanced decline of the precursor pool in the

i^6ade-treated protonema. The level of protease activity did not change during the several hours of the cytokinin treatment, so the enhanced precursor incorporation could not be a consequence of a decrease in the protein breakdown and had to be interpreted as a result of an increased protein synthesis (Spychała, 1980).

Further experiments were carried out with an in vitro system of protein synthesis using tRNAs from wheat germ, dialyzed wheat germ supernatant and polyribosomes isolated from a control and i^6ade-treated protonema, respecitvely. The incorporation of [^{14}C]-leucine in this system was inhibited considerably by cycloheximide and only slightly by chloramphenicol which indicated that the 80S ribosomes were mainly involved in the protein synthesis. The activity of polyribosomes from a cytokinin-treated protonema was considerably higher. After 12 h of i^6ade treatment, the stimulation amounted to 140%. As shown by a sucrose density gradient centrifugation, no essential differences were induced by the cytokinin in the level of polyribosomes, suggesting that the higher capacity of the ribosomal preparation after the cytokinin treatment of the protonema was not due to an increased formation of polyribosomes, but rather to their increased activity. In contrast to the results of Klämbt (1976), obtained with an in vitro system prepared from tobacco pith and young corn shoots, i^6ade added directly to the incubation mixture with protonemal polyribosomes did not increase the rate of protein synthesis, which indicated that the activation of polyribosomes was not a direct action, but required an intact cell environment.

It is rather difficult to speculate on the possible mechanism of the polyribosome activation. However, some data indicate that it is related more to the level of translation than to that of transcription. The levels of polyribosomes were insensitive to actinomycin D, used in a concentration that inhibited RNA synthesis by about 50%, and the antibiotic did not affect the in vivo incorporation of [^{14}C]-leucine into proteins in control (cytokinin-free) conditions, which demonstrated its independence of a de novo synthesis of RNA (Gwóźdź, 1979). Also no stimulation of a total RNA synthesis could be found, even after a 4-h, cytokinin treatment of the protonema, whereas a stimulation of protein synthesis appeared as soon as after 30 min. On the other hand, the stimulatory effect of cytokinin on protein synthesis in vivo was to some extent sensitive to actinomycin D. A fractionation of RNA by polyacrylamide gel electrophoresis (PAGE) showed that after 1 h of the cytokinin treatment an increased incorporation appeared in the region of polydisperse RNA, considered as the mRNA region. A PAGE fractionation of proteins did not show an induction by i^6ade of a synthesis of new protein fractions, however, a differential stimulation or inhibition of precursor incorporation into some fractions was observed (Spychała, 1980). Therefore, a participation of a pool of newly made RNAs of high template efficiency can not be excluded from the process of the cytokinin-induced polyribosome activation. It is very possible that this activation is due to events appearing at both the translational and transcriptional levels. The latter process is the more possible that the cytokinin induce an important step in moss morphogenesis — the differentiation of gametophore buds in the protonema.

In cucumber cotyledons, the activity of ribosomal preparations in an in vitro sysstem of protein synthesis was also markedly increased after an incubation of isolated cotyledons with a cytokinin (kinetin). Examination of the centrifugation profiles, as well as of electron micrographs, showed that the level of monoribosomes decreased and that of polyribosomes increased in preparations from kinetin-treated cotyledons (Woźny and Gwóźdź, 1980). The results were similar to those obtained by Muren and Fosket (1977) for soybean cells of cotyledonary origin and by Klyachko et al. (1978) for pumpkin cotyledons. The stimulatory effect of kinetin on polyribosome formation was not a consequence of the kinetin-induced promotion of cell growth. In conditions of an osmotic stress (in the presence of 0.3 M mannitol) which completely suppressed the kinetin stimulation of growth of the cotyledons, the promotive effect of the regulator on the level of polyribosomes, as well as their activity in protein synthesis, were only slightly affected. Also the RNase activity did not differ much in the experimental variants which indicated that the increase in polyribosomes did not result from an inhibitory effect of kinetin on the synthesis or activity of these enzymes.

Several experiments were carried out to throw some light on the problem of the mechanism of the kinetin action in protein synthesis and polyribosome formation. Two possibilities were considered in particular: a stimulation of a de novo synthesis of mRNA, and an enhancement of the formation of the initiation complex of protein synthesis. An analysis of the incorporation of [^{3}H]-adenosine into fractions of the ribosomal preparations showed that in the monoribosomal fraction it was proportional to the level of this fraction, i.e., it was higher in the control and lower in the kinetin variant, whereas in the polyribosomal fractions it was similar in both variants, in spite of the fact that the level of the polyribosomes was considerably higher in the kinetin preparation. Some differences were found in the distribution of label in the polyribosomal fractions: in the kinetin variant there was some preferentiation in the region of lighter polyribosomes which might indicate a stimulation by kinetin of the combining of monoribosomes into polysomal configurations. A fractionation of RNA labeled with [^{3}H]-adenosine on an oligo (dT)-cellulose column showed that kinetin stimulated the precursor incorporation into poly (A)$^-$RNA, but there was practically no difference in the incorporation into the poly (A)$^+$RNA, a fraction thought to constitute at least a major part of mRNA. The inhibitors of transcription — actinomycin D and 6-methylpurine, as well as an inhibitor of polyadenilation — cordicepin — did not suppress the stimulatory effect of kinetin on the level of polyribosomes. It was interesting that in several experiments there was even some increase in polyribosomes in the presence of cordicepin. The results coincide with those of Muren and Fosket (1977) who similarly concluded that the cytokinin-induced increase in the polyribosome/monoribosome ratio in soybean cells was not a reflection of a hormone-induced increase in the rate of mRNA synthesis. A de novo synthesis of mRNA, as well as its post-transcriptional processing appear not to be a requisite for the kinetin stimulation of polyribosome formation also in cucumber cotyledons. On the other hand, some decrease in the stimulation by kinetin of the polyribosomal level was observed in the presence of 5-fluorouracil, an inhibitor of rRNA synthesis which might indicate a participation of kinetin in the synthesis of new ribosomes.

A strongest suppression of the kinetin effect on polyribosomes was brought about by aurintricarboxylic acid, an inhibitor of the initiation of protein synthesis, blocking the binding of the smaller ribosomal subunit to the mRNA. In the presence of this inhibitor, the level of polyribosomes was lowered to the control level. In this respect, our results differ from those of Tepfer and Fosket (1978) who concluded from indirect experiments, with double-labeled ribosomes and cycloheximide treatment, that the cytokinin did not change the initiation rate of polypeptide synthesis. Our results rather support a conclusion that the kinetin action in protein synthesis in cucumber cotyledons occurs at the translational level by increasing the affinity of ribosomes to the template (Gwóźdź, 1979).

As a result of kinetin treatment, no qualitative differences could be demonstrated in the protein spectrum after an analysis of the [^{35}S]-labeled product of the in vitro translation by means of slab PAGE and fluorography. However, some quantitative differences appeared in the spectrum. This suggested that kinetin could selectively stimulate the synthesis of particular proteins which might be essential for cell growth.

It appears that the cytokinin stimulation of growth of the etiolated cucumber cotyledons is at least partly a consequence of its direct effect on the apparatus of protein synthesis. This effect results in an enhanced combining of free monosomes into polysomal configurations, characterized by a high ability to protein synthesis. The observed stimulation of the synthesis of some proteins may be a condition for the appearance of the growth effect. Although no unequivocal answer was obtained as to the origin of the mRNA participating in the stimulation of polyribosome formation, the results suggest that it is independent of a de novo mRNA synthesis.

It is concluded from the data we have presented that the cytokinins can promote the translation process in plant cells either by an activation of polyribosomes or by increasing their number through an enhancement of the ribosomal binding into polysome configurations. The two ways of action need not occur at the same time in the same cell or tissue: in the moss protonema only an activation of polyribosomes could be shown, and in the cucumber cotyledons a stimulation of the polyribosome formation was demonstrated.

References

Berridge MV, Ralph RK, Letham DS (1970) The binding of kinetin to plant ribosomes. Biochem J 119: 75-84

Biemann K, Tsunakawa S, Sonnenbichler J, Feldmann H, Dütting D, Zachau HG (1966) Struktur eines ungewöhnlichen Nucleosids aus Serin-spezifischer Transfer-Ribonucleinsäure. Angew Chem 78: 600-601

Fittler F, Hall RH (1966) Selective modification of yeast seryl-tRNA and its effect on the acceptance and binding functions. Biochem Biophys Res Commun 25: 441-446

Fletcher RA, McCullagh D (1971) Cytokinin-induced chlorophyll formation in cucumber cotyledons. Planta 101: 88-90

Fox JE, Erion JL (1975) A cytokinin binding protein from higher plant ribosomes. Biochem Biophys Res Commun 64: 694-700

Gefter NL, Russell RL (1969) Role of modifications in tyrosine transfer-RNA: a modified base affecting ribosome binding. J Mol Biol 39: 145-157

Gorton BS, Skinner CG, Eakin RE (1957) Activity of some 6-substituted purines on the development of the moss Tortella caespitosa. Arch Biochem Biophys 66: 493-496

Gwóźdź, EA (1979) The role of cytokinins in protein biosynthesis (in Polish with English summ.) Thesis, A Mickiewicz Univ Poznań, Ser Biol 16

Hahn H, Bopp M (1968) A cytokinin test with high specificity. Planta 83: 115-118

Klämbt D (1976) Cytokinin effects on protein synthesis of in vitro system of higher plants. Plant Cell Physiol 17: 73-76

Klyachko NL, Yakovleva LA, Kulaeva ON (1973) Effect of treating isolated pumpkin cotyledons with 6-benzylaminopurine on the activity of their ribosome preparation/in Russian with English summ./. Dokl Akad Nauk SSSR 211: 1235-1238

Klyachko NL, Ananiev E, Kulaeva ON (1978) Bystraya otvetnaya reakcya beloksinteziruyshtschego aparata izolirovanych semiadoley tykwy na deistrvye fitogormonov. Dokl Akad Nauk SSSR 243: 1334-1336

Kulaeva ON (1967) Citokininy i ich fizjologicheskoe deistvye. Usp Sov Biol 63: 28-53

Matthysse AG, Abrms M (1970) A factor mediating interaction of kinins with the genetic material. Biochim Biophys Acta 199: 511-518

Mothes K, Engelbrecht L, Kulaeva O (1959) Über die Wirkung des Kinetins auf Stickstoffverteilung und Eiweissynthese in isolierten Blättern. Flora 147: 445-464

Muren RC, Fosket DE (1977) Cytokinin-mediated translational control of protein synthesis in cultured cells of Glycine max. J Exp Bot 28: 775-784

Narain A, Laloraya MM (1974) Cucumber cotyledon expansion as a bioassay for cytokinins. Z Pflanzenphysiol 71: 313-322

Rijven AHGC (1974) Initiation of polyphenylalanine synthesis and the action of cytokinins in fenugreek cotyledons. Nature (London) 252: 257-259

Roychoudhury R, Datta A, Sen SP (1965) The mechanism of action of plant growth substances: the role of nuclear RNA in growth substance action. Biochim Biophys Acta 107: 346-351

Skoog F (1954) Substances involved in normal growth and differentiation of plants. In: Abnormal and pathological plant growth. Brookhaven Symp Biol 6: 1-21

Spychała M (1980) Protein and RNA synthesis in a protonema of the moss Ceratodon purpureus during the cytokinin-induced process of bud formation. Thesis, A Mickiewicz Univ, Poznań, unpublished

Szweykowska A, Dornowska E, Cybulska A, Wasiek G (1971) The cell division response to cytokinins in isolated cell cultures of the protonema of Funaria hygrometrica and its comparison with the bud induction response. Biochem Physiol Pflanz 162: 514-525

Szweykowska A, Korcz I, Jaśkiewicz-Mrocczkowska B, Metelska N (1972) The effect of various cytokinins and other factors on the protonemal cell divisions and the induction of gametophores in Ceratodon purpureus. Acta Soc Bot Pol 41: 401-409

Takegami T, Yoshida K (1977) Specific interaction of cytokinin binding protein with 40 S ribosomal subunits in the presence of cytokinin in vitro. Plant Cell Physiol 18: 337-346

Tepfer DA, Fosket DE (1978) Hormone-mediated translational control of protein synthesis in cultured cells of Glycine max. Dev Biol 62: 486-497

Wollgiehn R (1961) Untersuchungen über den Einfluß des Kinetins auf den Nucleinsäure- und Proteinstoffwechsel isolierter Blätter. Flora 151: 411-437

Woźny A, Szweykowska A (1975) Effect of cytokinins and antibiotics on chloroplast development in cytoledons of cucumis sativus. Biochem Physiol Pflanz 168: 195-209

Woźny A (1978) Effect of cytokinins on chloroplast development in protonema of Ceratodon purpureus and in cotyledons of Cucumis sativus (in Polish with in English summ. Thesis, A Mickiewicz Univ, Poznań, Ser Biol 9

Woźny A, Gwóźdź EA (1980) The effect of cytokinin on the polyribosome formation in cucumber cotyledons. Biochem Physiol Pflanz 175: 476-480

Yakovleva LA, Klyachko NL, Kulaeva ON (1975) Absence of the effect of phytohormones and abscisic acid on protein synthesis in vitro (in Russian with English summ). Fiziol Rast 22: 856-858

Yakovleva LA, Klyachko NL, Kulaeva ON (1977) Action of 6-benzylaminopurine on the incorporation of ^{14}C-leucine into protein in the cell-free system from isolated pumpkin cotyledons (in Russian with English summ). Mol Biol 11: 868-876

Zachau HG, Dütting D, Feldmann H (1966) Nucleotidsequenzen zweier serinspezifischer Transfer-Ribonucleinsäuren. Angew Chem 78: 392-393

Cytokinin Action on Transcription and Translation in Plants

O.N. Kulaeva[1]

1 Introduction

Cytokinins are well known to activate the syntheses of total RNA and proteins in plant tissues sensitive to this phytohormone [1-6]. The aim of this paper is to discuss the following questions:

1. What is the mechanism of activation of total RNA biosynthesis in plant cells by cytokinin?
2. Is the activation of RNA synthesis, i.e., increase in ribosome number and mRNA amount in cells the only mode of cytokinin action on protein synthesis or can cytokinin in addition take part in the control of protein synthesis at posttranscriptional level?
3. If cytokinin participates in regulation of protein synthesis at posttranscriptional level, is this effect due to the primary cytokinin action at transcriptional level?

These aspects of cytokinin effects were investigated in our laboratory on detached barley leaves [7-11] and isolated pumpkin cotyledons [12-15].

2 Materials and Methods

Barley seedlings (cv. Viner) were grown in boxes with soil in a greenhouse. The first leaf was taken for every experiment.

For chromatin preparation the leaves were homogenized in Tris-HCl buffer (pH 8.0). After filtration through several layers of cheese cloth and then through miracloth the homogenate was centrifugated at $1000\,g$. The chromatin-containing precipitate was washed six times with buffer in the presence of Triton X-100, and twice with the buffer alone, then it was dissolved in buffer containing glycerine.

RNA-polymerase activity was measured in 150 µl of incubation medium including 5 µmol of Tris-HCl buffer (pH 8.0), 1 µmol $MgCl_2$, 1 µM 2-mercaptoethanol, 0.2 µmol nucleoside phosphates, (among them, [^{14}C]-ATP) and chromatin (20-100 µg DNA). The reaction was carried out for 20 min at 32°C and stopped by addition of an equal volume of cold solution of 0.9% sodium pyrophosphate in 10% TCA. The isolation and purification of nuclei in sucrose density gradient has been described in [9].

---

[1] K.A. Timiriazev Institute of Plant Physiology, Moscow, USSR

To measure of RNA polymerase I activity nuclei (20-150 µg of DNA) were incubated in the presence of the following compounds; Tris-HCl-50 µmol (pH 8.3), $MgCl_2$ - 8 µmol, GTP, CTP, UTP — 0.1 µmol each, [^{14}C]-ATP — 0.0029 µmol (Czechoslovakia, sp.act. 321 µCi/µM), the final volume of incubation medium being 0.5 ml. The reaction mixture for RNA polymerase II included the same components but pH was adjusted to 7.8 and $MgCl_2$ was substituted for 2 µmol $MnCl_2$; ammonium sulfate was added at final concentration of 0.25 M [9].

Pumpkin cotyledons were isolated from etiolated seedlings grown in darkness for various periods [13, 15]. The cotyledons were then exposed to water or 6-benzylaminopurine (BAP) at a concentration of 10 mg/l. The preparation of ribosomes and their fractionation in sucrose density gradient were described in [15]. The technique of in vitro protein labeling was communicated in [13].

3 Results and Discussion

3.1 Mechanism of Activation of RNA Synthesis by Cytokinin

The activation of RNA synthesis in detached leaves by cytokinin is well established [1-6]. In a series of investigations it was demonstrated that cytokinin activated the production of all RNA types, including rRNA, tRNA and some mRNA-containing fractions as well. The mechanism of total activation of RNA syntheses in leaves under the influence of cytokinin was unclear. We have investigated it on detached barley leaves which are very sensitive to exogenous cytokinin. Cytokinin retarded their senescence and stimulated RNA synthesis [16, 17]. Earlier it was shown that cytokinin increased chromatin-bound RNA-polymerase I activity in detached barley leaves [7, 18]. We then demonstrated that cytokinin promoted the activities of not only RNA-polymerase I, but RNA-polymerase II was well [9]. It was shown on nuclei isolated from cytokinin-treated leaves (Table 1). To exclude the effect of cytokinin on RNA-polymerase synthesis cytokinin was added in the next experiment at the stage of leaf homogenization. After that nuclei were isolated, purified by sucrose density gradient and RNA-polymerase activity was measured. Cytokinin addition at the leaf homogenization step markedly stimulated the activities of RNA-polymerase I and II (Table 2).

More native nuclei were prepared from protoplasts obtained from barley leaves. These nuclei were tested for activities of RNA-polymerases I and II (Table 3). Activation of RNA-polymerases took place after cytokinin addition either at the moment of protoplast homogenization or to pure nuclei twice purified in sucrose density gradient. It follows that cytokinin receptors are located inside the cells, as protoplasts and even pure nuclei are sensitive to this phytohormone [19].

The discussion of data obtained must also take into account the possibility of cytokinin action on the permeability of nuclear membranes for labeled ATP. To exclude this possibility the next series of experiments was carried out not with isolated nuclei but with preparation of chromatin. In these experiments BAP was added at the leaf homogenization step, then chromatin-containing RNA polymerase I was isolated and used as template and at the same time as an RNA polymerase source for RNA synthesis. Figure 1 shows that under these conditions cytokinin has markedly stimulated the activity of chromatin-bound RNA polymerase. The action of cytokinin depended

Table 1. RNA-polymerase activity in nuclei isolated from 10-day-old barley leaves incubated for 24 h on water or on solution (2 mg/l) of 6-benzylaminopurine (BAP)

Enzyme	BAP	Incorporation of [^{14}C]-AMP	
		cpm per 100 µg of DNA	%
RNA-polymerase I	−	6340	100
	+	9200	145
RNA-polymerase II	−	3670	100
	+	5725	156

Nuclei preparation and technique of measurement of RNA-polymerase I and II activities are described in [9]

Table 2. Effect of BAP (0.1 mg/l) added to leaf homogenization medium on RNA-polymerase activity of nuclei isolated from 10-day-old barley leaves

Enzyme	BAP	Incorporation of ^{14}C-AMP	
		cpm per 50 µg of DNA	%
RNA-polymerase I	−	1451	100
	+	2762	190
RNA-polymerase II	−	1710	100
	+	6287	367

Nuclei preparation and technique of measurements of RNA polymerase I and II activities are described in [9]

Table 3. Activation of nuclear RNA-polymerase by BAP

Mode of treatment with BAP	Percent activation	
	RNA polymerase I	RNA polymerase II
Added during homogenization of protoplasts	267	259
Added to nuclei[a] purified in sucrose density gradient	150	145

[a]Suspension of nuclei was incubated in BAP (0.1 mg/l) for 10 min. Then an extra amount of BAP was removed by centrifugation of nuclei in 1.2 M sucrose. Reaction mixture for RNA-polymerase I estimation contained 50 mM Tris-HCl buffer (pH 8.0), 5 mM $MgCl_2$, 100 mM $(NH_4)_2SO_4$, 10 mM dithiotriethol, 20% glycerine, 0.4 mM UTP; 0.4 mM GTP, 0.4 mM CTP, 0.09 mM [^{14}C]-ATP (2.5 µCi in the final volume of reaction mixture), and nuclei (10-15 µg of DNA). In the medium for RNA-polymerase II estimation $MgCl_2$ was replaced by 2 mM $MnCl_2$ and concentration of $(NH)_2SO_4$ was increased up to 250 mM (Details in [19])

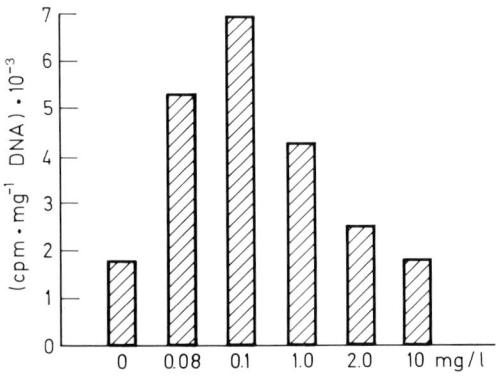

Fig. 1. Effect of 6-benzylaminopurine (BAP) on the activity of chromatin-bound RNA-polymerase. BAP was added to 10-day-old barley leaves during their homogenization in the course of chromatin isolation [8]. BAP-concentration (mg/l) is indicated under the columns. The reaction mixture for RNA polymerase estimation contained 33.5 mM Tris-HCl buffer, pH 8.0, 6.7 mM $MgCl_2$, 6.7 mM 2-mercaptoethanol, 1.34 mM M nucleoside triphosphates, including [^{14}C]-ATP, and chromatin (100 μg DNA). Incubation period 15 min at 30°C. (Details in [8])

strictly on its concentration. The optimal concentration in this case was 20 times lower than in the experiments with intact leaves. Activation of chromatin-bound RNA polymerase by cytokinin persisted after enzyme solubilization.

Hence cytokinin has modified the properties of RNA polymerase, although it is yet unclear whether cytokinin affected the enzyme directly or through its influence on chromatin to which RNA polymerase was bound.

It is important that there is a close relationship between cytokinin effect on activity of chromatin-bound RNA-polymerase and the structure of cytokinin molecules. The effect was observed only in the presence of natural cytokinins (zeatin, isopentenylaminopurine) and their highly active synthetic analogs (6–BAP, kinetin). No activity was revealed with adenine and some inactive analogs of 6-BAP (Table 4).

Some biological specificity was also evident in these experiments as BAP added during leaf homogenization affected chromatin-bound RNA polymerase activity only from cytokinin-sensitive 9-12 day-old leaves. Cytokinin had no effect in the experiments with cytokinin-insensitive younger and also old yellow leaves.

In all these experiments BAP was effective only when added at the leaf homogenization step. Addition of the phytohormone during the successive stages of chromatin purification or directly to the incubation medium for RNA synthesis was not effective. The data gave support to the assumption that some nonchromatin cytokinin acceptors or cofactors are necessary for its action on chromatin-bound RNA-polymerase. By analogy with animal steroid hormones we had assumed that supernatant contained some cytokinin receptors providing the effect on RNA-polymerase. This assumption was confirmed with the aid of affinity chromatography. Two fractions with high affinity for cytokinin were isolated from supernatant (1000 g) from 10-day-old leaves by means of column with epoxy-sepharose-bound 6-BAP. The addition of these fractions together with BAP to the incubation medium for chromatin-bound RNA-polymerase markedly increased the enzyme activity. This fraction alone or BAP alone had no effect on RNA-polymerase activity (Table 5).

The proteins with high affinity for cytokinin had been isolated already from an aquatic fungus *Achlya ambisexualis* [20], from tobacco leaves [21] and wheat germ [22-24]. But their function is yet not clear. By discussion of this problem we must take into account that not only the receptors but also some other proteins can possess high cytokinin-affinity, for example, all enzymes of cytokinin metabolism, as well as some hypothetical transport proteins and some proteins conjugating cytokinin.

Table 4. Effect of various cytokinins and their analogs on chromatin-bound RNA-polymerase activity from 10-day-old barley leaves

Substance[a]	Stimulation
Zeatin	+
6-isopentenilaminopurine	+
6-benzylaminopurine	+
Kinetin	+
Adenine	−
6-benzylthiopurine	−
Benzimidazol	−

[a]The substances were added to leaf homogenization medium [8] (For details see legend to Fig. 1)

Table 5. Effect of cytokinin-binding protein (CBP) and cytokinin on activity of chromatin-bound RNA polymerase from barley leaves

Treatment		Incorporation of [^{14}C]-ATP	
BAP 0.1 mg/l	CBP 5 µg	c.p.m. per 10 µg of DNA	%
−	−	1409	100
+	+	4620	328
−	+	1441	102
+	−	944	67

For details see legend to Table 1

A protein can be considered as a hormonal receptor only if it specifically binds the hormone and after that interaction, the receptor associated with the hormone induces a series of metabolic events in the cell which are necessary for the hormonal action. Activation of chromatin-bound RNA-polymerase by cytokinin-binding protein in the presence of cytokinin observed in our experiments gives evidence that this protein is really a receptor of cytokinin in barley leaves. On the basis of data obtained the following scheme of cytokinin action can be discussed (Fig. 2): Cytokinin is bound to some acceptors in cytoplasm of leaf cells, then acceptor-bound hormone penetrates the nucleus and affects the chromatin or chromatin-bound RNA-polymerase followed by RNA-polymerase activation. As a result, RNA and protein syntheses in cells are enhanced and, on the basis of this activation, the stimulation of physiological processes by the hormone is observed.

3.2 Effect of Cytokinin on Protein Synthesis

The effect of cytokinin on protein synthesis was investigated in more detail in our experiments on isolated pumpkin cotyledons [12-15]. They are very sensitive to exogenous cytokinin because the content of endogenous cytokinins of dry seeds has sharply decreased after cotyledon isolation [25].

An exogenous cytokinin, 6-benzylaminopurine (BAP), has stimulated cotyledon growth and development of subcellular structure providing cotyledon transformation

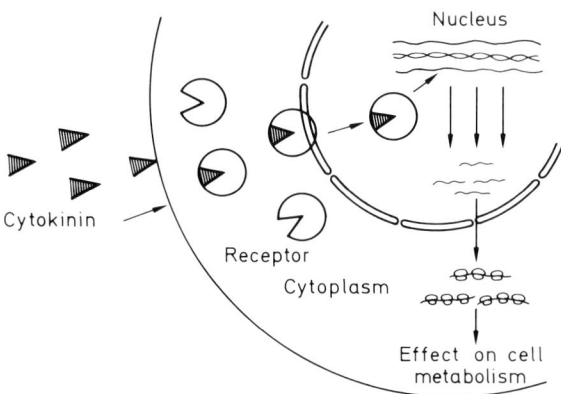

Fig. 2. A hypothetical model of cytokinin action on metabolism of leaf cells

from a storage organ to a green cotyledonary leaf [12, 26, 27]. Cytokinin promotes protein and fat mobilization, development of endoplasmic reticulum and mitochondria and development of proplastids to chloroplasts [12, 28, 29]. Accordingly cytokinin activates a number of enzymes, participating in a wide range of metabolic reactions in cotyledons [28-30]. We have shown these effects on endopeptidases providing degradation of the aleurone grain proteins, on pyrophosphatases capable of taking part in regulation of many syntheses, on malic enzyme participating in organic acid metabolism and on two photosynthetic enzymes, ribulosediphosphate carboxylase and phosphoenolpyruvate carboxylase [28, 30].

Stimulation by cytokinin of enzyme activities in pumpkin cotyledons is associated with protein synthesis because cycloheximide eliminated this cytokinin effect. Accordingly cytokinin promotes label incorporation into protein in isolated pumpkin cotyledons [31] (Fig. 3). This activation is partially due to increasing RNA synthesis in cotyledons which in its turn results in increasing "size" of protein-synthesizing machinery in the cells [14]. In addition the data obtained in our laboratory in 1973 [32] have shown that cytokinin not only increases ribosome number in the cells, but their protein-synthesizing activity as well. This conclusion was drawn from the experiments on protein synthesis in vitro by ribosomes isolated from cytokinin-treated and control cotyledons (Fig. 4). Cytokinin has markedly increased the rate of protein synthesis per unit of RNA, i.e., per ribosome. This result was obtained only with intact cytokinin-treated cotyledons. Cytokinin had no effect when it was added directly to the isolated ribosomes into the incubation medium for protein synthesis [33].

Fractionation of ribosomes in sucrose density gradient [15] has shown that BAP increased polysome/monosome ratio in cotyledons (Fig. 5). Hence cytokinin stimulated polysome formation in the cells. The higher percentage of polysomes in a ribosomal preparation from cytokinin-treated cotyledons explains the higher activity of these ribosomes in protein synthesis in vitro. The increase of polysome/monosome ratio can be detected in an hour after the cotyledon treatment by cytokinin (Fig. 6). It thus appears that polysome formation belongs to one of the most-rapid responses of cotyledons to cytokinin.

Polysome formation is known to occur either from de novo formed ribosomes and de novo synthesised mRNA, or from preformed ribosomes and mRNA stored as informosomes. To investigate the dependence of cytokinin-induced polysome formation

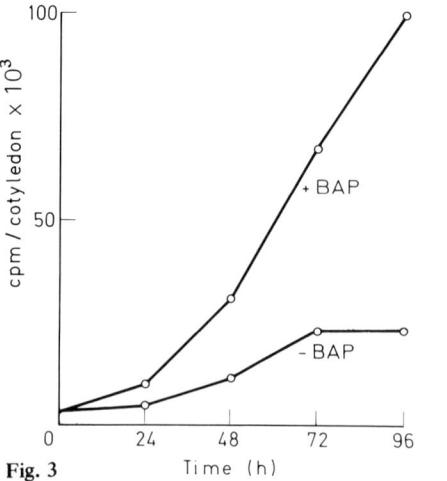 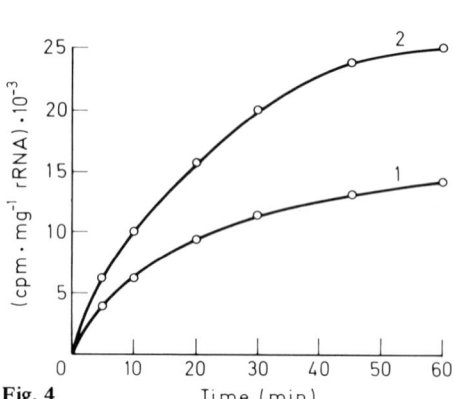

Fig. 3. [^{14}C]-leucine incorporation into proteins of pumpkin cotyledons. The cotyledons were isolated after 3-day-long seed germination on water in darkness. The isolated cotyledons were incubated in water or in *BAP* solution (10 mg/l) under light conditions [31]

Fig. 4. [^{14}C]-leucine incorporation into protein in vitro by ribosomes from excised pumpkin cotyledons. Cotyledons excised from etiolated 4-day-old seedlings were preincubated for 24 h in water in the dark, then incubated under illumination for 24 h in water (*1*) or in BAP (*2*) (10 mg/l) solution [32]

on RNA synthesis, cordycepine and α-amanitine were applied. Cordycepine is known to inhibit RNA synthesis and polyadenylation of mRNA. In our experiments cordycepine decreased the label incorporation into total RNA of cotyledons by 70%-75% but did not affect cytokinin-induced polysome formation. α-amanitin, an inhibitor of RNA-polymerase II, an enzyme providing mRNA synthesis, also has no effect on polysome formation (Table 6).

Hence hormone has induced polysome formation in cotyledons from preformed ribosomes and mRNA. It is an additional way of the cytokinin action on protein synthesis in the cell (Fig. 7). This process represents one of more rapid responses of cells to cytokinin.

This observation confirms the proposal by Fosket et al. of a posttranscriptional effect of cytokinin on translation [34].

On the basis of data discussed the conclusion can be drawn that cytokinin can take part in control of protein synthesis both at transcriptional and posttranscriptional levels.

It is important to discuss whether the posttranscriptional level of cytokinin action is independent of the transcriptional level. To consider this question the following data must be taken into account:

1. In our experiments with cordycepin and α-amanitin RNA syntheses in cotyledons diminished by 70%-80% and under these conditions cytokinin-induced polysome formation was not blocked. These results support the evidence that cytokinin stimu-

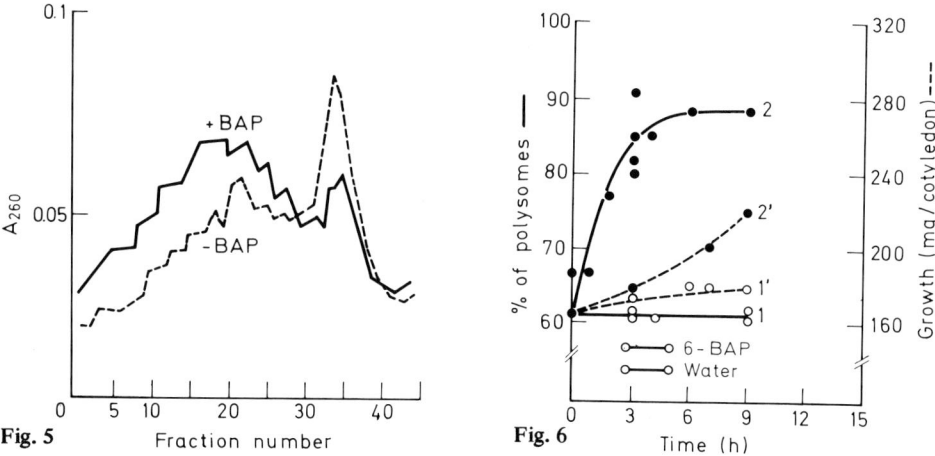

Fig. 5. Sedimentation profile of polysomes and monosomes from isolated pumpkin cotyledons in sucrose density gradient [15]. The cotyledons were isolated from 4-day-old etiolated seedlings, preincubated for 24 h in darkness on water and then incubated for 24 h in light on water or *BAP* (10 mg/l) solution [15]

Fig. 6. Effect of 6-benzylaminopurine (10 mg/l) on polysome percentage in ribosome preparation from isolated pumpkin cotyledons and on cotyledon growth. *Solid line* polysome percentage; *dotted line* growth; *open circles* cotyledons on water; *closed circles* cotyledons on *BAP* solution. Cotyledons were isolated from 4-day-old etiolated seedlings preincubated for 48 h in darkness on water and then incubated for 9 h in darkness on water or on *BAP* solution (10 mg/l). Procedures of ribosome preparation, fractionation in sucrose density gradient and measurement of polysome percentage are described in [15]

Table 6. Effect some inhibitors of RNA synthesis on cytokinin-induced formation of polysomes in isolated pumpkin cotyledons

Treatment		Polysomes, percentage of total ribosome content
Water		68
Cordycepine 50 mg/l		67
Cordycepine 400 mg/l		70
Benzylaminopurine	10 mg/l	83
Benzylaminopurine + cordycepine	10 mg/l + 50 mg/l	84
or	400 mg/l	86
Benzylaminopurine + α-amanitine	10 mg/l + 5 mg/l	84
or	20 mg/l	82

The cotyledons were isolated from 4-day-old etiolated seedlings and preincubated for 48 h in darkness on water, then transferred to various incubation media in darkness for 6 h. Ribosomes prepared from cotyledons were fractionated in sucrose density gradient (Details in [15])

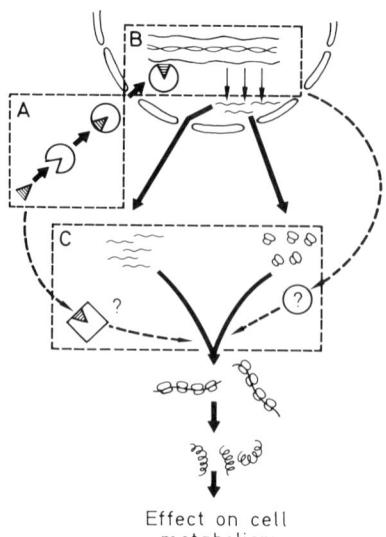

Fig. 7. A hypothetical model of cytokinin action on protein synthesis in cytokinin-sensitive plant cells. *A* interaction between cytokinin and receptor; *B* effect of cytokinin-receptor complex on transcription; *C* effect of cytokinin on polysome formation from preformed ribosomes and mRNA

lated polysome formation from preformed mRNA and ribosomes, but they do not exclude a possibility that 20%-30% of RNA synthesis was enough to provide the primary cytokinin action at transcriptional level.

2. Cytokinin-induced stimulations of polysome formation and label incorporation into RNA are detected simultaneously.

This fact also confirms a possibility of the primary effect of cytokinin at transcriptional level. Cytokinin can activate the synthesis of specific RNA's necessary for syntheses of some proteins controlling the translation of preformed mRNA's. Such a hypothetical model is demonstrated in Fig. 7.

Another possibility represents a direct, transcription-independent cytokinin effect on polysome formation.

In this case some cytokinin-binding proteins, associated with cytokinin, are supposed to induce polysome formation (Fig. 7). But at present it is not yet possible to favor one of these mechanisms of cytokinin action on polysome formation. It thus appears that we cannot now give a convincing answer to the question whether cytokinin effect at posttranscriptional level is independent of transcriptional level. To clarify this question is a very intriguing task for future investigations.

References

1. Wollgiehn R (1967) In: Symp Soc Exp Biol 21: 231
2. Osborne D (1962) Plant Physiol 37: 595
3. Kursanov AL, Kulaeva ON, Sveshnikova IN, Popova EA, Bolyakina YuP, Klyachko NL, Vorobyova IP (1964) Fiziol Rast 11: 838
4. Kulaeva ON, Fedina AB, Selivankina Sju, Kursanov AL (1967) Fiziol Rast 14: 972
5. Srivastava BJS (1968) In: Wightman F, Statterfield G (eds) Biochemistry and physiology of plant growth substances. Runge Press, Ottawa, p 1479

6. Kulaeva ON (1973) Cytokinins, their structure and function. Nauka, Moscow
7. Romanko EG, Selivankina Syu, Kuroyedov VA (1978) Fiziol Rast 25: 1199
8. Selivankina Syu, Romanko EG, Kuroyedov VA, Kulaeva ON (1979) Fiziol Rast 26: 41
9. Kulaeva ON, Selivankina Syu, Romanko EG, Nikolaeva MK, Nichiporovich AA (1979) Fiziol Rast 26: 1016
10. Selivankina Syu, Romanko EG, Kulaeva ON (1980) Fiziol Rast 27: 560
11. Kulaeva ON (1980) Stud Biophys 76: 167
12. Mikulovich TP, Khokhlova VA, Kulaeva ON, Svechnikova IN (1971) Fiziol Rast 18: 98
13. Yakovleva LA, Klyachko NL, Kulaeva ON (1977) Mol Biol 11: 868
14. Mikulovich TP, Wollgiehn R, Khokhlova VA, Neumann D, Kulaeva ON (1978) Biochem Physiol Pflanz 172: 101
15. Klyachko NL, Ananiev E, Kulaeva ON (1979) Physiol Veg 17: 501
16. Kulaeva ON, Selivankina Syu, Kuroyedov VA (1971) Fiziol Rast 18: 746
17. Selivankina Syu, Kuroyedov VA, Kulaeva ON (1972) Fiziol Rast 19: 508
18. Szweykowska A (1978) In: International conference on plant growth substances. Abstr, p 62, Prague
19. Burkhanova EA, Fedina AB, Dmitrieva GN, Kulaeva ON (1980) Dokl Acad Nauk 252: 488
20. Lejohn HB, Cameron L (1973) Biochem Biophys Res Commun 54: 1053
21. Takegami T, Joshida K (1977) Plant Cell Physiol 18: 337
22. Fox JE, Erion JL (1975) Biochem Biophys Res Commun 64: 694
23. Polya GM, Davis AW Planta (1978) 139: 139
24. Polya GM, Bowman JA (1979) Plant Physiol 64: 387
25. Engelbrecht L, Rybicka H, Kulaeva ON (1976) Biochem Physiol Pflanz 169: 317
26. Banerji D, Laloraya MM (1965) Naturwissenschaften 52: 349
27. Kursanov AL, Kulaeva ON, Mikulovich TP (1969) Fiziol Rast 16: 680
28. Karavaiko NN, Krawiarz K, Khokhlova VA, Kulaeva ON (1978) Fiziol Rast 25: 803
29. Khokhlova VA, Karavaiko NN, Podergina TA, Kulaeva ON (1978) Cytologia XX: 1033
30. Karavaiko NN, Ohmann E, Kulaeva ON (1975) Fiziol Rast 22: 903
31. Klyachko NL, Jakovleva LA, Kulaeva ON (1973) Fiziol Rast 20: 1219
32. Klyachko NL, Yakovleva LA, Kulaeva ON (1973) Dokl Acad Nauk 211: 1235
33. Yakovleva LA, Klyachko NL, Kulaeva ON (1975) Fiziol Rast 22: 856
34. Fosket DE, Tepfer DA (1978) In Vitro 14: 63

Covalent Insertion of N^6-Benzyladenine into rRNA Species of Tobacco Cells

J.-P. Jouanneau and B. Teyssendier de la Serve[1]

1 Introduction

The fact that cytokinins were recognized as adenine analogs early raised the question of their incorporation into plant RNA (Fox, 1966). The conversion of exogenous cytokinin bases to their corresponding nucleoside-5′-triphosphates was demonstrated later by Laloue et al. (1974); it provided biochemical support for the possible insertion of cytokinin nucleotides into polyribonucleotide chains.

Recently, two lines of experiments, performed with different methods by Skoog and coworkers and by our group, led to the evidence of direct incorporation of exogenous cytokinins into various RNA fractions. A detailed review on this subject has already been published (Péaud-Lenoël and Jouanneau, 1980). We reported previously that exogenous bzl^6Ade was covalently inserted into tobacco RNA and that covalently bound bzl^6Ade nucleotides did not arise from a transbenzylation reaction onto adenylic residues already present in preformed RNA transcripts (Jouanneau et al., 1977; Teyssendier de la Serve and Jouanneau, 1979). This result and other evidence provided by Walker et al. (1974) demonstrated that the mechanism of incorporation of bzl^6Ade into RNA was not related to the isopentenylation or other hypermodifications of adenine located adjacent to the 3′-end of the anticodon of defined tRNA molecules (for reviews see: Hall, 1970; Péaud-Lenoël and Jouanneau, 1980): i^6A was shown to be synthesized by the post-transcriptional transfer of the isopentenyl group of Δ^2-isopentenylpyrophosphate to the appropriate adenosyl residue in preformed tRNA (Kline et al., 1969).

The direct incorporation of exogenous bzl^6Ade and kinetin was further shown to occur in the tRNA fraction and in ribosomal RNA species of tobacco callus tissues and cell suspensions. The frequencies of incorporation according to different reports (Table 1) varied according to the experimental conditions but remained in the same order of magnitude for a given RNA fraction. For the calculation of values recorded in Table 1, it was assumed that bzl^6Ade nucleotide was inserted in all RNA molecules whereas the real number of acceptor RNA molecules remained unknown. Thereby, the estimated incorporation levels should be considered as minimal values. Compared to the frequencies of cytokinin incorporation in tRNA and 5S RNA, the levels of in-

---

Abbreviations: bzl^6Ade: N^6-benzyladenine; i^6A: N^6-(Δ^2-isopentenyl)adenosine

[1] Laboratoire de Biochimie Fonctionnelle des Plantes (ER 104), Faculté des Sciences de Luminy, 13228 Marseille Cedex 9, France

Table 1. Frequencies[a] of insertion of exogenous cytokinins into tobacco RNA fractions

Incubation conditions	RNA fractions			References
	25S rRNA	18S rRNA	4S + 5S RNA	
Callus tissues (3 weeks)				
0.067 μM bzlAde	$3.8 \ 10^5$	$3.8 \ 10^5$	$1.3 \ 10^6$	(Armstrong et al., 1976)
0.12 μM kinetin	$1.1 \ 10^5$	$1.1 \ 10^5$	$2.9 \ 10^5$	(Murai et al., 1977)
0.12 μM kinetin	$7 \ \ 10^4$	$6 \ \ 10^4$		(Murai et al., 1978)
Cell suspensions (10-16 h)				
0.1 μM bzlAde	$2.8 \ 10^5$	$2.8 \ 10^5$	$1.1 \ 10^6$	(Teyssendier de la Serve and Jouanneau, 1979)
0.4 μM bzlAde	$6.7 \ 10^4$	$6.7 \ 10^4$	$2.1 \ 10^5$	

[a] Number of bases for one inserted cytokinin

corporation into 18S and 25S rRNA were three to four times higher and very close together. When tobacco cell suspensions were incubated with 0.4 μM exogenous bzl^6-Ade (optimal concentration for cell division), the observed incorporation rates (Table 1) corresponded to the occurrence of one bzl^6Ade per twenty 25S rRNA and thirty five 18S rRNA molecules compared to one inserted bzl^6Ade per 2,800 tRNA molecules. 5.8S RNA was probably labeled at the same rate that 25S rRNA from which it was dissociated by denaturation; the insertion rates were found equal in 5S RNA and in the tRNA fraction (Teyssendier de la Serve and Jouanneau, 1979). We have no information on the levels of cytokinin incorporation into polydisperse RNA or organelle RNA fractions.

It appeared from the above experiments that the incorporation of exogenous cytokinins was not restricted to a single RNA species and that cytokinins were preferentially recovered from cytoplasmic 18S and 25S rRNA. To investigate further the specificity and the possible physiological significance of this incorporation, we considered as important to determine the localization of the cytokinin nucleotides into rRNA molecules and the mechanism of their insertion.

2 Possible Models for Cytokinin Incorporation into RNA

Apart from the evidence of a direct insertion process, the above experiments did not provide further information on the mechanism of incorporation. At least it might be assumed that the incorporation was an early process since high molecular weight RNA fractions, tentatively identified as rRNA precursors, were significantly labeled by bzl^6-Ade (Teyssendier de la Serve and Jouanneau, 1979).

By analogy with known events of RNA metabolism, several possible mechanisms might be assumed for cytokinin incorporation. For example, mismatches of non-complementary bases or base analogs are known to occur during transcription process (Chamberlin, 1974). On the other hand, 3′-polyadenylation and 5′-capping of mRNA or the CCA-end addition to tRNA are examples of post-transcriptional terminal addition of nucleotides to RNA molecules (Perry, 1976). Another post-transcriptional insertion mechanism is the transglycosylase mediated replacement of base G by a precursor of the base Q (a highly modified guanine derivative) in the first position of the anticodon of specific tRNA's (Okada et al., 1979). This base exchange results in the covalent insertion of the modified base Q precursor in the tRNA anticodons without cleavage of the phosphodiester bonds in the polynucleotide chains.

According to the above incorporation models, different locations of the cytokinin nucleotide into RNA molecules might be expected. The question was raised whether the incorporation took place
— at random or in a large number of distinct loci as the result of mismatching of complementary bases during transcription.
— or in one (or a small number of) specific sequence, presumably by a post-transcriptional insertion mechanism.

The sequence specificity of the cytokinin insertion was therefore investigated by studying the distribution of labeled bzl^{6}Ade in the oligonucleotide patterns of tobacco 25S and 18S rRNA species.

3 Bzl^6 Ade Nucleotide Distribution in Tobacco rRNA Sequences

3.1 Analysis of bzl^6 Ade Distribution in rRNA

In order to obtain rRNA and resulting oligonucleotides specifically labeled by the cytokinin nucleotide, an exogenous cytokinin precursor labeled on the side chain was utilized: for this purpose, [p-^{3}H]bzl^{6}Ade tritiated on the benzyl ring with high specific radioactivity (26 Ci/mmol) was synthesized according to Sussman and Firn (1976). Cytokinin-requiring tobacco cell suspensions were incubated in the presence of this cytokinin at the optimal concentration for cell division (0.4 μM), for 10h during the lag phase which precedes the hormone-induced mitotic response of cells: the supply of cytokinin, even restricted to this lag period, was previously shown to trigger one subsequent division cycle in control cell suspensions (Jouanneau and Tandeau de Marsac, 1973). RNA, specifically labeled by [p-^{3}H]bzl^{6}Ade, was extracted from the incubated cells, and rRNA were purified as described by Teyssendier de la Serve and Jouanneau (1979).

The procedure used for the analysis of rRNA oligonucleotides is outlined in Table 2. [p-^{3}H]bzl^{6}Ade-labeled 25S and 18S rRNA were separately submitted to digestion by pancreatic RNase or (and) RNase T1 or RNase U2 of different splitting specificities. Under controlled conditions, these endonucleases split phosphate ester bonds of polyribonucleotide chains, respectively between a pyrimidine or a guanine or a purine nucleoside-3′-P and the 5′-hydroxyl group of the adjacent nucleotide; they yield defined digestion products, which structures are recalled in Table 2. After the enzymatic digestion, a mild acid hydrolysis step was required to cleave the end cyclic- phosphodiester bond of mono- and oligonucleotides-2′, 3′-cyclic phosphate formed as inter-

Table 2. Analysis of oligonucleotides from [p-^{3}H]bzl^6 Ade-labeled rRNA

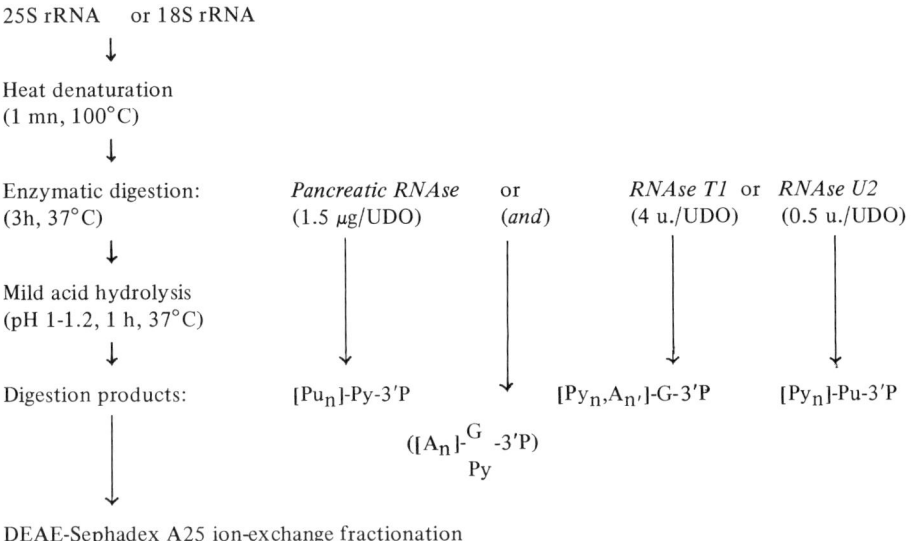

mediary reaction products by the above RNases. The resulting mixtures of mono- and oligonucleotides-3'-P were fractionated by DEAE-Sephadex A25 ion-exchange chromatography, according to their net charge, i.e., to their chain length.

3.2 Distribution of bzl^6 Ade in Oligonucleotide Patterns of 25S and 18S rRNA

The oligonucleotide patterns resulting from the digestion of labeled rRNA species by pancreatic RNase are shown in Fig. 1. Similar scannings of unlabeled oligonucleotides were obtained, allowing the separation of oligonucleotide families of common chain length. On each pattern, the radioactivity detecting the nucleotide of bzl^6 Ade was distributed in a series of distinct peaks. These peaks were eluted later than their unlabeled homologs, thus suggesting an over-binding of the bzl^6 Ade-containing oligonucleotides on the exchanger. However, radioactive peaks were associated to every oligonucleotide family of a given chain length with the exception of the mononucleotide fraction. The distribution of the bzl^6 Ade nucleotide throughout the series of oligonucleotides was also observed after the digestion of rRNA by RNase T1, by both pancreatic and T1 RNases or by RNase U2.

These results clearly demonstrated that the nucleotides of bzl^6 Ade were inserted in a large number of distinct sites and not in a unique sequence of any analyzed rRNA.

In the oligonucleotide patterns, the distribution of the labeled bzl^6 Ade did not exactly parallel the UV absorbance profiles: an odd variation of the abundance ratio of one radioactive peak to the relevant absorbance peak of same chain length was observed on each fractionation diagram (see Figs. 1, 3A and 4A). A striking variance was noticed at the level of the mononucleotide fraction of T1 or (and) pancreatic RNase digests which did not contain significant amounts of bzl^6 Adenosine-3'-P. On

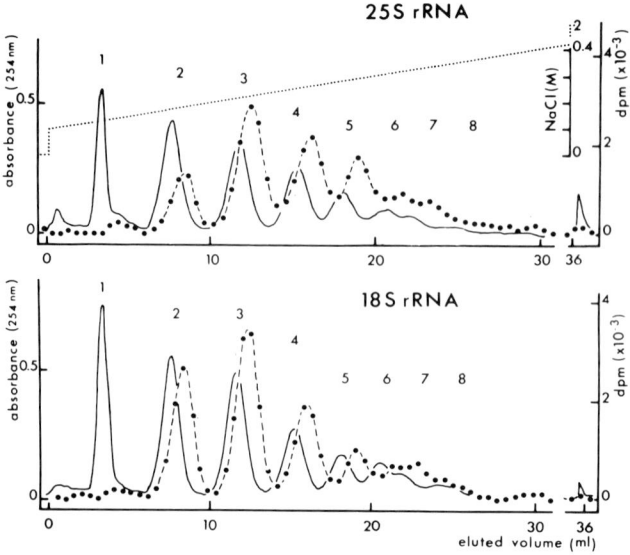

Fig. 1. Oligonucleotide patterns of [p-^{3}H]bzl^6 Ade-labeled 25S and 18S rRNA after digestion by pancreatic RNase.
25S and *18S rRNA* (6.5 A_{260} units, 5 10^4 dpm) were digested by pancreatic RNase as outlined by Table 2. Digests were fractionated on DEAE-Sephadex A25 columns (33 x 0.2 cm) by elution with a linear 0.1 to 0.45 M NaCl gradient (total vol.: 34 ml) containing 7 M urea and 10 mM Tris-Cl, pH 7.7 (Rushizky et al., 1964). *Solid lines* 254 nm absorbance of column effluents. *Dashed lines* [^{3}H]radioactivity in 0.54 ml fractions. *Dotted line* NaCl gradient concentration. *Numbers 1 to 8* mono- and oligonucleotide chain length (n)

the contrary, the mononucleotide fraction of RNase U2 digests was labeled by the bzl^6-Ade nucleotide (not shown). Thus it appeared likely that the internucleotidic bond adjacent to the 3′-P of the bzl^6 Ade nucleotide could be cleaved by RNAse U2, whereas it was resistant to hydrolysis by pancreatic or T1 RNases. Therefore, it was concluded that the bzl^6 Ade nucleotide could not occupy the 3′-terminal position of the radioactive oligonucleotides resulting from RNA digestion by pancreatic or T1 RNases.

3.3 Model of Distribution of bzl^6 Ade at Random in RNA Chains

The insertion of the bzl^6 Ade nucleotide at random in RNA chains was the simplest model to account for the observed distribution of radioactive oligonucleotides in 25S and 18S rRNA patterns. This model was based on the following assumptions:

1. Oligonucleotides are distributed according to a random sequence of the four usual bases A,U,C,G in RNA chains.
2. Bzl^6 Ade can occupy any position in the native RNA molecules.
3. After T1 or pancreatic RNase digestion, bzl^6 Ade is located in any position except the 3′-terminal position of any labeled oligonucleotide. According to this model, the specific radioactivity of an oligonucleotide family will vary as $\frac{n-1}{n}$ if n is the chain length of this family.

3.4 Comparison of the Experimental Distributions of Oligonucleotides to the Model of bzl^6 Ade Distribution

The above model provided for graphic analysis of the information derived from the oligonucleotide fractionations after pancreatic or (and) T1 RNase treatments. The experimental distributions of unlabeled and radioactive oligonucleotides recorded from a pancreatic RNase digest of 25S rRNA (see Fig. 1) have been compared to the relevant calculated model distributions: Fig. 2A shows the comparison between the experimental abundance of constituent mononucleotides contained in every oligonucleotide family, as recorded by its corrected absorbance, and the calculated values expected from a random distribution of the pancreatic RNase splitting sites U and C in RNA. A good agreement was observed between the experimental and the model distributions of oligonucleotide families in function of their chain length n (Fig. 2A). Such an agreement could be expected in so far as rRNA are long polynucleotide chains but this result did not preclude local fluctuations in the base sequence as compared to a random distribution of the four usual bases. At least, the agreement between the two distributions indicated that the conditions of enzymatic digestion by pancreatic RNase were optimal for the specific cleavage of RNA at U and C sites.

Since the experimental amounts of mononucleotides in every oligonucleotide family and the total radioactivity present in the oligonucleotide mixture were known, it was possible to calculate the radioactivity which would be assigned to a given family if bzl^6 Ade was randomly located at any position except the 3'-terminal position of the oligonucleotides. As shown by Fig. 2B, the experimental distribution of the label did not fit with the model diagram: a low radioactivity was noticed in the dinucleotide

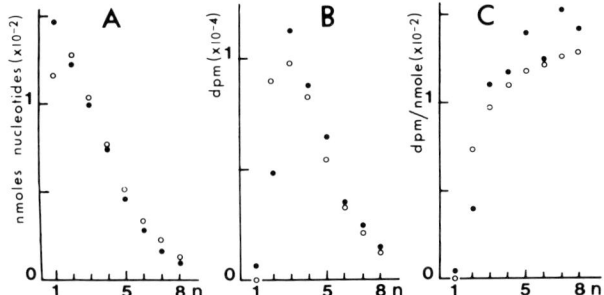

Fig. 2A-C. Comparison of the experimental distributions of oligonucleotides from a pancreatic digest of bzl^6 Ade-labeled 25S rRNA (see Fig. 1) with the model distributions calculated on the basis of a random insertion of bzl^6 Ade in RNA. **A, B** and **C.** *Abscissae* chain length (n) of oligonucleotide families. **A** *Ordinates* constituent mononucleotides (nmol) in each unlabeled oligonucleotide family; (•) experimental values; (o) theoretical values calculated according to the model of random distribution of the usual bases in a polyribonucleotide chain of infinite length with the same base composition as 25S rRNA (assumption no. 1 of model). **B** *Ordinates* bzl^6 Ade radioactivity in each oligonucleotide family; (•) experimental values; (o) calculated values of the radioactivity expected in each oligonucleotide family according to the assumptions no. 2 and 3 of the model. **C** *Ordinates* specific radioactivity of each oligonucleotide family; (•) experimental values (ratio of experimental radioactivity from diagram **B** to the experimental number of mononucleotides according to diagram **A**); (o) calculated values proportional to $\frac{n-1}{n}$ (ratio of calculated radioactivity from diagram **B** to the experimental number of mononucleotides of diagram **A**)

fraction whereas a higher radioactivity than expected was observed in oligonucleotides of higher chain length. Likewise, the specific radioactivity of oligonucleotide families was not proportional to the ratio $\frac{n-1}{n}$ predicted by the model of random distribution of bzl^6 Ade in RNA (Fig. 2C).

When the 18S rRNA oligonucleotides resulting from a pancreatic RNase digestion were fractionated, a specific radioactivity higher than predicted by the random insertion model was observed in the tri- and tetranucleotide fractions (not shown).

The oligonucleotide pattern of 25S rRNA digested by RNase T1 is shown by Fig. 3A. As expected from the different proportions of G and pyrimidine bases in 25S rRNA, the absorbance fractionation diagram was different from the scanning diagram after pancreatic RNase digestion. After RNase T1 digestion, the abundance of unlabeled oligonucleotides in each family was close to the abundance predicted from a random distribution of G in the 25S rRNA molecule (Fig. 3B). Again, the distribution of labeled bzl^6 Ade in the oligonucleotide series did not parallel the distribution of unlabeled oligonucleotides (Fig. 3A). When the experimental distribution of radioactivity was compared to the model of random insertion of bzl^6 Ade (Fig. 3C and D), a rather low specific radioactivity was observed in oligonucleotides of chain length 3-5; this low radioactivity was balanced by a specific radioactivity higher than expected at the level

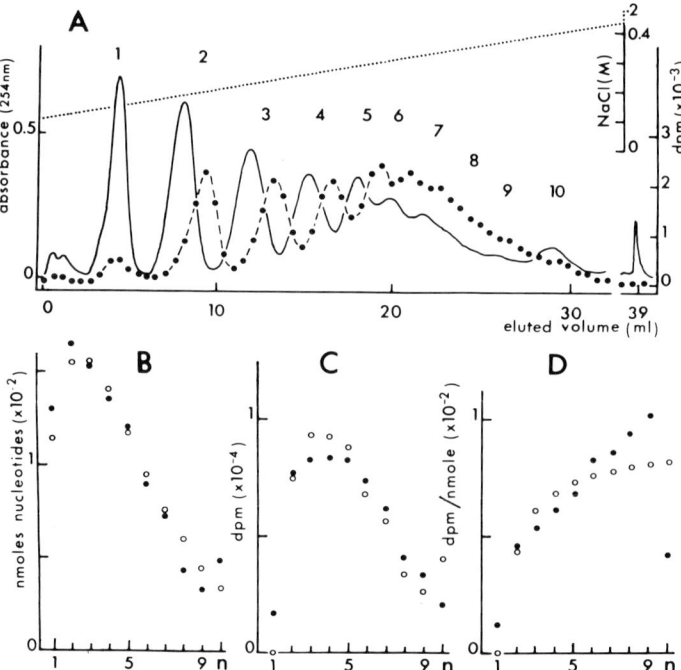

Fig. 3A-D. Oligonucleotide patterns resulting from the digestion of [p-^{3}H]bzl^6 Ade-labeled 25S rRNA by RNase T1. **A** DEAE-Sephadex chromatography of the digestion products of RNA (10 A$_{260}$ units, 5.8 10^4 dpm): see legend of Fig. 1; digestion by RNase T1: see Table 2. **B, C** and **D** Graphic analyses of the above oligonucleotide patterns according to the method described in legend of Fig. 2; *open circles* theoretical values, *black circles* experimental values

of oligonucleotides of chain length 6-9. An odd oligonucleotide distribution, as compared to the bzl^6Ade model distribution, was also obtained from the analysis of T1 RNase digests of 18S rRNA.

When the oligonucleotides resulting from the combined digestion of 25S rRNA by pancreatic and T1 RNases were fractionated (Fig. 4), bzl^6Ade radioactivity was found to be associated mainly to the dinucleotide fraction (Fig. 4C). The specific radioactivity of oligonucleotides of chain length larger than 2 was lower than was predicted from the model distribution (Fig. 4D).

According to the above analyses, the model was at variance with the experimental distributions of bzl^6Ade radioactivity in rRNA oligonucleotide fractions obtained after different selective enzymatic digestions. The first question was whether such a variance could be explained by an interference of the inserted bzl^6Ade with the activity or the splitting specificity of RNases. If the enhanced recovery of bzl^6Ade in high chain length oligonucleotides from T1 and pancreatic RNase digests was the result of impaired enzymatic digestions, it would likewise have been observed after the combined digestion of RNA by both enzymes. On the contrary, the specific radioactivity of trinucleotides and higher chain length oligonucleotides was smaller than predicted by the model distribution of bzl^6Ade (Fig. 4). Moreover, in the case of RNase T1 digests, it was observed that the odd distribution of bzl^6Ade was maintained after the use of a five to tenfold higher enzyme concentration than that used in the experiment of Fig. 3.

In view of these observations, it seems unlikely that the recorded variance in bzl^6-Ade distribution may result from incomplete enzymatic hydrolyses of RNA; this variance rather suggests that bzl^6Ade might be distributed in rRNA chains with an abundance correlated to the neighboring base sequence(s). The oligonucleotide structures favorable or unfavorable to the presence of bzl^6Ade nucleotide are summarized

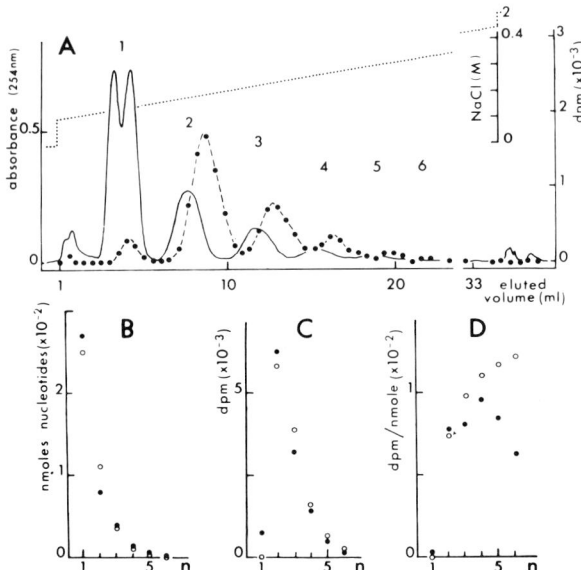

Fig. 4A-D. Oligonucleotide patterns of [p-^{3}H]bzl^6Ade-labeled 25S rRNA after combined digestion by mixed pancreatic and T1 RNases. A DEAE-Sephadex chromatography of the digestion products of RNA (4.2 A_{260} units, 1.2 10^4 dpm): see legend of Fig. 1; combined enzymatic digestion: see Table 2. B, C and D Graphic analyses of the results of the above oligonucleotide diagrams according to the method of Fig. 2; *open circles* theoretical values, *black circles* experimental values

in Table 3. The presence of bzl^6 Ade is favored in RNase T1 oligonucleotides of chain length higher than 6. If one considers the low radioactivity of the dinucleotide fraction of pancreatic RNase digests, bzl^6 Ade does not seem to be frequently inserted between two pyrimidine bases. On the contrary, its presence is favored in sequences Py-(Pu$_n$, BA)-Py traced by trinucleotides and oligonucleotides of higher chain length in the same digests. As suggested by the low radioactivity of oligonucleotides (A_n, BA)-$\frac{G}{Py}$ from combined T1 and pancreatic RNases digests, bzl^6 Ade appears to be rare in the neighborhood of A. Likewise, the neighborhood of G and a pyrimidine base might be considered as a propitious sequence to the presence of bzl^6 Ade nucleotide in RNA chains.

Table 3. Possible oligonucleotide sequences favorable or unfavorable to bzl^6 Ade [a] insertion into rRNA [b]

Enzymatic digestion	Favorable sequences	Unfavorable sequences
T1 RNase	Oligonucleotides $6 \leq n \leq 9$	Oligonucleotides $2 \leq n \leq 5$
Pancreatic RNase	Py-/ (Pu, BA)-Py/– or Py-/ (Pu$_n$, BA)-Py /–	Py-/ BA-Py /–
T1 + pancreatic RNases	$\left(\begin{array}{cc} G & G \\ -/BA-\\ Py & Py \end{array}\right)/-$	$\left(\begin{array}{cc} G & G \\ -/(A_n, BA)-\\ Py & Py \end{array}\right)/-$

[a] BA. [b] From analyses of oligonucleotide distributions of Figs. 2, 3 and 4
Bars represent enzymatic splitting sites according to the RNases specificities
In brackets: undetermined order of the bases

4 Conclusions

The question we attempted to answer in the present work was whether exogenous cytokinins were incorporated into selective nucleotide sequences of tobacco rRNA molecules. The oligonucleotide fractionations of partial RNase digests of bzl^6 Ade-labeled rRNA led to the conclusion that the cytokinin nucleotide was inserted in a large number of distinct sites and not in a unique sequence of 25S and 18S rRNA. Another conclusion of these experiments was that the observed distribution of bzl^6 Ade nucleotides in rRNA oligonucleotide fractions was at variance with a simplifying model of cytokinin incorporation at random in RNA sequences. This result was consistent with the statement that bzl^6 Ade nucleotides, although inserted in many distinct loci, might be located in rRNA chains with preference for some nucleotide sequences.

Our first attempt to correlate the abundance of inserted bzl^6 Ade to the neighboring base sequence was based on the analysis of the cytokinin distribution in the oligonucleotide series obtained from different selective enzymatic digestions of rRNA. This analysis suggested that a guanine and a pyrimidine base might be a neighborhood favorable to the presence of the bzl^6 Ade nucleotide in rRNA chains. More detailed study of rRNA oligonucleotide patterns than that we have achieved would be necessary in order to identify the nucleotide sequence(s) in which the incorporation of cytokinin is favored. A relevant question would be to determine whether the incorporation of bzl^6 Ade results from the addition of the cytokinin nucleotide to RNA polynucleotide chains or from the replacement of one preferential base or every base by the cytokinin.

The above results raised again the problem of the mechanism by which exogenous cytokinins were incorporated into RNA. As recalled in Section 2, the location of bzl^6-Ade in one or a few specific sequence(s) might be expected from the known mechanisms of post-transcriptional base insertion into RNA. Although the rRNA oligonucleotide patterns did not reveal such a specific location, bzl^6 Ade appeared likely to be distributed in rRNA chains more frequently in some sequences than in others. In this respect, Murai et al. (1978) reported that kinetin was more frequent in oligonucleotide fractions containing a terminal pyrimidine than in oligonucleotides with a terminal G prepared by combined pancreatic and T1 RNases digestion of rRNA. Whether the cytokinin incorporation in rRNA might result in part from a post-transcriptional insertion mechanism remains unknown.

On the other hand, the location of bzl^6 Ade in many distinct sites of rRNA chains was compatible with the statement that the cytokinin might be incorporated as the result of base mismatching during RNA transcription. If the incorporation resulted from transcriptional errors, it would remain to be determined for which base(s) is bzl^6 Ade mismatched and to explain by which mechanism the insertion rate could be correlated with the neighboring base sequence. In vitro experiments would be useful to answer the above questions and to identify the mechanism(s) of the incorporation of exogenous cytokinins into RNA.

Acknowledgments. This work was supported by the Centre National de la Recherche Scientifique (E.R. 104) and by the Délégation Générale à la Recherche Scientifique et Technique (contract no. 79-5-059). The authors are indebted to Dr. C. Péaud-Lenoël for collaboration and constant interest in this work. They acknowledge J.C. Gandar and M. Laloue for collaboration and M.C. Durand for competent technical assistance.

References

Armstrong DJ, Murai N, Taller BJ, Skoog F (1976) Incorporation of cytokinin N^6-benzyladenine into tobacco callus transfer ribonucleic acid and ribosomal ribonucleic acid preparations. Plant Physiol 57: 15-22

Chamberlin MJ (1974) Bacterial DNA-dependent RNA polymerase. In: Boyer PD (ed) The enzymes, vol X, 3rd edn. Academic Press, London New York, p 333

Fox JE (1966) Incorporation of a kinin, N^6-benzyladenine into soluble RNA. Plant Physiol 41: 75-82

Hall RH (1970) N^6-(Δ^2-isopentenyl)adenosine: chemical reactions, biosynthesis, metabolism and significance to the structure and function of tRNA. Prog Nucleic Acid Res Mol Biol 10: 57-86

Jouanneau JP, Tandeau de Marsac N (1973) Stepwise effects of cytokinin activity and DNA synthesis upon mitotic cycle events in partially synchronized tobacco cells. Exp Cell Res 77: 167-174

Jouanneau JP, Gandar JC, Péaud-Lenoël C, (1977) On the presence of benzyladenine nucleotide in the RNA of tobacco cells. Plant Sci Lett 9: 77-87

Kline LK, Fittler F, Hall RH (1969) N^6-(Δ^2-isopentenyl)adenosine. Biosynthesis in transfer ribonucleic acid in vitro. Biochemistry 8: 4361-4371

Laloue M, Terrine C, Gawer M (1974) Cytokinins: formation of the nucleoside-5'-triphosphate in tobacco and *Acer* cells. FEBS Lett 46: 45-50

Murai N, Taller BJ, Armstrong DJ, Skoog F, Micke MA, Schnoes HK (1977) Kinetin incorporated in tobacco callus ribosomal RNA and transfer RNA preparations. Plant Physiol 60: 197-202

Murai N, Armstrong DJ, Taller BJ, Skoog F (1978) Distribution of incorporated synthetic cytokinins in ribosomal RNA preparations from tobacco callus. Plant Physiol 61: 318-322

Okada N, Noguchi S, Kasai H, Shindo-Okada N, Ohgi T, Goto T, Nishimura S (1979) Novel mechanism of post-transcriptional modification of tRNA. Insertion of bases of Q precursors into tRNA by a specific tRNA transglycosylase reaction. J Biol Chem 254: 3067-3073

Péaud-Lenoël C, Jouanneau JP (1980) Presence and possible functions of cytokinins in RNA. In: Skoog F (ed) Proceedings in life sciences: 10th Int Conf Plant Growth Substances. Madison Wisconsin USA, 1979. Springer, Berlin Heidelberg New York, p 129

Perry RP (1976) Processing of RNA. Annu Rev Biochem 45: 605-629

Rushizky GW, Bartos EM, Sober HA (1964) Chromatography of mixed oligonucleotides on DEAE-Sephadex. Biochemistry 3: 626-629

Sussman MR, Firn R (1976) The synthesis of a radioactive cytokinin with high specific activity. Phytochemistry 15: 153-155

Teyssendier de la Serve B, Jouanneau JP (1979) Preferential incorporation of an exogenous cytokinin, N^6-benzyladenine, into 18S and 25S ribosomal RNA of tobacco cells in suspension culture. Biochimie 61: 913-922

Walker GC, Leonard NJ, Armstrong DJ, Murai N, Skoog F (1974) The mode of incorporation of 6-benzylaminopurine into tobacco callus transfer ribonucleic acid. A double labeling determination. Plant Physiol 54: 737-743

Section 5
Cytokinins and Chloroplast Development

Plastids and the Origin of Cytokinin Autonomy in Tobacco Tissue Cultures

M. Kamínek, V. Hadačová, and J. Luštinec[1]

1 Introduction

Investigation of the functions of individual plant cell structures gave interesting results relating to the relative autonomy of the plastids in the plant cell. It has been found that plastids contain their own DNA and corresponding protein synthetic apparatus. This apparatus differs from that in the cytoplasma in that the polypeptide are synthesized on 70S ribosomes. Some enzymes are formed only in the plastids and their synthesis is inhibited by antibiotics which act specifically on 70S ribosomes (see Schiff, 1973). The autonomy of plastids is limited and intracellular regulatory systems exist which coordinate their replication and development with that of the plant cell. In this respect it is of interest to note that the progress in experimental morphology in the 1920's inspired plant physiologists to study mechanisms which coordinated the development of organs in intact plants. The progress in molecular biology now enables the mechanisms responsible for the integrity of the plant cell to be investigated at the cellular level. We wish to report that cytokinins also participate in this intracellular regulatory system.

2 Effect of Cytokinins on Plastid Development

The role of cytokinins in chloroplast development was recently reviewed by Parthier (1979) who summarized their stimulating effects on chloroplast differentiation and chlorophyll synthesis in several plant systems including tobacco tissue and cell cultures. According to Parthier cytokinins exhibit a relatively low stimulation of the cytoplasmatic enzymes and a marked stimulation of plastid enzyme formation. In two syssystems, i.e., in some tobacco tissue and cell cultures (Stetler and Laetsch, 1965; Seyer et al., 1975) and in leaves of intact bean plants (Naito et al., 1979) exogenously applied cytokinins induced chloroplast differentiation and replication without affecting cell division. According to Lescure (1978) cytokinin stimulates specifically the synthesis of some plastid polypeptides in cultured tobacco cells during cytokinin-induced differentiation of chloroplasts.

Abbreviations: A_S — cytokinin autonomous strains induced by streptomycin; $A_{2,4-D}$ — -cytokinin autonomous strain induced by 2,4-D; D — cytokinin-and auxin-depent strain of tobacco callus.

[1] Institute of Experimental Botany, Czechoslovak Academy of Sciences, Praha 6, Czechoslovakia

Cytokinin affect biogenesis of some other plastid structures in addition to chloroplasts. We have investigated the differentiation of amyloplasts which is affected to great extent by phytohormones. An exellent material for such studies are stem pith explants of marrow-stem kale (*Brassica oleracea* L. var. *medullosa*). When these explants are cultured on agar medium containing only sugar their dry weight increases linearly with time and only a very limited amount of starch is formed. Addition of cytokinin to the medium stimulates, in explants derived from young plants, differentiation of amyloplasts and an accumulation of starch. This effect is further increased by the addition of an auxin (Luštinec et al., 1974).

When explants are derived from kale plants which have been grown either for a long time period under short day conditions or under weak illumination, stimulation of starch formation by auxin and cytokinin is very limited. However, when explants are excised from the stem in such a way that they contain vascular tissue, the starch is formed in the stem pith on agar medium containing sugar in the absence of cytokinin and auxin. In explants derived from young plants the vascular tissue can substitute for both auxin and cytokinin. In explants derived from old plants auxin and cytokinin cannot substitute for the effect of the vascular tissue. The results of these, and other experiments, indicate that cells located in and/or around vascular bundles produce or contain in addition to cytokinin and auxin some other as yet unidentified factor which stimulates plastid differentiation.

The fact that cytokinins affect not only cell division and differentiation but also plastid development indicates that they have several sites of action within the cell. Where more than one site exists, the final physiological effect of cytokinins is a product of the interaction of these activated systems. Competition between these systems, not only for metabolites but also for cytokinins, must be considered as a factor which plays an important role in the regulation of the processes which are under cytokinin control. Regardless of whether the primary site of cytokinin action is only in the cytoplasm or is also in the plastids (see Parthier, 1979), it is reasonable to assume that the cytokinin effects on plastids are accompanied by their utilization. In this way their concentration in the cytoplasm can be altered. The plastids may be, therefore, part of the system which regulates the endogenous level of cytokinins in plant cells.

3 Functions of Plastids in Hormonal Regulation of Growth

3.1 Origin of Cytokinin Autonomy and Change in Chlorophyll Formation in Tobacco Callus Tissue

The tobacco callus tissue derived from stem pith is an excellent tool for investigating the role of plastid functions in hormonal regulation. This tissue (1) requires cytokinin and auxin for growth, (2) this requirement can be lost following habituation or specific bacterial infection; (3) the ability of the tissue to form chlorophyll can be used as a marker to detect the presence of functional plastids in the cells.

We found that all three analyzed cytokinin-autonomous tissues (A_1, A_2, A_3) from various origins formed only a limited amount of chlorophyll (14%-36%) compared with the original cytokinin-dependent strain (D-strain) (Fig. 1, 5). The number of

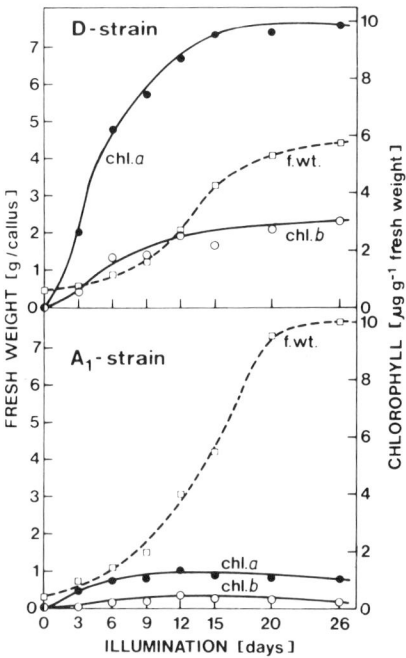

Fig. 1. The influence of photoperiod length on chlorophyl formation in 14-day-old cytokinin-dependent (*D*) and cytokinin-autonomous (A_1) strains of tobacco callus. (Kamínek and Luštinec, 1974a)

plastids per cell was reduced to a lesser extent (70%-83%) (Kamínek and Luštinec, 1974a). Vyskot and Novák (1977) investigated a collection of 20 tissue cultures derived from normal and chlorophyll-deficient plants of *Nicotiana tabacum* differing in both genotype and degree of ploidy. They found that the frequency of cytokinin habituation was much higher in cultures with reduced chlorophyll content. This relationship was confirmed both in tobacco tissue with chromosome-controlled chlorophyll deficiences and in tissues exhibiting plastid-regulated chlorophyll synthesis.

We did not succeed in eliminating the difference in chlorophyll formation between the cytokinin-dependent and cytokinin-autonomous tissues by modifying the auxin and cytokinin concentrations in the medium. The optimum concentration of naphtaleneacetic acid for chlorophyll formation in both tissues (5×10^{-7} M) was much lower than that for maximum tissue fresh weight (5×10^{-6} M). Chlorophyll synthesis in both tissues was stimulated by higher concentrations of kinetin (10^{-5} M) than are optima for their growth (10^{-6} M and 10^{-8} M for cytokinin-dependent and cytokinin-autonomous strains, respectively). It was not possible to reach the saturation point of kinetin for chlorophyll synthesis because tissue growth was completely inhibited at the highest kinetin concentration tested (Kamínek and Luštinec, 1974a). These data substantiate the reported inverse relation between chloroplast differentiation (chlorophyll formation) and tissue growth (Laetsch and Stetler, 1965, Kamínek and Luštinec, 1976). They indicate that the need for cytokinins for chloroplast differentiation and chlorophyll synthesis is much higher than that required for tissue growth. This is supported by the observations of Stetler and Laetsch (1965) and Seyer et al. (1975) that cytokinin-autonomous tobacco tissue and cell cultures require this hormone for chloroplast differentiation.

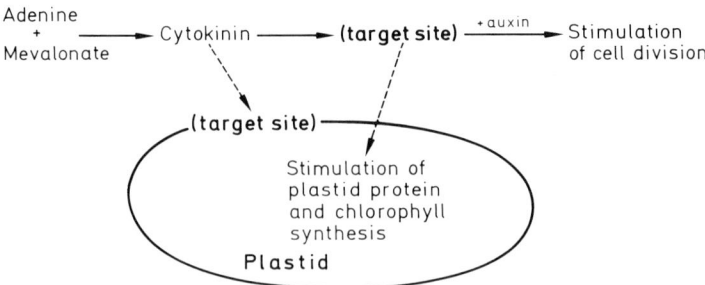

Fig. 2. The proposed action of cytokinins on growth and chlorophyll synthesis in tobacco callus tissue

On the basis of these results a hypothesis was proposed which, in an adapted form, is presented in Fig. 2. It is supposed that cytokinins are synthesized, at least in a limited amount, in both cytokinin-autonomous and dependent tissues. They are, however, extensively utilized for plastid growth and differentiation. In this way the concentration of free cytokinins in the cytokinin-dependent tissue, which contains functional plastids, is decreased and an exogenous source of cytokinin is necessary to maintain tissue growth. When some disorders in plastid function occur (cytokinin-autonomous strain) utilization of endogenous cytokinins for plastid growth and differentiation is reduced and their concentration in cytoplasm is sufficient to stimulate cell division (Kamínek and Luštinec, 1974a). These disorders in plastid function may also decrease the utilization of other substances required for plastogenesis which are now available to support tissue growth.

To date the explanation of cytokinin autonomy is based on the assumption that the growth substances synthesizing systems are activated in tumor cells (Braun, 1958). It has been reported that cytokinin-autonomous tissues incorporated radiolabeled adenine into free cytokinins (Burrows, 1978; Stuchbury et al., 1979) and an enzyme system was isolated and partially purified from similar tissue which catalyzes the formation of N^6-(Δ^2-isopentenyl)adenosine-5'-monophosphate from the appropriate precursors (Chen and Melitz, 1979). The lack of evidence for cytokinin biosynthesis in the cytokinin-dependent tissue (Einset and Skoog, 1973) cannot exclude the possibility that this tissue is capable of cytokinin synthesis which can be suppressed by cytokinin in culture medium.

3.2 Induction of Cytokinin Autonomy by Substances Which Interfere with Plastid Function

According to the above hypothesis the inhibition of cytokinin utilization associated with plastid development may increase the concentration of free cytokinins in the cytoplasm and thereby influence cell division. Streptomycin which inhibits protein synthesis on 70S ribosomes was used successfully to eliminate organized chloroplasts from *Euglena* cells (Ben-Shaul et al., 1972). In our experiments cytokinin-dependent tobacco tissue was grown for seven transfers on medium containing bleaching concentration of streptomycin (0.1 g · 1^{-1}) and then transferred to cytokinin- and streptomycin-free medium. The growth rate of the tissue was lower than that of the original cytokinin-dependent strain but rapidly increased during the successive transfers on the same medium, giving a fresh weight twice that of the dependent strain (Fig. 3). The new cytokinin-autonomous strain obtained after streptomycin treatment (A$_S$-strain) had a strict requirement for auxin and its growth was inhibited by kinetin. Its ability to synthesize chlorophyll was decreased to about one third of the D-strain value and the number of plastids per cell to 58% (Kamínek and Luštinec, 1974b).

The auxin, 2,4-dichlorophenoxyacetic acid (2,4-D), which is accumulated in plastids (Hallam and Sargent, 1970) having a disruptive effect on their structure (Hallam, 1970; Nadakavukaren and McCracken, 1977) had an effect similar to the effect of streptomycin. Witham (1968) reported that 2,4-D satisfied the requirements of tobacco and soybean callus for both auxin and cytokinin and Buiatti and Bennici (1970) found that this growth regulator increased the frequency of habituation in calli formed on *Nicotiana bigelovii* seedlings. As shown in Fig. 4 2,4-D induced callus formation on tobacco stem pith explants. Relatively low concentration of IAA enhanced this effect and decreased the optimum concentration of 2,4-D for callus formation. This suggests that 2,4-D affects two different processes and acts at least at two receptor sites in plant cells corresponding to its auxin and cytokinin-like effects. When applied alone 2,4-D induces callus formation at concentrations at which its cytokinin-like and auxin effects are balanced. The callus tissue induced by 2,4-D (A$_{2,4\text{-}D}$-strain) and kept for another two transfers on medium containing 10^{-6} M 2,4-D became cytokinin-autonomous and chlorophyll defficient (Kamínek and Luštinec, 1974c; Kamínek et al., 1977).

The ability of 2,4-D to induce cytokinin autonomy might be related to 2,4-D action in plants. Some effects of 2,4-D upon the growth, structure, and biochemical responses of plant cells are very similar to those caused by cytokinins as, e.g., induction of plant tumors and fasciations, initiation of cell division in parenchym cells of matured plant tissues, increase of RNA and protein synthesis and decline of RNA de-

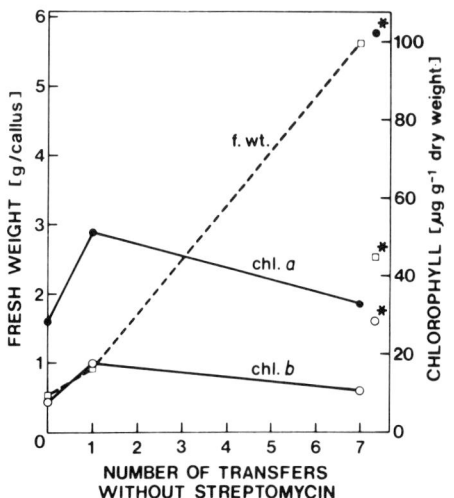

Fig. 3. The effect of transferring tobacco callus from kinetin- and streptomycin-containing medium to medium free of these substances on fresh weight and chlorophyll content. The tissue was grown for 14 days in darkness and then transferred for the another 14 days in continuous light. The chlorophyll content and fresh weight of the cytokinin-dependent tissue is shown in the *right* (*). (Kamínek and Luštinec, 1974b)

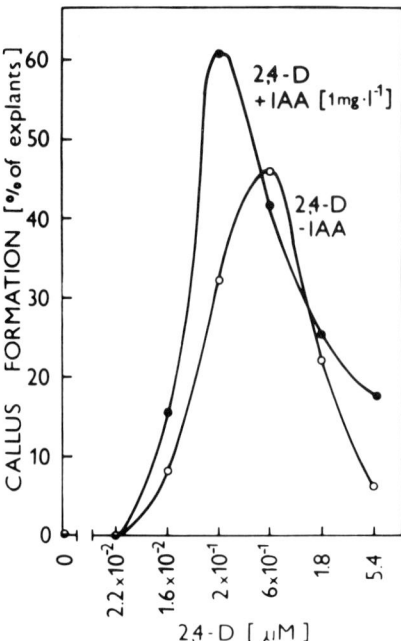

Fig. 4. Effect of *2,4-D* and *IAA* on callus formation in tobacco stem pith explants. (Kamínek et al., 1977)

gradation in excised tissues (see Ashton and Grafts, 1973). This indicates that some of the 2,4-D effects may be the result of increased cytokinin concentration in plant cells.

Streptomycin and 2,4-D are not the only substances which induce cytokinin autonomy. Bednar and Linsmaier-Bednar (1971) described the similar effect of some morfactins and aminofluorenes in tobacco callus tissue. Recently Mok et al. (1979) reported that N,N'-diphenylurea induced cytokinin autonomy in *Phaseolus lunatus* tissue cultures. It would be interesting to test whether these substances affect the plastid apparatus in a manner similar to streptomycin and 2,4-D.

4 Biological and Biochemical Properties of Cytokinin-Autonomous Tissues

The transition of cytokinin-dependent tobacco callus tissue to a cytokinin-autonomous one is associated with other changes, some of which are related to plastid functions. We have investigated properties of several tobacco callus strains differing in their requirement for auxin and cytokinin.

4.1 Chlorophyll Synthesis

This was very limited in all cytokinin-autonomous strains (Fig. 5). When grown under standard conditions the chlorophyll content of autonomous tissues does not exceed 40% of that of the D-strain. It is possible that this value is a consequence of the degree of plastid disorder which is necessary to achieve cytokinin autonomy.

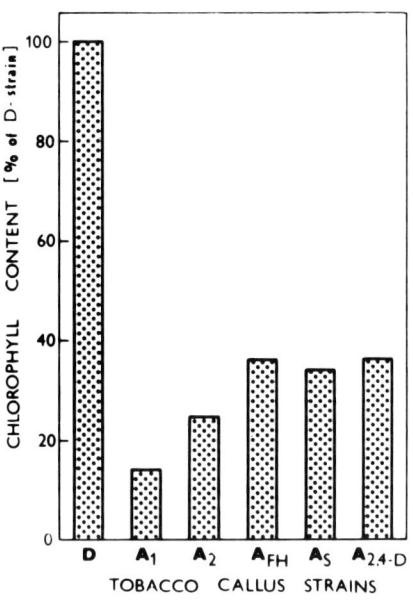

Fig. 5. The relative chlorophyll content in cytokinin-dependent (D), cytokinin-autonomous ($A_1, A_2, A_3, A_S, A_{2,4}\text{-}D$) and fully habituated ($A_{FH}$) strains of tobacco callus (*Nicotiana tabacum* L. cv. Wisconsin 38). The tissue was grown 14 days in the darkness and then transferred to continuous light for 10 days

4.2 Organogenetic Potency

This is very limited or completely destroyed in tissues with altered requirements for cytokinin (Kamínek et al., 1977). The loss of the ability to form organs by these tissues has some relation to Maliga's (1973) observation that streptomycin suppresses not only chlorophyll synthesis but also shoot formation in streptomycin-sensitive tobacco callus.

4.3 Streptomycin Resistance

There was greater resistance to streptomycin in cytokinin-autonomous tissues induced by streptomycin and 2,4-D as compared to the D-strain (Kamínek et al., 177). About 40% reduction in tissue fresh weight was obtained in the $A_{2,4}$ D and A_S-strain at streptomycin concentrations approximately 2 x and 4 x higher, respectively, than in the D-strain. The resistance of $A_{2,4}$-D-strain indicates that the resistance is not due to the appearance of an enzyme which inactivates streptomycin or interferes with its action in plant cells. One possible explanation could be that some streptomycin target structures are absent in plastids found in the two cytokinin-autonomous strains.

4.4 Isozyme Patterns

If the hypothesis that the state of cytokinin-autonomy is accompanied by the disruption of some plastid functions, it is likely this is accompanied by changes in plastid enzymes. We compared the isozyme patterns of several enzymes from tobacco callus tissues differing in their requirements for auxin and cytokinin (Fig. 6) (Kamínek et al.,

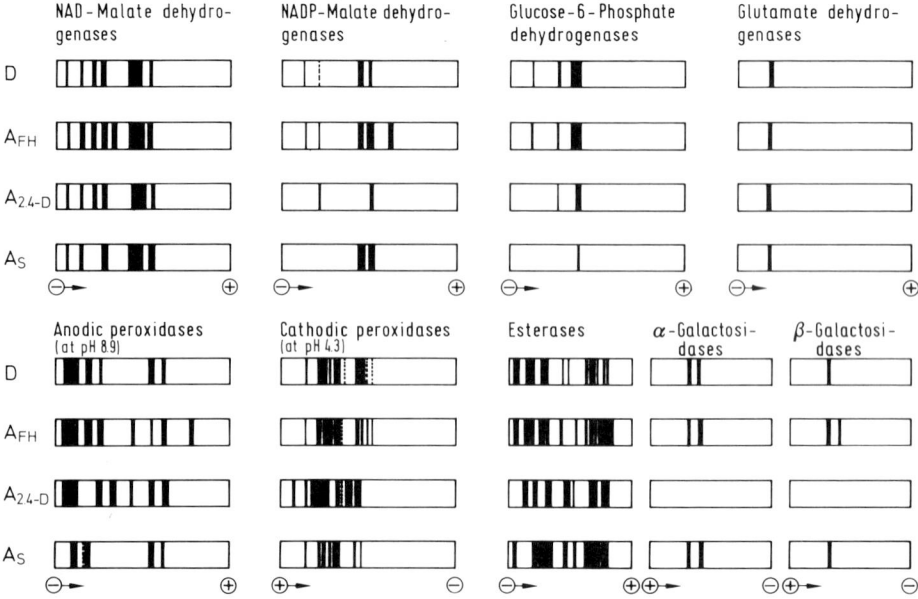

Fig. 6. Disc electrophoresis patterns of malate dehydrogenase, glucose-6-phosphate dehydrogenase, glutamate dehydrogenase, peroxidase, esterase and galactosidases from cytokinin-dependent (D), cytokinin-autonomous (A_S and $A_{2,4-D}$) and fully habituated (A_{FH}) strains of tobacco callus. The relative activities of enzymes are indicated by the *intensities of the bars*. (Kamínek et al., in press)

in press). Of interest are the isozymes of NAD- and NADP-dependent malate dehydrogenase (NAD-MDH and NADP-MDH). The NADP-MDH was found in maize leaf chloroplasts and in proplastids of dark-grown tobacco cells. However, very limited NAD-MDH activity was detected in these two organelles (Hatch and Slack, 1969; Washitani and Sato, 1977a, b). There are four NADP-MDH isozymes in the D-strain; this is reduces to two in the two cytokinin-autonomous strains. The number of NAD-MDH isozymes was much less affected by cytokinin habituation, one isozyme was missing in the A_S-strain.

Glucose-6-phosphate dehydrogenase (G-6-PDH) has been found in both plastid and cytoplasmatic fractions from various plants including dark-grown tobacco cells (Washitani and Sato, 1977b). Four isozymes of this enzyme were found in the D-strain. This was reduced to two and one in the cytokinin autonomous A_S- and $A_{2,4-D}$-strains, respectively, the relative activity of the remaining isozymes was significantly decreased. The total activity of this enzyme in the A_S-strain was reduced by 50% when compared to the original D-strain (Hadačová et al., 1975).

The remaining enzymes investigated are known to be mainly localized in the non-plastid cell fractions (peroxidases, esterases, α- and β-galactosidases). Their total number in the cytokinin-autonomous tissue was either decreased or not changed. The exception was an additional anodic peroxidase in $A_{2,4-D}$-strain.

The fully habituated strain (A_{FH}), which was both auxin- and cytokinin-autonomous, differed from cytokinin-autonomous strains. It contained either the same or an increased number of isozymes with respect to the D-strain. These results show that where changes in the number of isozymes occur their number is decreased in the cytokinin-autonomous strains and increased in the fully habituated strain. These findings support the possibility that the induction of cytokinin autonomy is accompanied by limitations in the synthesis of specific plastid proteins while some metabolic pathways are probably activated during the habituation of tissue for both auxin and cytokinin.

4.5 Regenerated Plants

Regenerated plants derived from tissue cultures with altered requirements for growth substances are useful tools for investigating changes that are of a heritable nature. We succeeded regenerating plants only from callus of $A_{2,4}$-D- and A_D-strains. Plants regenerated from $A_{2,4}$-D-strain differed from those differentiated from D-strain in the following characters: (1) decreased apical dominance (formation of thin, crawling, and branching stems), (2) formation of narrow thick leaves of xeromorphic character, (3) frequent fasciation of stem and petioles, (4) sterility and (5) loss of the cytokinin autonomy in explants derived from stem pith.

Despite the loss of the potential for expressing cytokinin autonomy it is evident that induction of cytokinin autonomy is accompanied by significant changes in other properties. These properties are, in the case of tumor tissues, dependent on the strain of *Agrobacterium tumefaciens* that incites the original tumor (Einset and Cheng, 1979). It is quite possible that the different chemical agents capable of inducing cytokinin-autonomy affect in different ways the retention of autonomy in regenerated plants.

5 Concluding Remarks

The proposed "negative" regulation of free cytokinins in tobacco callus cells by their utilization during cytokinin-induced plastid differentiation and growth may be criticized from several aspects. The evidence presented is based largely on experiments with greening plant tissues. One could argue that in dark-grown tissues utilization of cytokinins associated with plastid development is very limited. However, as shown above, cytokinins are required for amyloplast differentiation in tissues grown in darkness on sugar-containing medium. We do not suggest that disorders in plastid function and/or structure are the only factor responsible for cytokinin autonomy. It seems quite possible that there are different balances between the rates of cytokinin biosynthesis, utilization and degradation in the different autonomous tissues which make the solution of this problem more difficult. The evidence (Einset and Cheng, 1979) that the tumor characters depend upon the *Agrobacterium* strain supports such an opinion.

Plastids may be not the only structures which affect cytokinin utilization in plant cell. Recently Miller (1979) found that soybean mitochondrial preparations devoid of intact cells respond to cytokinins and that the mitochondria may contain a site for cytokinin action. This suggests that also other structures in plant cells may be involved in the regulation of free cytokinins. Stimulation of the processes which are associated with one cell structure may therefore affect other processes in the cell which also require cytokinins. The existence of such an inverse relationship has been proved already for the stimulation of tobacco tissue and cell culture growth and the differentiation of chloroplasts (Laetsch and Stetler, 1965; Seyer et al., 1975; Kamínek and Luštinec, 1976).

References

Ashton FM, Grafts AS (1973) Mode of action of herbicides. Willey-Interscience Publ. New York London Sydney Toronto

Bednar TW, Linsmaier-Bednar EM (1971) Induction of cytokinin-independent tobacco tissues by substituted fluorenes. Proc Natl Acad Sci USA 68: 1178-1179

Ben-Shaul YR, Silman R, Ophir I (1972) Effects of streptomycin on the ultrastructure of plastids in *Euglena*. Physiol Veg 10: 255-268

Braun AC (1958) A physiological basis for autonomous growth of the crown-gall tumor cell. Proc Natl Acad Sci USA 44: 344-349

Buiatti M, Bennici A (1970) Callus formation and habituation in *Nicotiana* species in relation to the specific ability for differentiation. Acad Naz Lincei Rend Cl Sci Fis Mat Nat Ser VIII 48: 261-269

Burrows WJ (1978) Incorporation of ^{3}H-adenine into free cytokinins by cytokinin-autonomous callus tissue. Biochem Biophys Res Commun 84: 743-748

Chen C-M, Melitz D (1979) Cytokinin biosynthesis in a cell free system from cytokinin-autotrophic tobacco tissue cultures. FEBS Lett 107: 15-20

Einset J, Cheng A (1979) Regeneration of tobacco plants from crown gall tumors. In Vitro 15: 704-708

Einset JW, Skoog F (1973) Biosynthesis of cytokinins in cytokinin-autotrophic tobacco callus. Proc Natl Acad Sci USA 70: 658-660

Hadačová V, Kamínek M, Luštinec J (1975) Glucose-6-phosphate dehydrogenase in tobacco callus strains differing in their growth and their requirement for auxin and cytokinin. Biol Plant 17: 448-451

Hallam ND (1979) Effect of 2,4-dichlorophenoxyacetic acid and related compounds on the fine structure of primary leaves of *Phaseolus vulgaris*. J. Expt Bot 21: 1031-1038

Hallam ND, Sargent JA (1970) The localization of 2,4-D in leaf tissue. Planta 94: 291-295

Hatch MD, Slack CR (1969) NADP-specific malate dehydrogenase and glycerate kinase in leaves and evidence for their location in chloroplasts. Biochem Biophys Res Commun 34: 589-593

Kamínek M, Luštinec J (1974a) Reduced chlorophyll synthesis in cytokinin-autonomous strains of tobacco callus. Z Pflanzenphysiol 73: 65-73

Kamínek M, Luštinec J (1974b) Induction of cytokinin-autonomy and chlorophyll-deficiency in tobacco callus tissue by streptomycin. Z Pflanzenphysiol 73: 74-81

Kamínek M, Luštinec J (1974c) Induction of cytokinin autonomy in tobacco stem pith tissue by 2,4-dichlorophenoxyacetic acid. In: Schreiber K, Schütte HR, Sembdner G (eds) Biochemistry and chemistry of plant growth regulators. Acad Sci GDR Halle (Saale), pp 307-310

Kamínek M, Luštinec J (1976) The effect of sugars on inverse relation between the growth and chlorophyll synthesis in tobacco tissue. Biol Plant 18: 384-388

Kamínek M, Hadačová V, Luštinec J (1977) Some biological and biochemical properties of the cytokinin-autonomous strains of tobacco callus tissue induced by 2,4-D and streptomycin. In: Novàk F (ed) Use of tissue cultures in plant breeding, Czech Acad Sci, Prague, pp 577-592

Kamínek M, Hadačová V, Luštinec J (in press) Origin of cytokinin-autonomy and changes in specific proteins in tobacco callus tissue. Biol Plant

Laetsch WM, Stetler A (1965) Chloroplast structure and function in cultured tobacco tissue. Am J Bot 52: 798-804

Lescure AM (1978) Chloroplast differentiation in cultured tobacco cells: in vitro protein synthesis efficiency of plastids at various stages of their evolution. Cell Differ 7: 139-152

Luštinec J, Hadačová V, Kamínek M (1974) The effect of various cytokinins and auxins on starch formation in kale and tobacco explants. In: Schreiber K, Schütte, HR, Sembdner G (eds) Biochemistry and chemistry of plant growth regulators. Acad Sci GDE, Halle (Saale), pp 311-314

Maliga P, Brezinovits A, Márton L (1973) Streptomycin-resistant plants from callus culture of haploid tobacco. Nature (London) New Biol 244: 29-30

Miller CO (1979) Cytokinin inhibition of respiration by cells and mitochondria of soybean, Glycine max (L.) Merrill. Planta 146: 503-511

Mok MC, Armstrong DJ, Mok DW (1979) Induction of cytokinin autonomy by N-N'-diphenylurea in tissue cultures of *Phaseolus lunatus* L. Proc Natl Acad Sci USA 76: 3880-3884

Nadakavukaren MJ, McCracken DA (1977) Effect of 2,4-dichlorophenoxyacetic acid on the structure and function of developing chloroplasts. Planta 137: 65-69

Naito K, Tsui H, Hatakeyama I (1979) Benzyladenine-induced increase in DNA content per cell, chloroplast size, and chloroplast number per cell in intact bean leaves. J Exp Bot 30: 1145-1151

Parthier B (1979) The role of phytohormones (cytokinins) in chloroplast development. Biochem Physiol Pflanz 174: 173-214

Schiff JA (1973) The development, inheritance, and origin of the plastid in *Euglena*. Adv Morphog 10: 265-312

Seyer P, Marty D, Lescure AM, Pèaud-Lenoel C (1975) Effect of cytokinin on chloroplast cyclic differentiation in cultured tobacco cells. Cell Differ 4: 187-197

Stetler DA, Laetsch WM (1965) Kinetin-induced chloroplast maturation in cultures of tobacco tissue. Science 149: 1387-1389

Stuchbury LM, Palni LM, Horgan R, Wareing PF (1979) The biosynthesis of cytokinins in crown-gall tissue of *Vinca rosea*. Planta 147: 97-102

Vyskot B, Novák F (1977) Habituation and organogenesis in callus cultures of chlorophyll mutants of *Nicotiana tabacum* L. Z Pflanzenphysiol 18: 117-125

Washitani I, Sato S (1977a) Studies on the function of proplastids in the metabolism of in vitro cultured tobacco cells. I. Localization of nitrite reductase and NADP-dependent glutamate dehydrogenase. Plant Cell Physiol 18: 117-125

Washitani I, Sato S (1977b) Studies on the function of proplastids in the metabolism in vitro cultured tobacco cells. III. Source of reducing power for amino acid synthesis from the nitrite. Plant Cell Physiol 18: 1235-1241

Witham FH (1968) Effect of 2,4-dichlorophenoxyacetic acid on the cytokinin requirement of soybean cotyledon and tobacco stem pith callus tissues. Plant Physiol 43: 1455-1457

Influence of Cytokinin on Plastid Biogenesis in Rye Leaves

J. Feierabend[1]

1 Introduction into the Experimental System

Cytokinins are known to promote the multiplication and growth of higher plant cells (for review see Kende, 1971). Besides this general growth-promoting action they appear to exert some specific influence on the differentiation of chloroplasts, which has been the subject of our own research, as well as on the stability of chloroplasts during leaf senescence (for review see Parthier, 1979). The specific quality of the plant response to cytokinin is, however, not universally and exclusively determined by the hormone, but primarily depends on the physiological conditions, the specific stage of differentiation, and the genetic competence of the individual target tissue, so that even in different organs or tissues of the same plant cytokinin can be employed as a trigger to initiate the expression of quite opposing types of response. Thus, in the shoot organs of rye seedlings, the experimental system with which we have been working, cytokinins promote the growth of the leaves but, in complete contrast to their well-known senescence-retarding action in leaves, enhance the onset of senescence in the coleoptiles, including the degradation of nucleic acids and protein (de Boer and Feierabend, 1978; Sossinka and Feierabend, 1978). In order to respond to an exogenously applied hormone, the tissue must not only be genetically competent but must also have a deficiency of this hormone. While a kinetin treatment clearly stimulated the growth of rye leaves at 22°C, it had, for instance, no effect on their protein and enzyme formation at 2°C (Feierabend, 1970a). Under the low temperature conditions the endogenous cytokinin supply appeared to be saturating, and only above about 17°C does it seem to become limiting so that the leaves respond to an exogenous application of kinetin. In order to obtain a safe and maximal response an experimental system should be, as much as possible, depleted of the hormone under investigation. The major endogenous source of cytokinins for the shoot organs appear to be the roots (Torrey, 1976). Therefore, we have generated cytokinin-depleted shoots by early excision of the seedling roots and experimentally raised the cytokinin level for studying its action by the application of kinetin to the rootless seedlings at the age of two days. In order to separate and discriminate the effects of light which were very similar to those of cytokinin, both promoting leaf growth and plastid differentiation and enhancing coleoptile senescence (Feierabend, 1969; de Boer and Feierabend, 1978; Sossinka

---

[1] Botanisches Institut, Fachbereich Biologie, J.W. Goethe-Universität, Siesmayerstraße 70, 6000 Frankfurt/Main, FRG

2 Influence of Cytokinin on Leaf Growth

Growth of etiolated rye leaves raised at 22°C is to some extent controlled by the cytokinin supply (Fig. 1). Several different growth parameters, such as the accumulation of total amino nitrogen, protein, nucleic acids or the ribosome content, were rather uniformly and nonspecifically affected. They were lowered in cytokinin-depleted leaves and to a fairly similar extent increased after application of kinetin, the kinetin-induced increases amounting to 70%-100%, relative to the cytokinin-deficient condition (de Boer and Feierabend, 1978). The greatest difference was in the DNA content. The percentage of cytoplasmic ribosomes bound as polysomes increased only slightly, if at all, after treatment of cytokinin-depleted leaves with kinetin (de Boer and Feierabend, 1978; Feierabend and de Boer, 1978). Therefore, the cytokinin action on leaf growth cannot be clearly attributed to any exclusive or predominant effect on either DNA replication, transcription or translation. Since DNA replication was most affected, this might possibly be considered to be under a main cytokinin control, thereby uniformly limiting the growth rate and all other parameters. However, there are also several lines of evidence indicating that either the availability or the capacity and rate of translocation of substrates from the endosperm to the leaves were the essential growth-limiting steps determined by the cytokinin level. The cytokinin-dependent differences in the protein content were always strictly related to the total amino nitrogen content and those of the amino acid incorporation to the total amino acid uptake in the leaves. Irrespective of the cytokinin level, protein accounted for 80% of the amino nitrogen content and an equal percentage of the radioactive amino acid uptake was incorporated into protein in etiolated leaves, whereas in light both ratios were significantly increased (de Boer and Feierabend, 1978). The amino nitrogen content and the rate and amount of amino acid uptake varied according to the cytokinin level (Fig. 1; de Boer and Feierabend, 1978). It is, however, still uncertain at which level substrate transport into the leaf was most importantly controlled by cytokinin. The amino acid uptake capacity seemed to be affected (de Boer and Feierabend, 1978). In a system as complicated as intact seedlings, correlative relationships between different organs may, however, also be involved, representing a serious disadvantage for the analysis of a specific hormone action. Thus, it is quite conceivable that already the breakdown of the reserves in the endosperm was accelerated under the influence of cytokinin, as reported for watermelon cytoledons by Longo et al. (1979). In addition, cytokinin enhanced the onset of protein breakdown in the coleoptiles and the translocation of amino acids from the coleoptiles to the leaves (Fig. 1) and might through such a correlative phenomenon advance the development of the leaves at the cost of the coleoptiles (de Boer and Feierabend, 1978; Sossinka and Feierabend, 1978). Therefore, it is at present not yet possible to identify the cytokinin-dependent event in the translocation of substrates. An essential cytokinin control of substrate mobilization or translocation would, however, explain the nonspecific nature of the hormone influence on leaf growth.

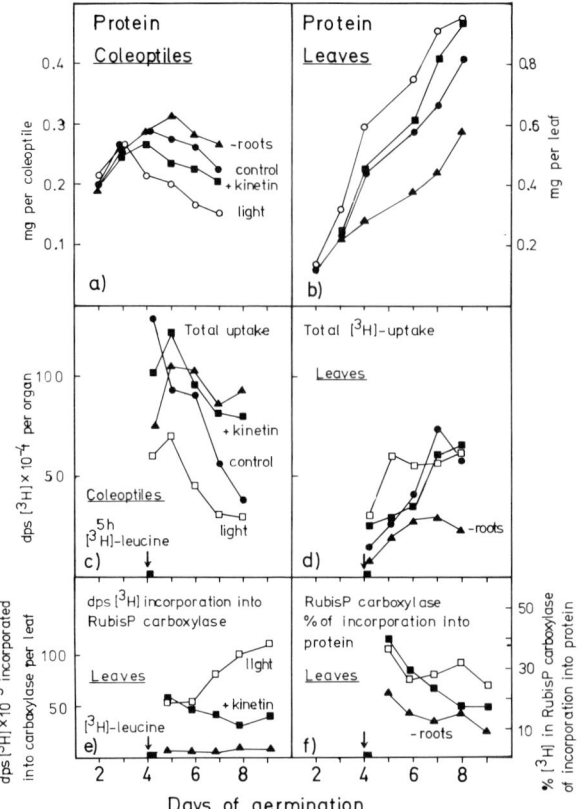

Fig. 1a-f. Protein content, uptake of [^{3}H]-leucine, and [^{3}H]-leucine incorporation into ribulosebisphosphate carboxylase in coleoptiles or leaves of rye seedlings. Radioactivity incorporated into ribulosebisphosphate carboxylase was determined after double precipitation with a monospecific rabbit antiserum to the carboxylase and goat antiserum to rabbit γ-globulin and corrected for radioactivity in precipitates with control serum. In c-f the shoots of 4-day-old etiolated seedlings were exposed for 5h to a solution of L-[^{3}H]-leucine. ● control dark; ○ control white light; ▲ seedlings roots removed on day 1, darkness; ■ seedling roots removed on day 1, application of kinetin on day 2 of germination in darkness; □ seedling roots removed on day 1, application of kinetin on day 2 of germination in darkness, white light after the end of the 5h labeling period on day 4. (Modified from data of de Boer and Feierabend, 1974 and 1978, Sossinka and Feierabend, 1978, and Feierabend, 1979)

3 Influence of Cytokinin on Chloroplast Differentiation and Enzyme Formation

3.1 Specificity and Characterization of the Phenomenon

In several higher plant systems capable of greening, such as developing leaves, cotyledons, or tissue cultures, cytokinins were found to exert a stronger and much more specific effect on chloroplast differentiation and chloroplast enzyme development

Influence of Cytokinin on Plastid Biogenesis in Rye Leaves

than on growth in general. In the etiolated rye leaves where we have studied the cytokinin control of chloroplast enzyme development (Feierabend, 1969; de Boer and Feierabend, 1974; Feierabend and de Boer, 1978) the activities of several photosynthetic enzymes were much more decreased after lowering the endogenous cytokinin supply by removal of the seedling roots, relative to intact plants, than growth in general (e.g., dry weight, amino nitrogen content, protein, nucleic acids) or than several nonphotosynthetic enzymes, such as glucose-6-phosphate dehydrogenase or mitochondrial enzymes. While the increases induced by the application of kinetin to derooted seedlings were only 70%-100% (de Boer and Feierabend, 1974, 1978) for parameters of general growth, the kinetin-induced increases of chloroplast enzyme activities amounted to 200%-600% (Fig. 2a; Feierabend and de Boer, 1978). For the best investigated example, ribulose bisphosphate carboxylase, it has been determined that the kinetin-induced changes in activity reflected corresponding increases of the amount of the enzyme protein (Fig. 2b; Feierabend and de Boer, 1978).

Adenosine, auxin, or gibberellic acid were not able to substitute for kinetin in the derooted seedlings (Fig. 2a; Feierabend 1969, 1970b). For some but not all photosynthetic enzymes the activities were so much increased by the kinetin treatment that they reached similarly high activities in darkness as in control seedlings in light (Fig. 2a;

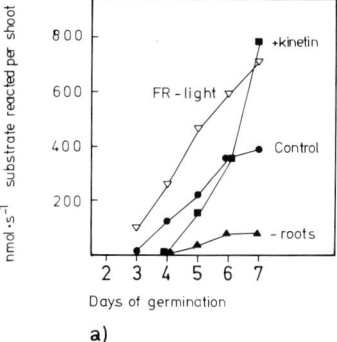

a)

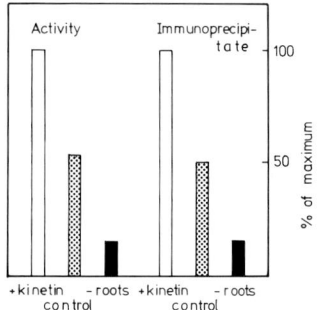

b)

Fig. 2a, b. a Development of ribulosebisphosphate carboxylase activity in the leaves of rye seedlings. ● control, dark; ▽ control in continuous far-red light; ▲ seedling roots excised on day 1 of germination, darkness; ■ seedling roots excised on day 1, application of kinetin on day 2 of germination in darkness. b Comparison of the activity of ribulosebisphosphate carboxylase and the amounts of precipitate obtained with a monospecific antiserum to the enzyme in extracts of leaves of 7-day-old dark-grown rye seedlings. Roots were excised on day 1 of germination (*−roots*), kinetin was applied on day 2 to derooted seedlings (*+ kinetin*). (Modified from data of Feierabend, 1969 and Feierabend and de Boer 1978)

Feierabend, 1969; Feierabend and de Boer, 1978). When kinetin-treated etiolated plants were, in addition, exposed to light treatments the activities were, however, still further increased. A detailed analysis of the cooperation of cytokinin and of light, acting through the phytochrome system, revealed that they were acting in a multiplicative way. Each of the different rates of increase of the photosynthetic enzyme activities determined by the different cytokinin levels in darkness (derooted, kinetin-treated, or intact control plants) could be further increased by a virtually constant factor through the light treatment. From this it may be concluded that both factors were acting independently via different mechanisms and without interaction (Feierabend, 1969).

Besides the photosynthetic enzymes of the chloroplasts, peroxisomal enzymes engaged in photorespiratory glycolate metabolism which is closely related to photosynthesis were also preferentially affected by changes in the cytokinin levels. The kinetin-induced increases, relative to cytokinin-depleted leaves, were 200%-300% (de Boer and Feierabend, 1974).

This demonstrates that not only individual chloroplast enzymes were specifically affected, but that cytokinin appeared to trigger the expression of a whole developmental pattern leading to the photosynthetic competence of the leaf.

3.2 Mode of Action

Mainly two chloroplast enzymes have been studied to characterize the mechanism of cytokinin action: NADP-dependent-glyceraldehydephosphate dehydrogenase which is synthesized on cytoplasmic ribosomes (Feierabend, 1979) and ribulosebisphosphate carboxylase, consisting of two types of subunits of which the small ones are synthesized on cytoplasmic 80S ribosomes and coded on nuclear DNA while the large ones are synthesized on chloroplast 70S ribosomes and coded on chloroplast DNA (for further literature see Feierabend, 1979; Parthier, 1979). A comparison of the development of these enzymes after kinetin treatments at different developmental stages showed that their time of appearance in the young leaf was not affected by cytokinin but only their rate of increase (Feierabend, 1979). From this it follows that cytokinin was not capable of turning on individual enzymes but only of controlling their quantitative expression and had, for this purpose, to await the appropriate competence of the leaf tissue.

Regarding the mechanism of the cytokinin control of chloroplast enzymes we have asked the following further questions: (a) Why is the chloroplast differentiation so much more sensitive to cytokinin than leaf growth in general? (b) At which molecular level is the expression of chloroplast proteins controlled? and (c) is the cytokinin affecting the chloroplast directly or only acting through the nucleo-cytoplasmic system?

The greater sensitivity of chloroplast constituents to changes in the cytokinin level seems to indicate that more cytokinin is needed for the maximal expression of the plastid activities than for growth. This is best illustrated by some tissue cultures which are able to multiply at a low, or without any, exogenous cytokinin source but require a higher exogenous cytokinin concentration for greening (Stetler and Laetsch,

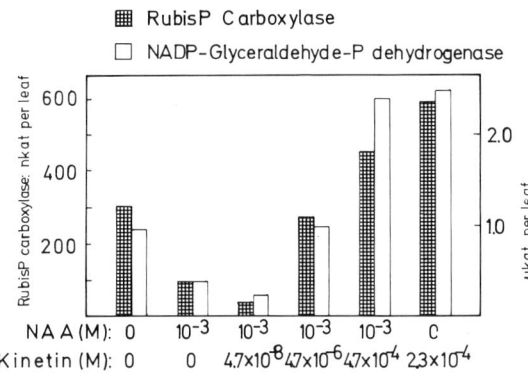

Fig. 3. Activity of ribulosebisphosphate carboxylase and NADP-glyceraldehyde-phosphate dehydrogenase in the leaves of 6-day-old etiolated rye seedlings. Application of *NAA* and *kinetin* to intact seedlings at the concentrations indicated on day 2 of germination. (Modified from Feierabend, 1970b)

1965; Lescure, 1978). A conceivable — while so far only speculative — model to explain such a greater cytokinin requirement has been presented (Feierabend, 1970b) by demonstrating an antagonistic action of the synthetic auxin α-naphthalene acetic acid (NAA) and kinetin. NAA was preferentially suppressing the development of ribulose-bisphosphate carboxylase (Fig. 3) as well as that of NADP-glyceraldehydephosphate dehydrogenase, and kinetin was capable of antagonistically overcoming the inhibition by NAA so that the enzyme activities were not related to the absolute concentration of one of the growth regulators but rather to their ratio to each other (Feierabend, 1970b). This is of course only an artificial model since the auxin concentrations used were unphysiologically high. But in young embryonic cells and leaves other endogenous inhibitors might conceivably occur, exerting some selective suppressing effect on plastid development which would need to be overcome by an appropriately high concentration of cytokinin before plastid differentiation can proceed at maximum rate. The natural occurrence, involvement and possible nature of such inhibitory factors antagonistic to cytokinin has not yet been established in our system. Control systems involving antagonistically acting factors have, however, been described to occur, e.g., in the regulation of shoot development in tissue culture or of apical dominance through auxin or ethylene and cytokinin (Skoog and Miller, 1957; Wickson and Thimann, 1958; Burg and Burg, 1968).

Most information regarding the molecular mechanism of the cytokinin control are available for the ribulosebisphosphate carboxylase. In this case it is clear that the formation of the enzyme protein is controlled by the cytokinin supply but it is not yet sufficiently understood at which level. The formation of ribulosebisphosphate carboxylase was extremely sensitive to actinomycin D in cytokinin-depleted leaves. The kinetin-induced increase was, however, only partially inhibited (Feierabend and de Boer, 1978). Such results are, however, ambiguous and hardly allow any safe conclusions about the necessity of transcription or the stability or new formation of mRNA, since RNA synthesis of the rye leaves was only slightly affected in the presence of actinomycin D (de Boer and Feierabend, 1978). There are, however, several lines of evidence indicating that cytokinin induces an increased synthesis of this enzyme. The kinetin-induced increases of the ribulosebisphosphate carboxylase activity were totally inhibited by chloramphenicol, an inhibitor of chloroplast protein synthesis, as well as by cycloheximide, an inhibitor of cytoplasmic protein synthesis (Feierabend, 1970b). In addition, a higher percentage of labeled leucine was incorporated into the enzyme protein in kinetin-treated than in cytokinin-depleted leaves (Fig. 1; Feierabend, 1979), whereas under both conditions an almost identical proportion of the total uptake of the radioactive amino acid was incorporated into protein (de Boer and Feierabend, 1978). Following a period of labeling with [^{3}H]-leucine the ratio of radioactivity incorporated in ribulosebisphosphate carboxylase decreased again, both in cytokinin-depleted and in kinetin-treated leaves, when they were kept in darkness, indicating that turnover of the enzyme protein occurred and was not arrested by the kinetin treatment. Only light was able to stabilize incorporation into ribulosebisphosphate carboxylase (Fig. 1; Feierabend, 1979). This is in accord with the observation of Peterson et al. (1973) who found that in barley leaves turnover of the ribulosebisphosphate carboxylase was virtually absent in light but significantly increased in darkness. In summary, the data of amino acid incorporation suggest that cytokinin regulated the accumulation of ribulosebisphosphate carboxylase in developing rye leaves through an increased synthesis and, unlike the cytokinin action in leaf senescence, not through decreased degradation. By contrast, the senescence-delaying action of cytokinin in detached leaves is at present more attributed to a retardation of protein degradation rather than to a stimulation of protein synthesis (see Kende, 1971; Martin and Thimann, 1972).

The fact that a preferential cytokinin-induced enhancement was observed also for such chloroplast enzymes which are known to be translation products of 80S ribosomes and to some extent even for peroxisomal enzymes, clearly demonstrates that cytokinin did not exclusively act on the genetic system or on biosynthetic events of the chloroplast alone but also on the nucleo-cytoplasmic system and the differentiation of the leaf cells as a whole (de Boer and Feierabend, 1974; Feierabend and de Boer, 1978). It appears that cytokinins do not specifically induce any single photosynthetic enzyme but rather enhance, by acting as "permissive factors" in the sense of Fosket and Tepfer (1978), the realization of a whole and integrated differentiation pattern, leading to the characteristic organization of photosynthetic mesophyll cells, that has already been initiated on the basis of the leaf's genetic program during the course of prior development making the leaf competent for its expression. In this context, synthetic events in the chloroplast seem, however, to be even more sensitive to a cytokinin control than those in the cytoplasm. This is illustrated by the fact that the cytokinin-dependent increase of ribulosebisphosphate carboxylase was far greater than that of any other chloroplast enzyme of cytoplasmic origin and seems to correlate with a general selective enlargement of the capacity and stimulation of the activity of plastid protein synthesis under the influence of cytokinin. The size of the plastid protein synthesis machinery seems to become particularly reduced in cytokinin-depleted leaves. The plastids remained much smaller and the amount of plastid 70S ribosomes (as monitored by the content of 16S rRNA per DNA) was unproportionally low (Feierabend and de Boer, 1978). Under the influence of kinetin the 70S ribosome content and the proportion of 70S, relative to 80S, ribosomes (as measured by the ratio of (23S + 16S) (23S + 18S) rRNA were strikingly increased, in parallel with an increase in the size of the plastids which might, in addition, be accompanied by an increase of plastid numbers as described by Naito et al. (1979). An additional cytokinin-dependent increase of the proportion of plastid polyribosomes (as measured by the percentage of 16S rRNA found in the total polyribosome fraction) seems to indicate a cytokinin-induced increase of the activity of the plastid protein synthesis machinery which is not to any comparable extent seen for the cytoplasmic polyribosomes (Feierabend and de Boer, 1978). The strong and unproportional deficiency of the plastid protein synthesis in cytokinin-depleted etiolated rye leaves is further illustrated by a striking accumulation of small subunits of ribulosebisphosphate carboxylase, presumably resulting from an insufficient synthesis of large subunits under these conditions (Feierabend and de Boer, 1978), since the latter have to be produced in the plastids. The cytoplasmic synthesis of the small subunits seemed to be less impaired. However, in etiolated rye leaves, also kinetin treatment obviously did not sufficiently reinforce plastid protein synthesis to balance the plastidic production of large, with the cytoplasmic synthesis of small, subunits of the carboxylase. Only in light did the excess of small subunits disappear. While light did not augment the proportion of 70S ribosomes within the cell, it increased the 70S polyribosome proportion and herewith presumably the activity of the chloroplast protein synthesis much more strikingly than kinetin in darkness (Feierabend and de Boer, 1978). It is not yet understood whether cytokinin and light can directly affect the activity of plastid protein synthesis or whether their action has to be mediated through the nucleo-cytoplasmic system.

Acknowledgment. Support of our research by the Deutsche Forschungsgemeinschaft is greatly appreciated.

References

Boer J de, Feierabend J (1974) Comparison of the effects of cytokinins on enzyme development in different cell compartments of the shoots organs of rye seedlings. Z Pflanzenphysiol 71: 261-270

Boer J de, Feierabend J (1978) Comparative analysis of the action of cytokinin and light on the growth of rye leaves. Planta 142: 67-73

Burg SP, Burg EA (1968) Ethylene formation in pea seedlings, its relation to the inhibition of bud growth caused by indole-3-acetic acid. Plant Physiol 43: 1069-1074

Feierabend J (1969) Der Einfluß von Cytokininen auf die Bildung von Photosyntheseenzymen in Roggenkeimlingen. Planta 84: 11-29

Feierabend J (1970a) Proteinsynthese und Enzymbildung in Keimlingen bei niedrigen Wachstumstemperaturen und ihre Beziehungen zum Cytokininhaushalt. Z Pflanzenphysiol 62: 70-82

Feierabend J (1970b) Characterization of cytokinin action on enzyme formation during the development of the photosynthetic apparatus in rye seedlings. Planta 94: 1-15

Feierabend J (1979) Role of cytoplasmic protein synthesis and its coordination with the plastidic protein synthesis in the biogenesis of chloroplasts. Ber Dtsch Bot Ges 92: 553-574

Feierabend J, Boer J de (1978) Comparative analysis of the action of cytokinin and light on the formation of ribulosebisphosphate carboxylase and plastid biogenesis. Planta 142: 75-82

Fosket DE, Tepfer DA (1978) Hormonal regulation of growth in cultured plant cells. In Vitro 14: 63-75

Kende H (1971) The cytokinins. Int Rev Cytol 31: 301-338

Lescure AM (1978) Chloroplast differentiation in cultured tobacco cells: in vitro protein synthesis efficiency of plastids at various stages of their evolution. Cell Differ 7: 139-152

Longo GP, Pedretti M, Rossi G, Longo CP (1979) Effect of benzyladenine on the development of plastids and microbodies in excised watermelon cotyledons. Planta 145: 209-217

Martin C, Thimann KV (1972) The role of protein synthesis in the senescence of leaves. I. The formation of protease. Plant Physiol 49: 64-71

Naito K, Tsuji H, Hatakeyama I, Ueda K (1979) Benzyladenine-induced increase in DNA content per cell, chloroplast size, and chloroplast number per cell in intact bean leaves. J Exp Bot 30: 1145-1152

Parthier B (1979) The role of phytohormones (cytokinins) in chloroplast development. Biochem Physiol Pflanz 174: 173-214

Skoog F, Miller CO (1957) Chemical regulation of growth and organ formation in plant tissues cultured in vitro. Symp Soc Exp Biol 11: 118-131

Sossinka J, Feierabend J (1978) Influence of cytokinin and light on nucleic acid and protein metabolism of senescing coleoptiles. Biochem Physiol Pflanz 173: 505-513

Stetler DA, Laetsch MW (1965) Kinetin induced chloroplast maturation in cultures of tobacco tissue. Science 149: 1387

Torrey JG (1976) Root hormones and plant growth. Ann Rev Plant Physiol 27: 435-459

Wickson M, Thimann KV (1958) The antagonism of auxin and kinetin in apical deminance. Physiol Plant 11: 62-74

Light and Fusicoccin as Tools for Discriminating Among Responses of Cotyledons to Cytokinins

C.P. Longo, G.P. Longo, M. Lampugnani, G. Rossi, and O. Servettaz[1]

1 Introduction

Cytokinins promote the development of excised cotyledons. The most important mechanisms stimulated by this class of hormones are growth by cell expansion, breakdown of storage material and development of the three main classes of cell organelles: plastids, mitochondria, and microbodies (Longo et al., 1979a, b). The overall result of this wide range of responses is an accelerated transformation of the cotyledon from a storage organ into a photosynthetic organ. Such a transformation normally occurs during germination in light (Gruber et al., 1970; Kagawa et al., 1973).

We have tried a classification of this wide array of responses. We were able to show that the responses of excised watermelon cotyledons to benzyladenine (BA) can be divided into two classes on the basis of concentration requirements (Longo et al., 1979c). We called "class 1" responses those that are not saturated by a high concentration of BA (10^{-4} M) and "class 2" responses those that show a sharp optimum at 10^{-5} M BA concentration. The distinction between the two classes seems to be a neat one: there are very few effects whose allocation is uncertain. "Class 1" responses include growth, development of mitochondria and breakdown of reserve lipid while "class 2" responses are all related to the development of plastids and microbodies. "Class 2" responses are mimicked by white light while "class 1" responses are not.

Since irradiation with white light induces in cotyledons many responses that are similar to those obtained with cytokinins, it may be of some interest to study the action of the two factors when applied simultaneously. Such studies of interactions between two factors can give a preliminary clue whether a linkage between their mechanisms of action has to be expected or not. According to Bühler et al. (1978), the following theoretical possibilities are given:
a) Additive interaction. The response caused by the two factors applied simultaneously is identical with the sum of the responses caused by the two factors applied separately. In this case two completely different mechanisms of action have to be expected.
b) Multiplicative interaction. The response caused by the two factors applied simultaneously is a defined fraction of the response obtained with one factor irrespective of the magnitude of the response. Such an interaction means that the two factors are acting over the same pathway, but independently of each other and at different sites.
c) Any other outcome means direct interference between the two factors (one possibility is that one factor modifies the magnitude of the other).

---

[1] Centro di Studio del CNR per la Biologia Cellulare e Molecolare delle Piante, Università degli Studi di Milano, Istituto di Scienze Botaniche, via G. Colombi, 60, 20133 Milano, Italy

The interaction between BA and light can also give some preliminary information whether "class 2" responses to BA are all related to each other or not. If BA induces all of these responses through a common mechanism there should exist only one pattern of interaction between BA and light, regardless of the particular type of response.

We used the fungal toxin fusicoccin (FC) as another tool for analyzing the relationships between the different responses of cotyledons to BA. FC is known to stimulate cell expansion in cucurbit cotyledons (Marré et al., 1974), but other effects of the toxin on this material have not yet been reported, to our knowledge. If "class 1" responses drive "class 2" responses as has been suggested (Bewley and Black, 1978), then FC should be expected to mimick "class 2" responses as well.

2 Materials and Methods

Watermelon (*Citrullus vulgaris* Schrad., cv. Fairfax) seeds were imbibed for 24 h at 27°C in the dark. The cotyledons were then excised from the embryo axis and cultured for 5-6 days in Petri dishes on filter paper saturated with water or with the growth regulator solution. The cotyledons were grown either in continuous darkness or in continuous white light. Sylvania Gro-Lux fluorescent lamps were used as light source; energy fluence rate was 950 $\mu W/cm^2$. All other experimental details are described in former papers (Longo et al., 1979a, c).

3 Results and Discussion

Figure 1 shows the effect of a simultaneous treatment with light and BA on growth and on hydroxypyruvate reductase (HPR) level. Stimulation of HPR is a typical "class 2" response to BA while stimulation of growth is a "class 1" response. Although light is not active in itself on cotyledon expansion it could be expected to modify the growth response to BA. This is, however, not the case: growth in presence of BA is unaffected by light. This insensitivity to light has proved to hold also for other "class 1" responses; e.g., lipid breakdown (not shown).

HPR activity is enhanced both by BA and light. The type of interaction between the two factors is multiplicative. This sort of interaction holds over a wide range of BA concentrations including the optimal one (10^{-5} M) as shown in Fig. 2.

The multiplicative interaction between light and BA is common to a whole group of effects, all of which are related to the development of plastids and peroxisomes (see Table 1).

The multiplicative interaction of BA and light on plastids and peroxisomes allows the following deductions. (a) It seems very unlikely that light acts on the development of these organelles by raising endogenous cytokinin levels. If this were true, one would expect little or no enhancement by light when the response is fully saturated with BA. (b) It is likely that BA and light act on the same sequence of events that leads to development of plastids and peroxisomes, but at different sites and independently from each other. (c) There may be a common control of plastid and peroxisome development in the greening cotyledon. Such a common control (through light, endogenous

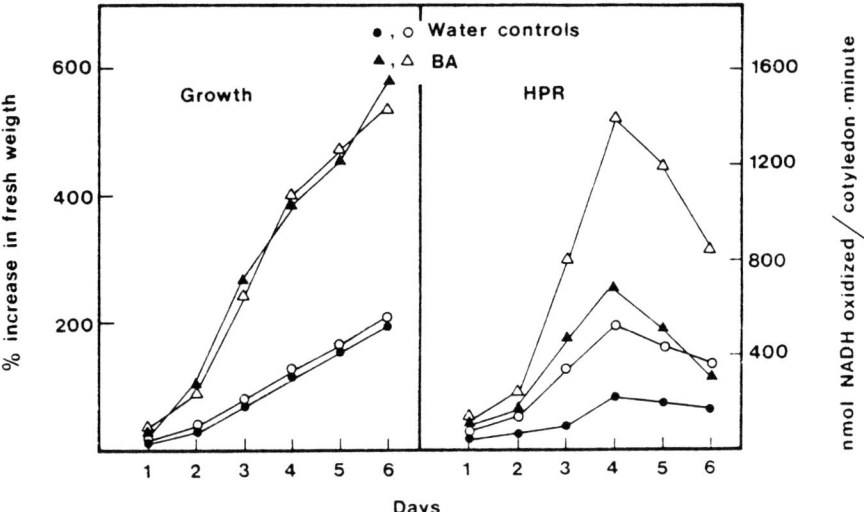

Fig. 1. Action of light and *BA* on growth and hydroxypyruvate reductase (*HPR*) level of excised watermelon cotyledons. The cotyledons were treated with 10^{-4} M BA for 4 h immediately after excision. This treatment has the same effect as a continuous exposure to 10^{-5} M BA. *Full symbols* dark; *empty symbols* light. The time of excision of the cotyledons was taken as day zero

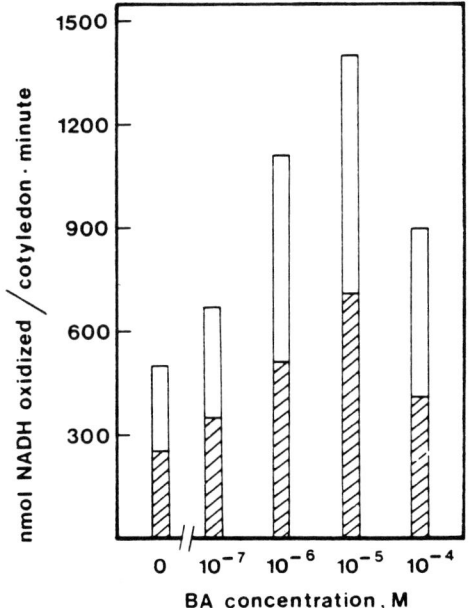

Fig. 2. Stimulation by BA of hydroxypyruvate reductase level in light and in darkness as a function of BA concentration. Enzyme activities were measured 4 days after excision

Table 1. Effect of light on some enzyme activities of watermelon cotyledons in presence and in absence of BA. The cotyledons were treated with 10^{-4} M BA immediately after excision. They were then grown either in continuous light or in continuous darkness. Enzyme activities were measured 4 days after excision. The results are reported as ratios between enzyme activity per cotyledon in the light and the corresponding activity in darkness

Enzyme activity	Activity in light/Activity in darkness	
	Water control	BA
Ribulose 1,5-diphosphate carboxylase	1.43	1.49
NADPH-glyoxylate reductase	1.96	1.75
Catalase	0.80	0.76
Hydroxypyruvate reductase	2.15	2.02
Glycolate oxidase	2.12	2.05

cytokinins, etc.) seems reasonable in view of the close functional integration between the two types of organelles.

The study of interactions between BA and light has thus contributed to put in a sharper focus the differences between the two classes of responses to the hormone. Light does not modify "class 1" responses, but it interacts multiplicatively with BA on all "class 2" responses tested.

The distinction between the two classes of responses suggests that BA may act on cotyledons by at least two completely distinct mechanisms. Alternatively it is possible that BA induces some "primary response" that, in turn, evokes all other responses. Some authors think that the second hypothesis may be more correct. According to them, cytokinins act on cotyledons just by stimulating cell expansion. This type of growth involves synthesis of new cell wall material from reserve lipids. The conversion of lipids to cell wall material creates an internal sink inside the cotyledons: it is this sink that represents the driving force for all other responses (Bewley and Black, 1978).

According to this hypothesis the growth response to BA (a "class 1" response) should entrain all "class 2" responses. We have used the fungal toxin fusicoccin (FC) to test whether this hypothesis was correct. FC promotes growth by cell expansion in a wide range of plant materials including cotyledons (Marré at al., 1974). The growth promotion is believed to occur through a single well-defined mechanism, i.e., activation of proton extrusion by a pump presumably located in the plasmalemma (Marré, 1979). If expansion growth really provided the driving force for the other responses, then FC should not only stimulate growth, but duplicate the other responses to BA as well. Figure 3 shows that this hypothesis is not correct. Treatment with FC strongly stimulates expansion in watermelon cotyledons (the magnitude of the response is about the same as for treatment with BA), but fails to stimulate carotenoid synthesis,

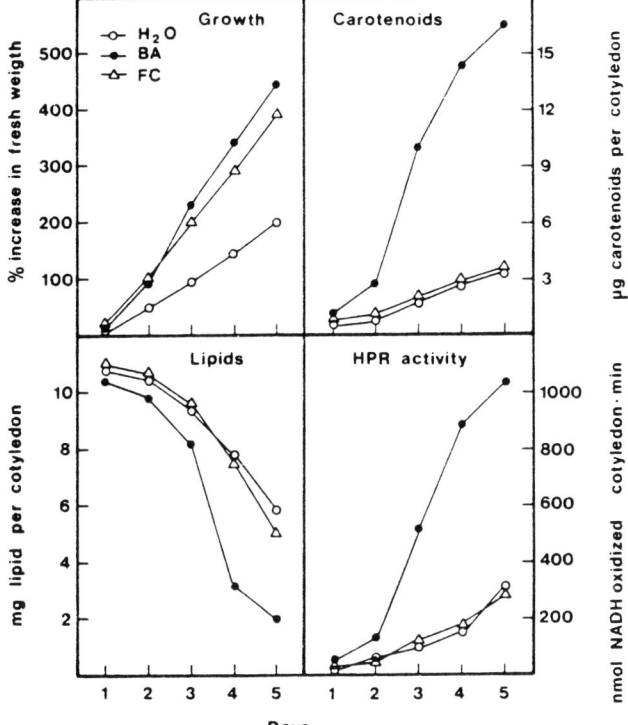

Fig. 3. Effect of *BA* and fusicoccin (*FC*) on growth, carotenoid level, lipid breakdown and hydroxypyruvate reductase (*HPR*) activity. The cotyledons were grown in the dark in continuous presence of the relevant growth regulator. Concentrations of BA and FC were 10^{-5} M

lipid breakdown and HPR activity. Also the level of cytochrome oxidase (a "class 1" response to BA) is unaffected (not shown). On the other hand FC and BA seem to stimulate growth through the same basic mechanism. In both cases a correlation between H^+ extrusion and growth responses has been shown (Marré et al., 1974). Simultaneous treatment of cotyledons with BA and FC yields a less than additive growth response as should be expected if the two factors act via the same mechanism (C.P. Longo, unpubl. data).

The experiments with FC suggest that, at least in watermelon cotyledons, cell expansion does not provide the driving force for organelle development. The toxin even failed to mimick all "class 1" responses to BA since lipid breakdown and cytochrome oxidase activity were hardly stimulated.

4 Conclusion

The evidence obtained through experiments with light and FC does not support the hypothesis of a multiple cytokinin action mediated by a single primary response (e.g., proton extrusion). A more likely model for describing the action of BA on cotyledons seems to be the following:

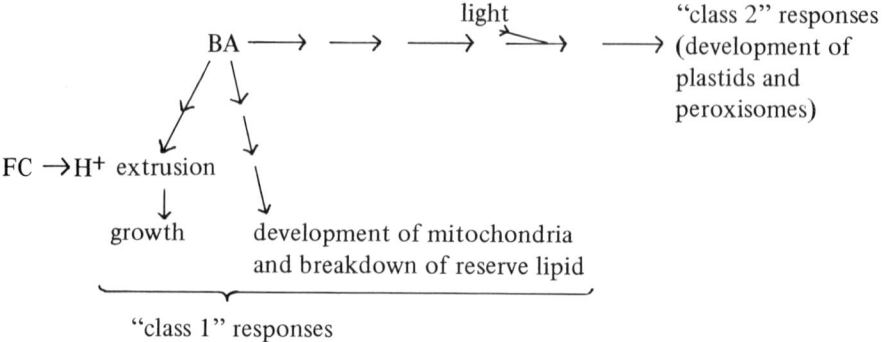

This model was intended only to accommodate the present results of our work. The actual action mechanism of BA will probably turn out to be even more complex. It is quite possible, as an example, that lipid breakdown and development of mitochondria are two completely unrelated responses.

References

Beevers H (1979) Microbodies in higher plants. Ann Rev Plant Physiol 30: 159-186
Bewley JD, Black M (1978) Physiology and biochemistry of seeds in relation to germination, vol. 1. Springer, Berlin Heidelberg New York
Bühler B, Drumm H, Mohr H (1978) Investigations on the role of ethylene in phytochrome-mediated photomorphogenesis. Planta 142: 109-117
Gruber PJ, Trelease RN, Becker WM, Newcomb EH (1970) A correlative ultrastructural and enzymatic study of cotyledonary microbodies following germination of fat-storing seeds. Planta 93: 269-288
Kagawa T, Mc Gregor DI, Beevers H (1973) Development of enzymes in cotyledons of watermelon seedlings. Plant Physiol 51: 66-71
Longo GP, Pedretti M, Rossi G, Longo CP (1979a) Effect of benzyladenine on the development of plastids and microbodies in excised watermelon cotyledons. Planta 145: 209-217
Longo GP, Pedretti M, Rossi G, Longo CP (1979b) Benzyladenine stimulates the development of mitochondria in watermelon cotyledons. Plant Sci Lett 14: 213-223
Longo GP, Lampugnani MG, Servettaz O, Rossi G, Longo CP (1979c) Evidence for two classes of responses of watermelon cotyledons to benzyladenine. Plant Sci Lett 16: 51-57
Marré E (1979) Fusicoccin: a tool in plant physiology. Annu Rev Plant Physiol 30: 273-288
Marré E, Colombo R, Lado P, Rasi-Caldogno F (1974) Correlation between proton extrusion and stimulation of cell enlargement. Effects of fusicoccin and cytokinins on leaf fragments and isolated cotyledons. Plant Sci Lett 2: 139-150

Studies on the Action of Cytokinin and Light in RNA Synthesis of Pumpkin Cotyledons by Autoradiography

D. Neumann[1] and W.A. Khokhlova[2]

1 Introduction

Although the plastids are autonomous to a certain degree in their RNA and protein metabolism, close connections exist with the nucleo-cytoplasmic compartment. Only a certain proportion of plastid proteins is coded for by chloroplast DNA and is synthesized within the organelle. Many plastid components are under nuclear control (Parthier et al., 1975). The dramatic effects of cytokinin on the differentiation of plastids in excised pumpkin cotyledons (Khokhlova et al., 1971; Khokhlova, 1977; Kulaeva et al., 1973) encourages us to study the role of phytohormones in the regulation of transcription in the two compartments.

Khokhlova (1977) has shown that in excised pumpkin cotyledons the differentiation of plastids is stimulated by 6-benzylaminopurine (BAP) in the dark and in the light. The volume of the plastids, as well as the formation of the membrane system, are increased by the action of BAP. The hormone accelerates the normal differentiation process, like many other processes in different plant systems. The stimulation in the structural development of the plastids is correlated with accelerations in the synthetic pathways of many of their constituents (refs. see Parthier, 1979). We have studied by means of autoradiography whether the effect of the hormone in stimulating plastid development is controlled on the nucleo-cytoplasmic level of RNA synthesis.

---

[1] Institute of Plant Biochemistry, Academy of Sciences of the GDR, 401 Halle, Weinberg, GDR
[2] Timiriazev Institute of Plant Physiology of the Academy of Sciences of the USSR, Botanicheskaya, ul 35, 127106 Moscow, USSR

2 Materials and Methods

Cotyledons were isolated from dry seeds of pumpkin (*Cucurbita pepo* L.) and placed in petri dishes on distilled water or 6-benzylaminopurine solution (10 mg/l). The dishes were put in a damp air-conditioned chamber at 28°C in complete darkness or in continuous light (fluorescent tubes, 2500 lux).

After 24, 48, 72, and 96 h, small pieces of the cotyledons were incubated with [^{3}H]-5-uridine (spec. act. 10 C/mmol) for 2 h. The specimens were fixed with 1% OsO_4 according to Kellenberger at al. (1958). Controls were incubated with RNase (0.1 mg/l) for 4 h at 37°C after 15 min OsO_4-fixation. Specimens were dehydrated in a graded series of acetone and embedded in Durcupan ACM. For light microscopic investigations, semithin sections (1-2 μm) were stained with azure II/methylene blue. For autoradiography ultrathin sections were covered with a monolayer of Ilford L 4 emulsion following the method of Caro and van Tubergen (1962). After 2 months exposition, the grids were developed in Kodak D 19b (2 min) and stained with Pb.

3 Results and Discussion

Cotyledons grown in the dark on water for 96 h contain small proplastids embedded in a large number of spherosomes. Labeling of these structures is very rare (Fig. 1), only a distinct incorporation of [^{3}H]-uridine into the nuclei is observed (Fig. 2). BAP stimulates the development of plastid structures in the dark and the incorporation of uridine into plastids and also into nuclei (Figs. 3, 4).

Cotyledons grown in continuous light show a normal plastid differentiation and labeling of the organelles after [^{3}H]-uridine treatment (Fig. 5, 6). In these cotyledons BAP stimulates plastid differentiation. The size of the plastids, their membrane system and the labeling after [^{3}H]-uridine incorporation is increased in comparison with water-treated cotyledons (Figs. 7, 8). In our investigations, light and the hormone cause a synergistic effect on the labeling pattern. Cycloheximide and CAP have no effect on the incorporation of [^{3}H]-uridine into plastids and nuclei after 20 h preincubation with the inhibitors.

The development of plastids and plastid RNA synthesis is strongly inhibited by ABA (10 mg/l; Fig. 9). Labeling of nuclear RNA seems to be stimulated by ABA (Fig. 10). Simultaneous action of ABA and BAP "neutralized" the effect of ABA on differentiation and RNA synthesis of plastids (Figs. 11, 12). In all experiments RNase controls exhibit no specific labeling. The single silver grains arise from the background of the nuclear emulsion.

One of the disadvantages of the method used is the difficulty in the quantitative evaluation of the results. However, quantitative determinations of the RNA content and synthesis confirm the autoradiographic results (Mikulovich et al., 1978). Total RNA content as well as plastid and cytoplasmic ribosomal RNA amounts increase after BAP treatment. Similar results were obtained by Kinoshita et al. (1979) with cucumber cotyledons.

In our material the increase in plastid and nuclear RNA contents reflect not only hormone-stimulated RNA synthesis but may also be caused by cell division processes.

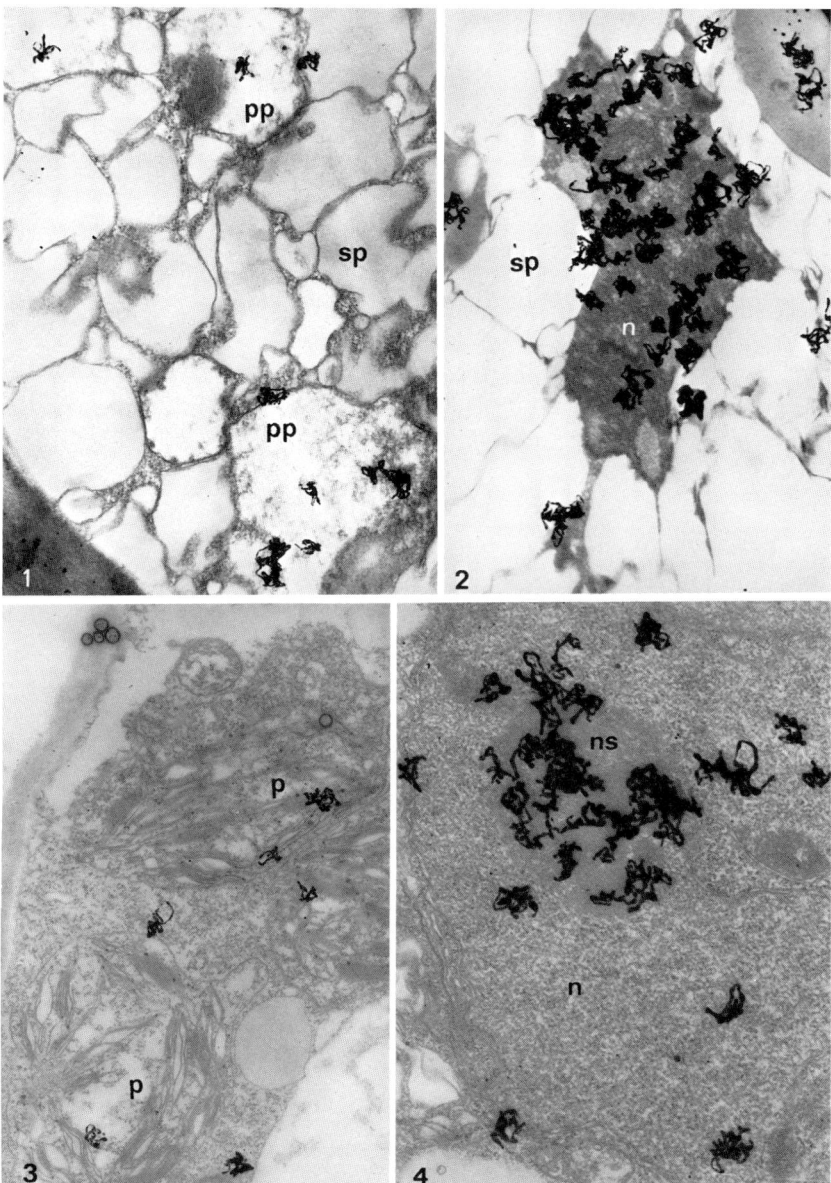

Figs. 1-4. Incorporation of [^{3}H]-uridine into the RNA of plastids and nuclei in the dark. After 96 h growth in darkness on water or BAP, the material was incubated for 2 h with [^{3}H]-uridine. Fixation: OsO_4, proplastid (*pp*), plastid (*p*), spherosome (*sp*), nucleus (*n*), nucleolus (*ns*)

Fig. 1. Proplastids of water-treated cotyledons in the dark; 12,000 x

Fig. 2. Nucleus of water-treated cotyledons in the dark; 12,000 x

Fig. 3. Plastids of BAP-treated cotyledons in the dark; 16,000 x

Fig. 4. Nucleus of BAP-treated cotyledons in the dark; 24,000 x. Note the stimulation of the structural development by BAP in darkness

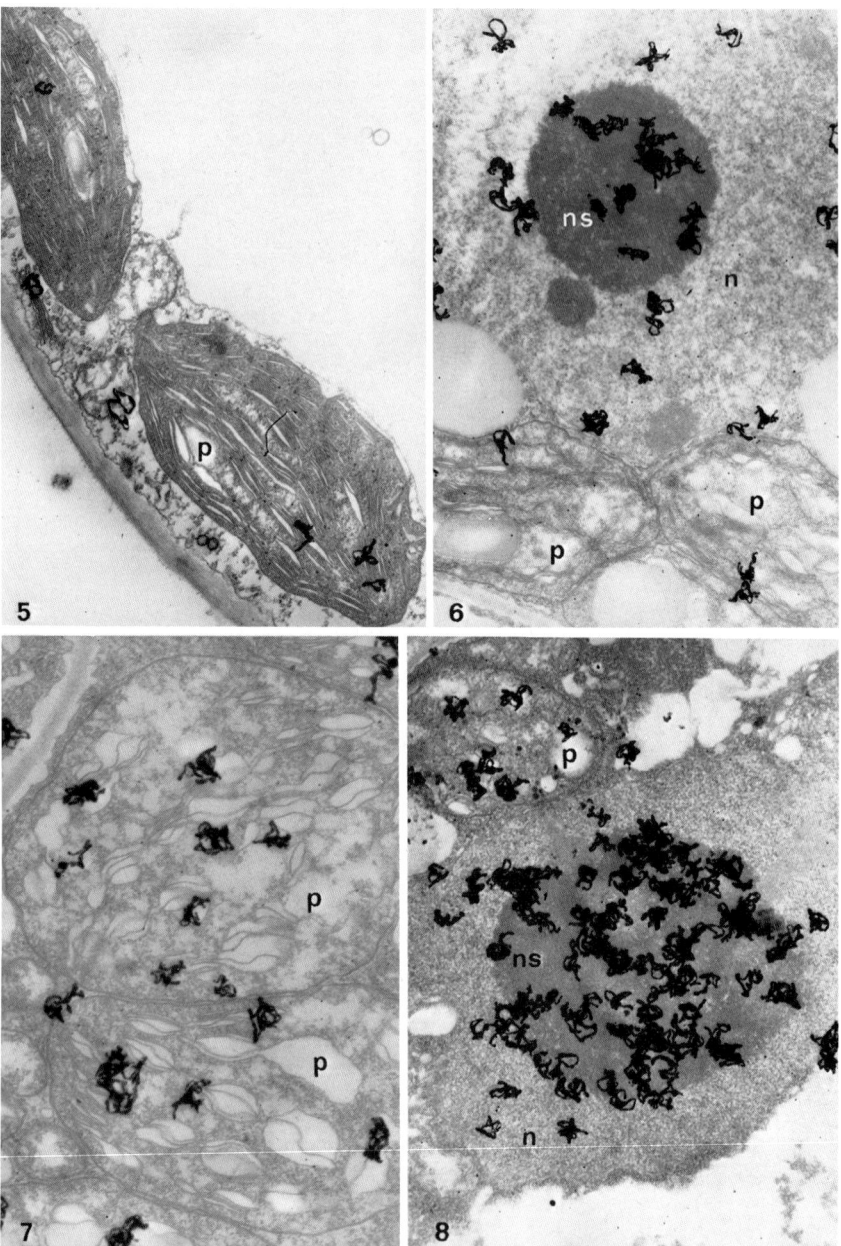

Fig. 5-8. Incorporation of [^{3}H]-uridine into the RNA of plastids and nuclei in the light. After 96 h growth in the light on water or BAP, the material was incubated for 2 h with [^{3}H]-uridine. Fixation: OsO_4

Fig. 5. Plastids of water-treated cotyledons in the light; 16,000 x

Fig. 6. Nucleus of water-treated cotyledons in the light; 18,000 x

Fig. 7. Plastids of BAP-treated cotyledons in the light; 24,000 x

Fig. 8. Nucleus of BAP-treated cotyledons in the light; 16,000 x. Note the increased labeling in the BAP-treated plastids

Studies on the Action of Cytokinin and Light in RNA Synthesis of Pumpkin Cotyledons 271

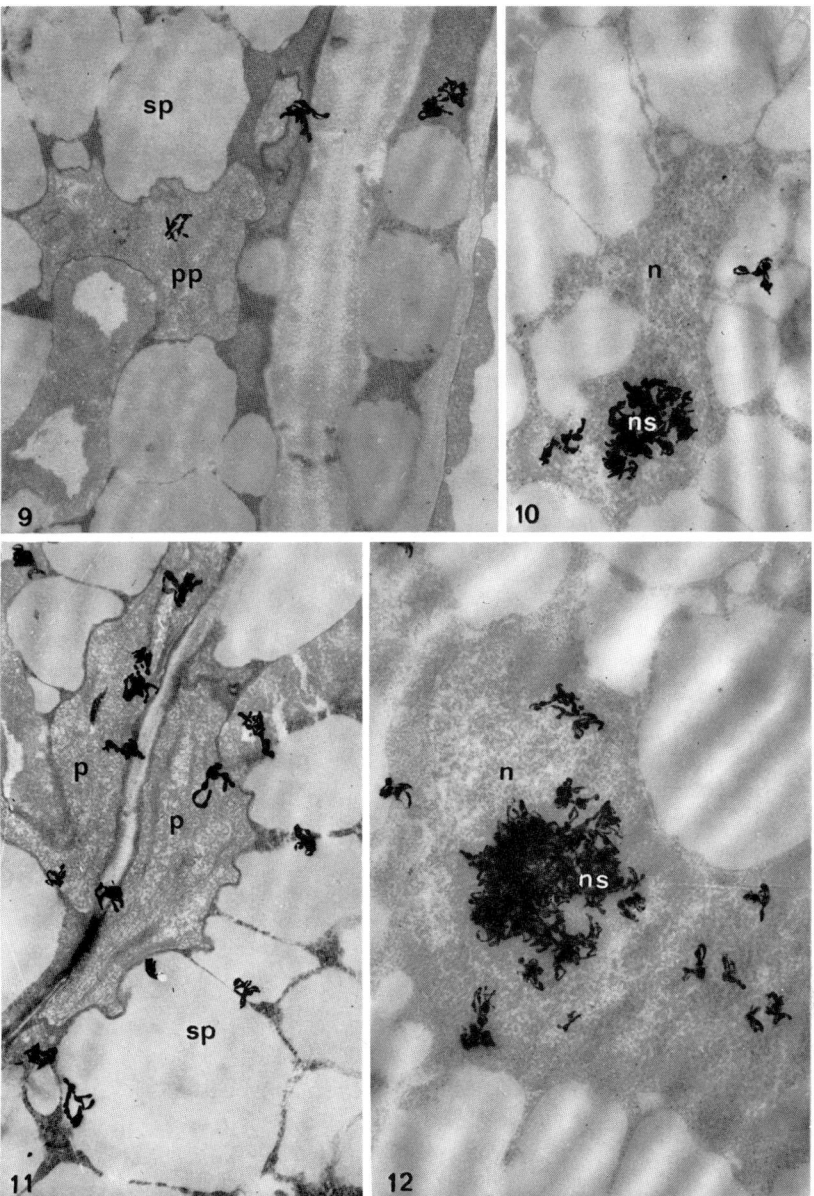

Fig. 9-12. Incorporation of [^{3}H]-uridine into RNA of plastids and nuclei under the influence of ABA. After 60 h growth in darkness on water, the material was preincubated for 3 h with ABA and subsequently with [^{3}H]-uridine for 2 h.
Fig. 9. Incorporation of [^{3}H]-uridine into plastids of ABA-treated cotyledons. 20,000 x
Fig. 10. Incorporation of [^{3}H]-uridine into the nucleus of ABA-treated cotyledons. 20,000 x
Fig. 11. Incorporation of [^{3}H]-uridine into plastids of ABA + BAP-treated cotyledons. 20,000 x
Fig. 12. Incorporation of [^{3}H]-uridine into the nucleus of ABA + BAP-treated cotyledons. 20,000 x.
Note the inhibition of plastid RNA labeling by ABA and the compensation of this effect by BAP

The number of palisade parenchyma cells in BAP-treated cotyledons and the number of labeled nuclei in light microscopic autoradiographs after [^{3}H]-thymidine incorporation have increased in comparison to water controls. This indicates that the hormone-promoted growth of the cotyledons is the result of cell elongation plus cell division.

The stimulation of RNA synthesis of plastids by BAP can be discussed on the basis of the following considerations:

1. BAP acts on the immediate precursor pools for RNA? Experiments have been done in which isolated cotyledons cannot take up external precursors. This excludes an influence by BAP on the external uptake system. However, we cannot exclude an effect on the changes in distribution of RNA precursors within and between intracellular pools.
2. The regulation of the plastid RNA synthesis is influenced on the nucleo-cytoplasmic level by stimulation of plastid RNA polymerase synthesis? The labeling experiments after preincubation with cycloheximide and chloramphenicol show no effect on [^{3}H]-uridine incorporation into plastid and nuclear RNA. A regulation on the nucleo-cytoplasmic level seems to be unlikely according to these experiments.
3. The BAP stimulated RNA synthesis is caused by a regulation of RNA polymerase activity on the plastid level? We have no experimental evidence supporting this idea directly, but some indications point in this direction: In isolated pumpkin cotyledons, BAP influences plastid differentiation at different levels by promoting the development of the membrane system, by increasing the size of chloroplasts, chlorophyll content and the activity of photosynthetic enzymes with lag phases of about 24 h. These different actions led to the conclusion that BAP does not affect the activity of the plastid RNA polymerase in a direct way but influences it via more common BAP action at other levels. Some indications for this assumption came from investigations of the ABA action on RNA synthesis in isolated pumpkin cotyledons. Sussex et al. (1975) and Walbot et al. (1975) showed with other systems similar inhibition effects of ABA on RNA synthesis and growth. ABA seems to be a specific counterpart in the action of BAP.

Many results concerning (a) primary action(s) of ABA point to effects at the membrane level in connection with changes in the compartmentation of ions (Milborrow, 1974). The significance of such changes in ion concentrations for regulation of enzyme systems is not understood at the moment. In general the activity of enzymes depends strongly from particular ions, e.g., the activity of RNA polymerase I is regulated by divalent cations (Mg, Mn) (Grossmann and Seitz, 1979) and the cytokinin-dependent betacyanin synthesis is stimulated by monovalent cations but inhibited by divalent cations (Elliott, 1979). The situation will be more complicated by the fact that subsequently with ion movements the precursor pools can be changed. It seems possible that those effects also influence possible actions by cytokinin on transcription and translational levels (Trewavas 1976).

The various biological responses to cytokinins, the antagonistic effects to ABA and first indications in the literature (Elliott, 1979; Sonka, 1976; Göring and Mardanov, 1976) led to the conclusion that one site of action of cytokinins is, may be, the cell membranes.

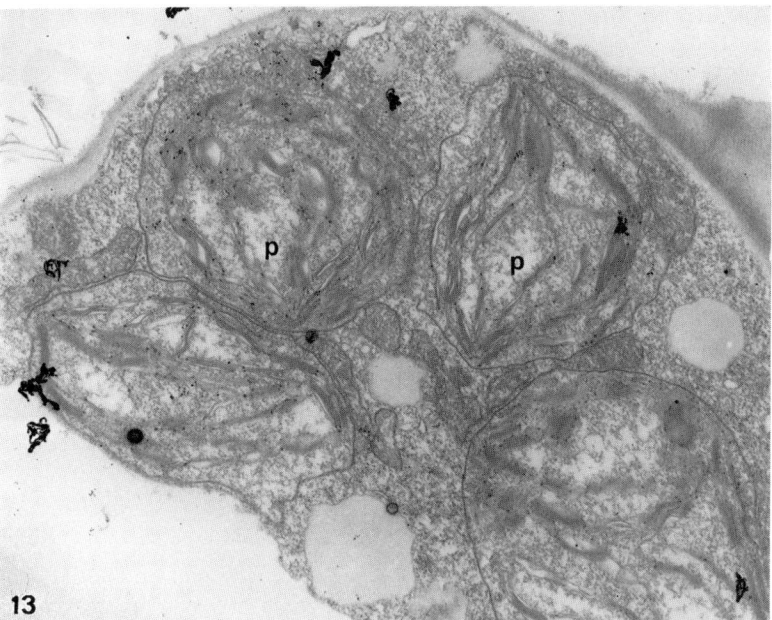

Fig. 13. RNase control after 96 h growth with BAP in the light and labeling with [^{3}H]-uridine for 2 h. 16,000 x

References

Caro LG, van Tubergen RP (1962) High resolution autoradiography. J Cell Biol 15: 173-199
Elliott DC (1979) Ionic regulation for cytokinin-dependent betacyanin synthesis in Amaranthus seedlings. Plant Physiol 63: 264-268
Göring H, Mardanov AA (1976) Beziehungen zwischen K$^+$/Ca^{++}Verhältnis im Gewebe und der Wirkung von Zytokininen in höheren Pflanzen. Biol Rundsch 14: 177-189
Grossmann K, Seitz U (1979) RNA polymerase I from higher plants. Evidence for allosteric regulation and interaction with nuclear phosphatase activity controlled NTP pool. Nucleic Acid Res 7: 2015-2029
Kellenberger E, Ryter A, Sechaud J (1958) Electron microscope study of DNA-containing plasms. II Vegetative and mature phase DNA as compared with normal bacterial nucleoids in different physiological states. J Biophys Biochem Cytol 4: 671-678
Khokhlova WA (1977) The effect of cytokinin on plastid formation in excised pumpkin cotyledons in the light and in the dark. Fiziol Rast 24: 1189-1192
Khokhlova WA, Sveshnikova IN, Kulaeva ON (1971) The effect of phytohormones on the formation of the chloroplast structures in isolated pumpkin cotyledons. Cytologia 19: 1023-1033
Kinoshita I, Katagiri K, Tsuji H (1979) Effects of benzyladenine and light on changes in various RNA species in etiolated cucumber cotyledons. Plant Cell Physiol 20: 707-713
Kulaeva ON (1973) Cytokinins, their structure and function. Nauka, Moscow
Kulaeva ON, Erkeev MK, Khokhlova WA, Sveshnikova IN (1972) Effect of phytohormones on physiological processes in isolated pumpkin cotyledons. Fiziol Rast 19: 1023-1033
Mikulovich TP, Wollgiehn R, Khokhlova WA, Neumann D, Kulaeva ON (1978) Synthesis of plastid and cytoplasmic ribosomal RNAs in isolated pumpkin cotyledons. I Effect of cytokinin and light. Biochem Physiol Pflanz 172: 101-110

Milborrow BU (1974) The chemistry and physiology of abscisic acid. Annu Rev Plant Physiol 25: 259-307

Parthier B (1979) The role of phytohormones (cytokinins) in chloroplast development. Biochem Physiol Pflanz 174: 173-214

Parthier B, Krauspe R, Munsche D, Wollgiehn R (1975) The biogenesis of chloroplasts. In: Harborne JB, van Sumere CF (eds) The chemistry and biochemistry of plant proteins. Academic Press, London New York, p 167

Sonka J (1976) Biphasic regulation of transport through plant cell membranes by kinetin and its possible relation to direct transport between source and sink. Experientia 32: 1010-1011

Sussex J, Clutter M, Walbot U (1975) Benzyladenine reversal of abscisic acid inhibition of growth and RNA synthesis in germinating bean axes. Plant Physiol 56: 575-578

Trewavas AJ (1976) Plant growth substances. In: Bryant JA (ed) Molecular aspects of gene expression in plants. Academic Press, London New York, pp 249-298

Walbot V, Clutter M, Sussex J (1975) Effects of abscisic acid on growth, RNA metabolism and respiration in germinating bean axes. Plant Physiol 56: 570-574

Plastogenesis and Cytokinin Action. Cytokinin and Light Interactions in Plastid Enzyme Formation of Detached *Cucurbita* Cotyledons

B. Parthier[1], S. Lerbs[1], and N.L. Klyachko[2]

1 Introduction

Cytokinins enhance growth and intracellular differentiation of isolated plant organs or cells. Of particular interest is the light-induced greening process in response to hormone addition (Stetler and Laetsch, 1965; Feierabend, 1969; Fletcher and McCullagh, 1971; Khokhlova et al., 1971; Mikulovich et al., 1971), i.e., the interaction of light and cytokinins in plastogenesis, especially in the formation of photosynthetically active chloroplasts. The question arises as to how this differentiation process is primarily controlled by cytokinins, and if so, in which way and at which level hormone and light interact.

There is ample evidence that both light and cytokinins promote plastid gene expression (Harvey et al., 1974; Frosch et al., 1976; Smith et al., 1977; Mikulovich et al., 1978; Romanko et al., 1976; Feierabend and de Boer, 1978; Klyachko et al, 1979; Kinoshita et al., 1979; Lampugnani et al., 1980; Roussaux et al., 1980; for further refs cf. Parthier, 1979). However, their primary sites and modes of action are difficult to localize, since functional chloroplasts are the products of both plastid and nuclear genes. Induction or stimulation of chloroplastogenesis by light and cytokinin, therefore, can reflect involvement in various steps of transcription and translation in the plastid an nucleo-cytoplasmic compartments, as shown in the oversimplified scheme of Fig. 1. At the molecular level neither of the actions has been elucidated in detail, nor are the interactions of the two signal systems understood.

In order to learn more about the involvement of cytokinins in light-induced chloroplastogenesis we have studied the effect of 6-benzylaminopurine and white light on activity and formation of two types of plastid enzymes, RuBPCase and aminoacyl-tRNA synthetases. The latter enzyme group has been chosen, because each organelle isoenzyme species which binds a certain amino acid to its cognate isoacceptor tRNA species exists as a counterpart isoenzyme in the cytosol (Weil, 1979). Both isoenzymes are synthesized on cytoplasmic polyribosomes (Parthier, 1973). Comparison, during

---

Abbreviations: BAP, 6-benzylaminopurine; Leu-RS, leucyl-tRNA synthetase; RuBPCase, ribulose-1,5-bisphosphate carboxylase

[1] Institute of Plant Biochemistry, Academy of Sciences of the GDR, Halle, GDR
[2] Institute of Plant Physiology, Academy of Sciences of the USSR, Moscow, USSR

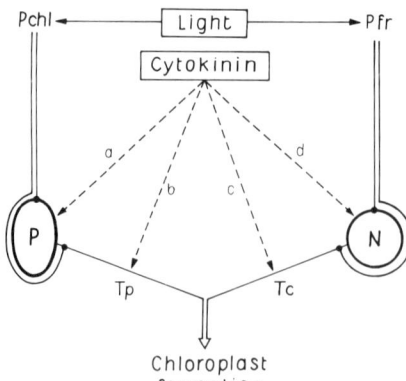

Fig. 1. Possible sites of action and interaction of light and cytokinin in chloroplast biogenesis. P plastid; N nucleus; *Pchl* protochlorophyll(ide); *Pfr* phytochrome; T_p plastid translation system; T_c cytoplasmic translation system. Cytokinin influence on: *a* plastid gene activation and/or transcription; *b* plastid translation (or post-translation); *c* cytoplasmic translation (or post-translation); *d* nuclear gene activation and/or transcription

chloroplastogenesis, of the two isoenzyme activities should discriminate a subsequent action of cytokinins in enzyme compartmentation from a possible stimulation of enzyme synthesis.

RuBPCase consists of two biogenetically different types of subunits. The large subunits are encoded and synthesized inside the plastid, while the small ones are formed on cytoplasmic 80S ribosomes as precursor polypeptides, which are processed to the final size during transmembrane import into the chloroplasts (Ellis, 1977; Chua and Schmidt, 1979). Assembly of the two types of subunits takes place in the stroma compartment and results in the functional holoenzyme. Studies on the influence of cytokinins on the biosynthesis and the activity of this enzyme, as a major plastid component, should help to discriminate between the site(s) of action of the hormone in plastogenesis inside or outside the organelles, or in both plastid and nucleo-cytoplasmic compartments.

Several data of this paper correspond with our results in two previous papers (Klyachko and Parthier, 1980; Klyachko et al., 1980), from which technical and experimental details can be obtained. Here we deal with the kinetics of enzyme activities of excised pumpkin cotyledons cultivated under various BAP and light regimes (Fig. 2). Preliminary data on RuBPCase polypeptide synthesis in BAP-treated cotyledons and in cell-free translation systems are also reported.

2 Results and Discussion

2.1 Stimulation of Aminoacyl-tRNA Synthetase Activities

Plastid-specific Leu-RS of pumpkin cotyledons can be separated from the cytoplasmic isoenzyme using hydroxyapatite chromatography (Klyachko and Parthier, 1980). We have extended the chromatographic separation procedure also for 10 other pumpkin aminoacyl-tRNA synthetases. The elution pattern of enzyme activities are demonstrated in Fig. 3.

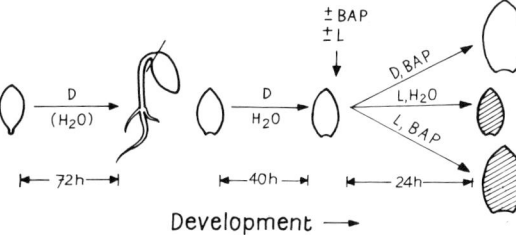

Fig. 2. Outline of the experimental processing and treatments of *Cucurbita* cotyledons as used in this paper, beginning with seed germination (*left*). D darkness; L light; *hatched areas* green cytoledons

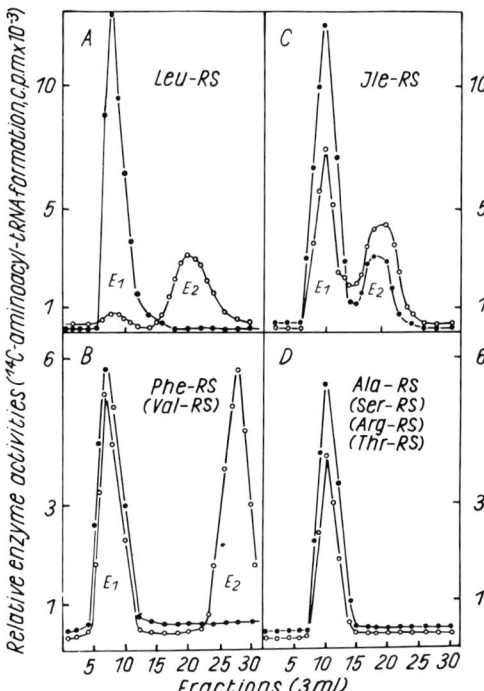

Fig. 3A-D. Hydroxyapatite chromatography of various aminoacyl-tRNA synthetases from green pumpkin cotyledons. 10 g cotyledons were homogenized with mortar and sand in 0.01 M phosphate buffer containing 1 mM Mg acetate, 0.005 M β-mercaptoethanol (pH 7.6). The homogenate was centrifuged at 100 · g and 25,000 g, and an aliquot of 60 mg protein was applied to a 15 x 1 cm hydroxyapatite column. Elution was performed with a linear 0.01 to 0.3 M potassium phosphate gradient containing 1 mM Mg acetate. Three ml fractions were collected and assayed for enzyme activities with either tRNA from *Anacystis nidulans* (●—●) or cytoplasmic tRNA from *Euglena gracilis* W_3 BUL mutant (○—○), and the respective [^{14}C]-labeled amino acids (Assay conditions see Klyachko and Parthier, 1980). The enzyme species written in *brackets* show similar elution behavior but do not necessarily elute with the same fractions as the enzyme species written on *top*

It seems obvious that Leu-RS isoenzymes are exceptional in terms of tRNA specificity. The first eluted enzyme activity (cf. Fig. 3A, E 1) aminoacylates tRNA^{Leu} from *Anacycstis nidulans* (substitutes chloroplast tRNA^{Leu}: Parthier and Krauspe, 1974), but almost does not charge cytosol tRNA prepared from the bleached *Euglena* mutant W_3BUL or rat liver. On the other hand, these two tRNA preparations contain tRNA^{Leu} species which are exclusively aminoacylated by the second peak of activity (E2). We have shown previously that E 1 isoenzyme is of organellar origin, while E2 represents the cytosol Leu-RS isoenzyme (Klyachko and Parthier, 1980).

Certain aminoacyl-tRNA synthetase isoenzyme species probably aminoacylate tRNA of prokaryotic and eukaryotic origin (Fig. 3B, peak E 1; Fig. 3C, both peaks). Other isoenzymes may not be separated by the chromatographic method used (Fig. 3D). These findings are in contrast to our observations with *Euglena* enzymes, when the majority of aminoacyl-tRNA synthetases could be separated by hydroxyapatite chromatography (Krauspe and Parthier, 1973). Thus, determination of plastid and cytoplasmic isoenzyme activities in the post-ribosomal supernatant of the cotyledon homogenate is possible for Leu-RS but fails for the other tested aminoacyl-tRNA synthetases.

Fig. 4 shows the results for plastid (A) and cytosol (B) Leu-RS isoenzyme activities during 24 h of development of BAP-treated, etiolated or illuminated pumpkin cotyledons. The hormone increases the activities per cotyledon and likewise the specific activities (now shown) or both isoenzymes (curves a vs. c; b vs. f). Plastid Leu-RS activity appears more enhanced during cotyledon development than that of the cytosol isoenzyme. Although additive to the BAP stimulation, light plays a subordinate role, especially for cytosol Leu-RS activity. In this case the small stimulatory effect may be the consequence of a general metabolic activation by light.

Light stimulation is more pronounced for plastid Leu-RS (Fig. 4A, curves a vs. b; c vs. f), thus suggesting a specific light action in addition to the cytokinin-stimulated enzyme synthesis in the cytoplasmic compartment. However, we cannot exclude that this phenomenon is the result of a general promotion of light-induced chloroplast differentiation and enlargement, possibly also due to an increase in chloroplast number per cotyledon. Nevertheless BAP-stimulated increase in plastid Leu-RS activity dominates over light stimulation, since the rate of increase becomes markedly higher, if BAP is added 6 h after illumination of the cotyledons (Fig. 4, curves d) than illumination of dark-grown cotyledons which had been pretreated for 6 h with BAP (Fig. 4, curves e).

In conclusion, the activity of the Leu-RS isoenzyme partitioned within the plastid but synthesized at cytoplasmic ribosomes is efficiently controlled by cytokinins during plastogenesis. A general master control in the nucleocytoplasmic compartment probably determines enzyme quantity by de novo synthesis. The observed stimulation by light might reflect mostly, if not exclusively, secondary effects in turn of the light-dependent chloroplast differentiation and enlargement, processes stimulated by an increase of the plastid translation machinery including aminoacyl-tRNA synthetases.

The results obtained for the aminoacylating enzymes are almost identical with the effects of BAP and light on fresh weight increase (Fig. 5A; Klyachko et al., 1980) or other cytoplasmic enzyme activities (Romanko et al., 1976). In contrast, the influence on chlorophyll accumulation is qualitatively different. Here light is a more efficient promoter than BAP (Fig. 5B), clearly indicated in the parallel rates of chlorophyll accumulation in BAP-treated cotyledons illuminated at 0 or 6 h after BAP treatment (curves f vs. e). This observations point to a light-independent effect of BAP in chlorophyll or chromoprotein biosynthesis. It is likely that cytokinins regulate the activities of protochlorophyll(ide) synthesizing enzymes, especially enhancing the amount of δ-aminol evulinate synthetase (as shown with sunflower cotyledons, Ford et al., 1979). This enzyme is characterized by a high turnover (Nadler and Granick, 1970) and thus represents a key enzyme in the chlorophyll biosynthesis pathway. Its activity may be regulated via phytohormone-controlled turnover, a process independent of the light regime control of later steps in chlorophyll biosynthesis.

Plastogenesis and Cytokinin Action, Cytokinin and Light Interactions in Plastid Enzyme. 279

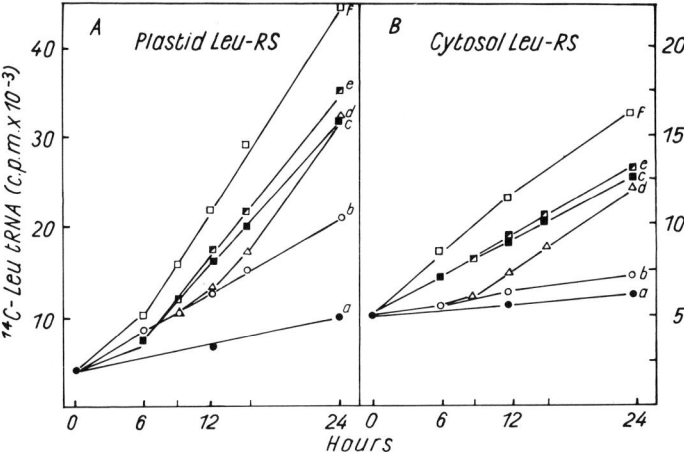

Fig. 4A, B. Interaction of BAP and light on plastid and cytosol leucyl-tRNA synthetase activities in detached pumpkin cotyledons. **A** plastid Leu-RS ([^{14}C]-Leu binding to total *Anacystis* tRNA); **B** cytosol Leu-RS ([^{14}C]-Leu binding to *Euglena* cytoplasmic tRNA). After 72 h of germination the cotyledons have been detached and cultivated on water for 40 h in darkness. Then they were selected for mean sizes and transferred to Petri dishes with water in darkness (*a*, ●), water in light (*b*, ○), BAP in darkness (*c*, ■) or BAP in light (*f*, □). Other cotyledons are treated with BAP in darkness and illuminated after 6 h (*e*, ▨) or were illuminated on water and after 6 h were transferred to BAP in light (*d*, △). Enzyme activities assayed according to Klyachko and Parthier (1980) are expressed as synthesized [^{14}C]-Leu-tRNA, calculated per cotyledon. Concentration of the BAP solution 10 mg/liter; light 3200 lux

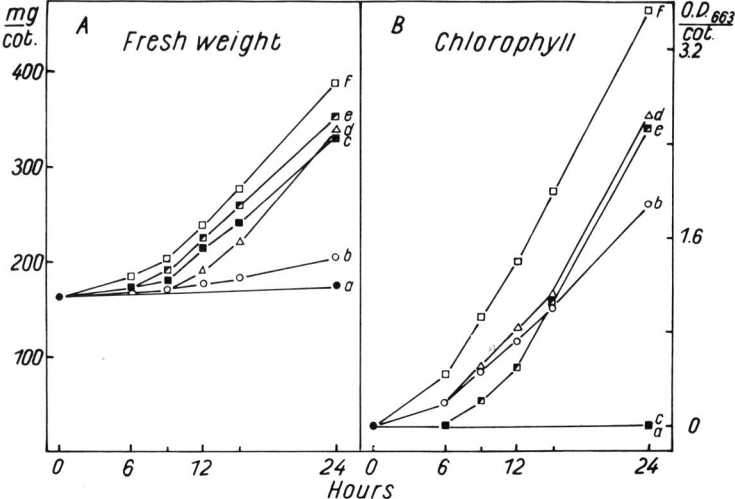

Fig. 5A, B. Interaction of BAP and light on fresh weight (**A**) and chlorophyll accumulation (**B**) in detached cotyledons. Symbols and experimental details are the same as in Fig. 4. Aliquots of the crude homogenates have been used to estimate chlorophyll concentration as optical density at 663 nm. All calculations per cotyledon

2.2 Stimulation of RuBPCase Activity

The increase of RuBPCase activity in detached, dark-grown or illuminated pumpkin cotyledons treated with BAP is illustrated in Fig. 6. Etiolated cotyledons grown on water show a low level of enzyme activity; its increase is insignificant within the experimental period. During the first 30 h after cytokinin treatment of dark-grown cotyledons, the enzyme activity remains almost constant but thereafter rapidly rises. Forty eight h after BAP-treatment the carboxylase activity in BAP-treated cotyledons is about five fold higher than in water-treated ones. These data are typical for cotyledons excised after 72 h of germination and immediately exposed to BAP solution. If cotyledons, after detachment, are transferred to water for 40-48 h followed by BAP treatment, the delay in BAP response is reduced to about 6-9 h (cf. also Figs. 4 and 7). This phenomenon is suggested to reflect the different amounts of endogenous cytokinins in the tissues and, consequently, the growing sensitivity toward exogenous cytokinin with increasing time after detachment.

Exposure of BAP-treated tissues to light results in a lag-free increase in the rate of RuBPCase enzyme activity (Fig. 6). This light response seems not to be dependent on the differentiation stage of the cotyledon or on the duration of BAP-treatment, respectively. In order to learn more about the interaction between light and BAP on the RuBPCase level, we have studied the changes in enzyme activity in cotyledons subjected to varying treatment periods with BAP and light, respectively.

On exposure to light and simultaneous BAP treatment, RuBPCase activity increases with a markedly higher rate than in both illuminated control and BAP-treated darkened cotyledons (Fig. 7A). The lag phase of 3-6 h in illuminated cotyledons is shorter than that observed for BAP treatment. The reason is unclear; it might be due to the different times the signals need to reach their site(s) of action. Illumination of BAP-pretreated cotyledons results in an additional, substantial and linear increase in enzyme activity without lag phase (Fig. 7B). Illumination for 6 h of dark-grown cotyledons and subsequent BAP treatment produces no significant increase in RuBPCase activity (at least not in the following hours), in comparison with water-treated, illuminated cotyledons (Fig. 7C). This indicates that light has promoted the cells to a stage of maximal RuBPCase activity, which cannot be further increased by BAP. Finally, illumination for 6 h followed by darkness and concomitant BAP addition demonstrates that the hormone preserves enzyme activity at a level as high as the activity in continuously illuminated controls. Redarkened water-treated tissues soon slow down their increase in RuBPCase activity (Fig. 7D).

These results on the kinetics of RuBPCase activity demonstrate that (1), the combination of light and BAP always yields the highest activities, probably a reflection of both accelerated etioplast differentiation by BAP and chloroplast formation induced by light; (2), elevation of RuBPCase activity triggered by light does not substantially increase further upon subsequent BAP treatment; (3), the lag phase in the rate of light-induced RuBPCase activity seems not to be altered in etioplasts which have been processed to differentiate after BAP pretreatment in darkness; (4), the lowered increase of RuBPCase activity, observed in illuminated control cotyledons after return to darkness, can be prevented by BAP either due to maintaining the rate of enzyme synthesis in darkness or decreasing possible dark-dependent enzyme degradation. RuBPCarbo-

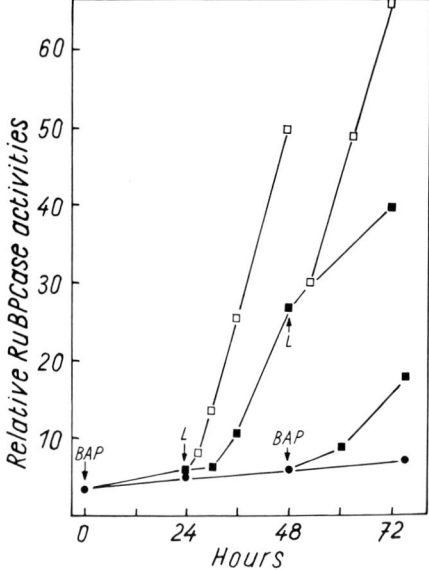

Fig. 6. RuBPCase activities in dark-grown or illuminated, *BAP*-treated detached cotyledons. Enzyme activities assayed according to Klyachko et al. (1980) are expressed as relative units (c.p.m. [^{14}C]-3-phosphoglycerate formed per cotyledon) parenthesis after cotyledon. Symbols are the same as in Fig. 4

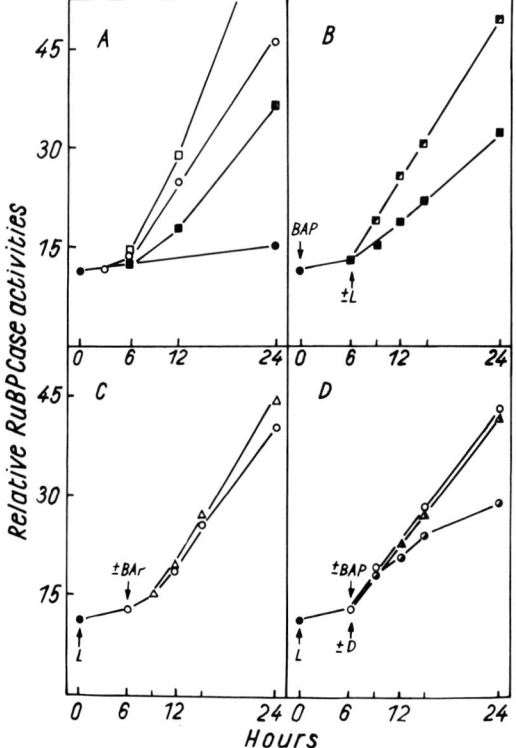

Fig. 7A-D. Interaction of *BAP* and light on RuBPCase activity during growth and greening of detached cotyledons. **A** simultaneous *BAP* treatment and illumination of detached cotyledons incubated in darkness on water for 40 h; **B** as in **A** but *BAP* treatment 6 h prior to illumination; **C** as in **A** but illumination 6 h before *BAP* treatment; **D**, as in **A** but illumination 6 h before *BAP* treatment with concomitant redarkening of the cotyledons. Symbols: ▲, illuminated cotyledons on water are transferred to *BAP* and darkness after 6 h; ⊙, illuminated cotyledons on water are transferred to darkness after 6 h. Other symbols and details are the same as in Fig. 4

xylase is degraded when barley leaves (Peterson and Huffaker, 1975) or *Lemna* plants (Tobin and Suttie, 1980) are put into darkness.

In conclusion, the comparison of cytosol and plastid enzyme activities under various light and cytokinin regimes, in accordance with similar findings of other authors (e.g., Lampugnani et al., 1980), suggests that BAP interacts with light in the developing and greening cotyledons in various ways: (1), no interaction (no specific light effect) in synthesis (activity) of constituents which are produced and localized in the cytosol; (2), both cytokinin and light stimulate synthesis (activity), but the hormone effect clearly dominates over the light effect, e.g., in plastid-located components of cytoplasmic origin; (3), cytokinin and light synergistically promote synthesis (activity) of plastid constituents synthesized within the organelle; however, light plays a dominating role.

It is still difficult to draw conclusions about similarities or differences in the actions of the two signal systems on the biosynthesis and functionalization of plastid enzymes, since quantitative and qualitative aspects are difficult to separate in this complicated multistep process. At present we can only assume that cytokinins and light control the biosynthetic pathway of plastid constituents, but the molecular sites of action might be different from each other.

2.3 Stimulation of RuBPCase Synthesis

This general suggestion derived from enzyme activity measurements should be confirmed and specified by direct studies on de novo synthesis of RuBPCase, since increase in activity can reflect either synthesis of one or both subunits in different cell compartments, or activation of the existing holoenzyme, including post-translational modifications, subunit assembly etc. To discriminate between the two possibilities it is necessary to estimate the amount of synthesized enzyme protein in comparison with changes in activity. Using monospecific antibodies raised against native *Euglena* RuBPCase as a tool for the quantitative determination of RuBPCase large subunits of pumpkin cotyledons, Klyachko et al. (1980) have shown that the increase in RuBPCase activity corresponds with the increase in labeling of immunoprecipitates, suggesting de novo synthesis of the enzyme during growth and greening of the cotyledons.

The two inhibitors of plastid and cytoplasmic protein synthesis, chloramphenicol and cycloheximide, have been used to block subunit synthesis at plastid or cytoplasmic ribosomes, respectively, thus arresting holoenzyme formation by the limitation of synthesis of either type of subunit. The results obtained are summarized in Fig. 8. They indicate that increase of RuBPCase activity of BAP-treated cotyledons, within the experimental period, is due to de novo synthesis of the enzyme (A versus B), even in darkness (panels 3 and 4). Amino acid incorporation into large subunit polypeptide and BAP-dependent increase of enzyme activity are completely blocked by chloramphenicol. Cycloheximide does not suppress the BAP-induced enhancement in activity, however, de novo polypeptide synthesis is inhibited between 30% and 80%, compared with untreated controls.

The data obtained in the presence of cycloheximide could mean that RuBPCase large subunits are synthesized, to a limited extent, in the absence of simultaneous synthesis of the small subunits in BAP-treated cotyledons. Hence, small subunits might be present in excess before inhibitor addition, resulting in unaffected rates of holoenzyme formation. These results clearly demonstrate an uncoupling in synthesis of large and small RuBPCase subunits, respectively, in BAP-treated cotyledons. This is probably not a specific BAP effect since similar observations have been reported in other developing and greening plant tissues (Criddle et al., 1971; Roy et al., 1978; Feierabend and Wildner, 1978; Barraclough and Ellis, 1979; Tobin and Suttie, 1980). This control by individual subunit pools seems to be the rule for higher plants, while in *Chlamydomonas* a strong correlation in the synthesis of both subunits was observed (Iwanij et al., 1975).

In which way do cytokinin and light mediate this process of intracellular biogenetic cooperation? There are a number of possibilities summarized by Parthier (1979). Feierabend and de Boer (1978) obtained highest accumulation of free RuBPCase small subunits in dark-grown, cytokinin-depleted rye seedlings, while light with or without addition of kinetin reduced this amount in favor of de novo formation of the holoenzyme. These findings in combination with our own kinetic results on enzyme activity (Fig. 7) and biosynthesis (Fig. 8) suggest that the limiting step in RuBPCase synthesis in the dark phases of plastogenesis seems to be de novo synthesis of large subunits. Light and to a less marked extent cytokinins accelerate this plastid-specific event. This is in accordance with an increased content of plastid polyribosomes after illumination or kinetin treatment (Feierabend and de Boer, 1978). The observation of cytoplasmic (Short et al., 1974; Feierabend, 1979) or total cell polysome formation (Klyachko et al., 1979) as an early response to cytokinin action strongly indicates that the cytoplasmic translation apparatus may be likewise involved in BAP-stimulated holoenzyme formation.

These findings point to translation regulation of enzyme synthesis by cytokinin in general and for the plastid enzymes in particular. Allosteric modifications of ribosome proteins, protein phosphorylation especially of initiation factors, tRNA function or mRNA functionalization can be discussed as possible targets for cytokinin actions (cf. Burrows, 1978), but each of them is speculative as yet. In addition, the results available at present do not exclude a control of enzyme synthesis by cytokinin or light at the level of mRNA transcription. Light-induced stimulation of the synthesis of RuBPCase large subunits (Walden and Leaver, 1978) or small subunits (Tobin and Suttie, 1980) seems to be realized in the enhanced transcription of translatable mRNA's.

We tried to differentiate between the two levels of control by quantitative estimation of the available translatable mRNA in cell-free translation systems and subsequent product analysis. In a few preliminary experiments we have used both *E. coli* and wheat germ translation systems for the determination of the proportion of translatable RNA in total RNA fractions prepared from BAP- and light-treated cotyledons. The data in Table 1 provide evidence for a marked stimulation of incorporated amino acids into acid-precipitable material using total RNA prepared from BAP-treated tissue. RNA from illuminated cotyledons shows a less promoting effect than the RNA from BAP-treated cotyledons in darkness. Since the *E. coli* system translates selectively plastid mRNA (Ellis, 1977) and the wheat germ system prefers cytoplasmic poly(A)-RNA, our results tentatively suggest that both cell compartments contain a higher proportion of translatable RNA after BAP and light treatments, respectively, relative to nontreated cotyledons.

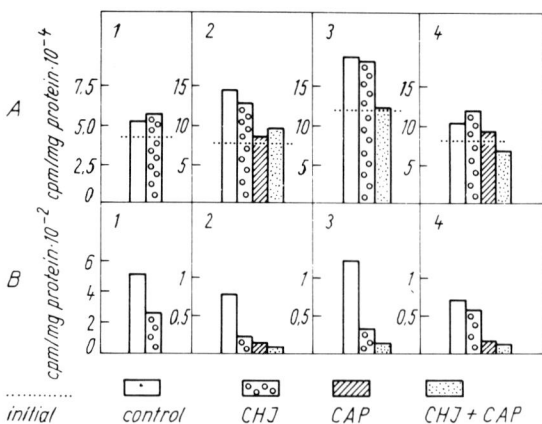

Fig. 8A, B. Effects of cycloheximide and chloramphenicol on activity (**A**) and synthesis (**B**) of RuBPCase in BAP-treated cotyledons. Dark-grown cotyledons were treated with 10 mg/l BAP for 30 h before cycloheximide (10 mg/l) or chloramphenicol (200 mg/l) were added 1 h prior to 4 h labeling with a mixture of [^{14}C]-amino acids. The cotyledons were homogenized and the post-ribosomal supernatants were used to estimate enzyme activity (**A**) or incorporation into RuBP-Case large subunit protein (**B**) as revealed by precipitation with antibodies raised against RuBP-Case large subunits from *Euglena gracilis*. Experiments carried out either in light (*1, 2*) or in darkness (*3, 4*). Initial activity as a reference was determined before the addition of the antibiotics. Controls: BAP-treated cotyledons without inhibitor addition. (Klyachko et al., 1980)

Table 1. Stimulation of [^{3}H]-leucine incorporation into polypeptides by total RNA from cotyledons in a cell-free *E. coli* or wheat germ system. 7.5 µg of total RNA prepared according to Leaver and Ingle (1971) have been added to cell-free translation systems from *E. coli* (Bottomley and Whitfeld, 1979) or wheat germs (Roberts and Paterson, 1973). Incubation 15 min at 30°C in 60 or 67 µl, resp., aliquots transferred to filters and processed for counting. Endogenous incorporation substracted

Cotyledon treatment (RNA source)	Net ^{3}H incorporated (c.p.m x 10^{-3})	
	E. coli S-30	Wheat germ S-30
0 h, darkness, water	6.2	20
24 h, darkness, water	5.9	20
9 h, light, water	6.7	32
24 h, light, water	7.9	38
9 h, darkness, BAP	9.6	40
24 h, darkness, BAP	11.7	42

Product analyses we have carried using SDS polyacrylamide gel electrophoresis and fluorography do not yet allow conclusions on alterations in RuBPCase subunit synthesis. Interestingly, few differences in the pattern of polypeptides synthesized in the wheat germ system could be discriminated using the RNA sources as indicated in Table 1. Hence, at least quantitative differences in translatable mRNA's exist in these treated cotyledons.

References

Barraclough R, Ellis RJ (1979) The biosynthesis of ribulose bisphosphate carboxylase. Uncoupling of the synthesis of the large and small subunits in isolated soybean leaf cells. Eur J Biochem 94: 165-177

Bottomley W, Whitfeld PR (1979) Cell-free transcription and translation of total spinach chloroplast DNA. Eur J Biochem 93: 31-39

Burrows WJ (1978) Cytokinins. Biochem Soc Trans 6: 1395-1400

Chua NH, Schmidt GW (1979) Transport of proteins into mitochondria and chloroplasts. J Cell Biol 81: 461-483

Criddle RS, Dau B, Kleinkopf GE, Huffaker RC (1971) Differential synthesis of ribulose diphosphate carboxylase subunits. Biochem Biophys Res Commun 41: 621-627

Ellis RJ (1977) Protein synthesis by isolated chloroplasts. Biochim Biophys Acta 463: 185-215

Feierabend J (1969) Der Einfluß von Cytokininen auf die Bildung von Photosyntheseenzymen in Roggenkeimlingen. Planta 84: 11-29

Feierabend J, Wildner G (1978) Formation of the small subunit in the absence of the large subunit of ribulose-1-5-bisphosphate carboxylase in 70S ribosome deficient rye leaves. Arch Biochem Biophys 186: 283-291

Feierabend J, de Boer J (1978) Comparative analysis on the action of cytokinin and light on the formation of ribulose-bisphosphate carboxylase and plastid biogenesis. Planta 142: 75-82

Feierabend J (1979) Role of cytoplasmic protein synthesis and its coordination with the plastidic protein synthesis in the biogenesis of chloroplasts. Ber Dtsch Bot Ges 92: 553-574

Fletcher RA, McCullagh D (1971) Cytokinin-induced chlorophyll formation in cucumber cotyledons. Planta 101: 88-90

Ford MJ, Alhadeff, M, Chapman JM, Black M (1979) A rapid and selective action of 6-benzylaminopurine on 5-aminolevulinate production in excised sunflower cotyledons. Plant Sci Lett 16: 397-402

Frosch S, Bergfeld R, Mohr H (1976) Light control of plastogenesis and ribulose bisphosphate carboxylase levels in mustard seedling cotyledons. Planta 133: 53-56

Harvey BMR, Lu BC, Fletcher RA (1974) Benzyladenine accelerates chloroplast differentiation and stimulates photosynthetic enzyme activity in cucumber cotyledons. Can J Bot 52: 2581-2586

Iwanij V, Chua NH, Siekevitz P (1975) Synthesis and turnover of ribulose bisphosphate carboxylase and of its subunits during the cell cycle of *Chlamydomonas reinhardi*. J. Cell Biol 64: 572-585

Khokhlova VA, Sveshnikova IN, Kulaeva ON (1971) The effect of phytohormones on the formation of chloroplast structure in isolated pumpkin cotyledons (russ.). Tsitologia (Moscow) 13: 1074-1079

Kinoshita I, Katagiri K, Tsuji H (1979) Effects of benzyladenine and light on changes in various RNA species in etiolated cucumber cotyledons. Plant Cell Physiol 20: 707-713

Klyachko NL, Parthier B (1980) Cytokinin control of aminoacyl-tRNA synthetases and ribulosebisphosphate carboxylase in developing and greening excised *Cucurbita* cotyledons. Biochem Physiol Pflanz 175: 333-345

Klyachko NL, Ananiev E, Kulaeva ON (1979) Effect of 6-benzylaminopurine and abscisic acid on protein synthesis in isolated pumpkin cotyledons. Physiol Veg 17: 607-617

Klyachko NL, Parthier B, Chayanova SS, Volodarsky AD, Kulaeva ON (1980) Ribulose-1,5 bisphosphate carboxylase synthesis in detached cytokinin-treated pumpkin cotyledons. Biochem Physiol Pflanz 175: 712-721

Krauspe R, Parthier B (1973) Chloroplast- and cytoplasm-specific aminoacyl-tRNA synthetases of *Euglena gracilis:* Separation, characterization and site of synthesis. Biochem Soc Symp 38: 111-135

Lampugnani MG, Martellini P, Servettaz O, Longo CP (1980) Interaction between benzyladenine and white light on excised watermelon cotyledons. Plant Sci Lett 18: 351-358

Leaver CJ, Ingle J (1971) The molecular integrity of chloroplast ribosomal ribonucleic acid. Biochem J 123: 235-243

Lord JM, Armitage TL, Merrett MJ (1975) Ribulose 1,5-diphosphate carboxylase synthesis in *Euglena*. II, Effect of inhibitors on enzyme synthesis during regreening and subsequent transfer to darkness. Plant Physiol 56: 600-604

Mikulovich TP, Khokhlova VA, Kulaeva ON, Sveshnikova IN (1971) Effect of 6-benzylaminopurine on isolated pumpkin cotyledons (russ.). Fiziol Rast 18: 98-106

Mikulovich TP, Wollgiehn R, Khokhlova VA, Neumann D, Kulaeva ON (1978) Synthesis of plastid and cytoplasmic ribosomal RNAs in isolated pumpkin cotyledons. II. Effect of cytokinin and light. Biochem Physiol Pflanz 172: 101-110

Nadler K, Granick S (1970) Controls on chlorophyll synthesis in barley. Plant Physiol 46: 240-246

Parthier B (1973) Cytoplasmic site of synthesis of chloroplast aminoacyl-tRNA synthetase in *Euglena gracilis*. FEBS Lett 38: 70-74

Parthier B (1979) The role of phytohormones (cytokinins) in chloroplast development. Biochem Physiol Pflanz 174: 173-214

Parthier B, Krauspe R (1974) Chloroplast and cytoplasmic transfer RNA of *Euglena gracilis*. Biochem Physiol Pflanz 165: 1-17

Peterson LW, Huffaker RC (1975) Loss of ribulose 1,5-diphosphate carboxylase and increase in proteolytic acticity during senescence of detached primary barley leaves. Plant Physiol 55: 1009-1015

Roberts BE, Paterson WM (1973) Efficient translation of tobacco mosaic virus RNA and rabbit globin 9S RNA in a cell-free system from commercial wheat germ. Proc Natl Acad Sci USA 70: 2330-2334

Romanko EG, Selivankina S Y, Ohmann E (1976) Effect of cytokinins on activity of a number of chloroplast and cytoplasmic enzymes (russ.). Fiziol Rast 23: 543-549

Roussaux J, Hoffelt M, Farineau N (1980) Interactions entre le cycloheximide et la 6-benzylaminopurine au cours du verdissement de cotylédons de concombre. Can J Bot 58: 1101–1110

Roy H, Costa KA, Adari H (1978) Free subunits of ribulose-1,5-bisphosphate carboxylase in pea leaves. Plant Sci Lett 11: 159-168

Short KC, Tepfer DA, Fosket DE (1974) Regulation of polyribosome formation and cell division in cultured soybean cells by cytokinin. J Cell Sci 15: 75-87

Smith H, Billett EE, Giles AB (1977) The photocontrol of gene expression in higher plants. Phytochem Soc Symp 14: 93-127

Stetler DA, Laetsch WM (1965) Kinetin-induced chloroplast maturation in cultures of tobacco tissue. Science 149: 1387-1389

Tobin E, Suttie JL (1980) Light effects on the synthesis of ribulose-1,5-bisphosphate carboxylase in *Lemna gibba* L. G-3. Plant Physiol 65: 641-647

Walden, R, Leaver CJ (1978) Regulation of chloroplast synthesis during germination and early development of Cucumis sativus. In Akoyunoglou G et al (eds) Chloroplast development. Elsevier/North Holland, Amsterdam, pp 251-256

Weil JH (1979) Cytoplasmic and organellar tRNAs in plants. In: Hall TC, Davies JW (eds) Nucleic acids in plants vol 1 CRC Press, Boca Raton Fa, pp 143-192

Cytokinin Action on RNA Synthesis in Chloroplasts

T.P. Mikulovich[1], E.R. Romanko[1], S. Yu-Selivankina[1], I.M. Kukina[1], and R. Wollgiehn[2]

1 Introduction

The influence of cytokinin on the formation and maintenance of the structure and function of chloroplasts is convincingly demonstrated by numerous investigations both on young developing leaves and cotyledons as well as mature or senescing leaves (Kulaeva, 1973; Parthier, 1979). However, little is known yet about the effect of cytokinin on the nucleic acid synthesis in plastids.

In excised pumpkin cotyledons developing from storage organ to a green photosynthesizing cotyledonary leaf, cytokinin influences plastid differentiation by promoting the development of the membrane system and the formation of grana, and by increasing the size of chloroplasts (Khokhlova et al., 1971; Mikulovich et al., 1971). The stimulation of prolamellar body formation in etioplasts was also observed (Khokhlova, 1977). Along with the action on structure of plastids, cytokinin activates chlorophyll accumulation (Mikulovich et al., 1971) and increase the activity of some photosynthetic enzymes (Karavaiko et al., 1975).

In senescing barley leaves kinetin retards the degradation of chloroplasts and prevents the decrease of chlorophyll content (Kulaeva, 1973).

We have chosen these two plant systems to investigate the cytokinin effect on transcription in chloroplasts.

2 Materials and Methods

2.1 Excised Pumpkin Cotyledons

Pumpkin seeds (*Curcurbita pepo* L. cv Mozoleyevskaya 49) were soaked in distilled water for 5 h and germinated on moist filter paper in a dark damp air-conditioned chamber at 28°C. After 4 days the cotyledons were excised, separated and placed on 10 ml of distilled water in petri dishes. All manipulations were conducted under dim green light. The excised cotyledons were incubated in total darkness at 28°C for 24 h. Next, the cotyledons were transferred to either 10 ml of 6-benzylaminopurine solution (BAP, 10 mg/l) or fresh distilled water (Kursanov et al., 1969). Half of the

---

[1] K.A. Timiriazev Institute of Plant Physiology of the Academy of Sciences of the USSR, Moscow, USSR
[2] Institute of Plant Biochemistry GDR, Halle/Saale, GDR

samples of each treatment was placed in continuous light (white fluorescent tubes, 2500 lux) at 28°C, while the other half was kept in total darkness. After 24 h the cotyledons were taken for analysis and were sterilized by dipping into 1% solution of calcium hypochlorid for 10 sec, then washed several times with sterile distilled water.

2.2 Detached Barley Leaves

The barley seedlings (cv Viner) were grown in soil-containing boxes in growth chamber. The first-formed leaf of 9-12 day-old plants were used for experiments.

2.3 Analytical Methods

Quantitative determination of RNA was assayed according to Wollgiehn and Parthier (1964).

Total RNA was extracted by phenol-chloroform method as described by Mikulovich et al. (1975).

Nucleic acids were fractionated by polyacrylamide gel electrophoresis according to Munsche and Wollgiehn (1973). The amounts of individual nucleic acids were estimated as described by Mikulovich et al. (1978).

For the labeling of nucleic acids, cotyledon discs (d = 10 mm) were sterilzed, floated on sterile water for 2 h, then incubated with $Na_2H^{32}PO_4$ (0.4 m Ci/40 discs) for 6 h in the light.

Determination of radioactivity was carried out after measurement of the optical density of the gels and drying them on filter paper. Autoradiograms of the dry gels were scanned at 600 nm and the integration values of the peaks were used as a relative measure of radioactivity.

Carotenoid content was determined as described by Metzner et al. (1965).

DNA was determined by the method of Burton (1956). To measure the rate of the Hill reaction chloroplasts were isolated in a medium consisting of 0.35 M NaCl and 0.05 M Tris-HCl buffer, pH 7.8 (in 10:1 ratio). The sediment was resuspended in 0.035 M NaCl. The rate of Hill reaction was measured as the rate of $K_3Fe(CN)_6$ reduction. The reaction mixture (1.5 ml) contained (μmol): $MgCl_2$-5, $K_3Fe(CN)_6$-2, Tris-HCl, pH 7.8-30 and chloroplast suspension (0.03-0.04 mg of chlorophyll). The irradiance was 200 Wm^{-2} for 3 min. Reaction was ceased by switching out the light and adding 10% trichloracetic acid. The Hill reaction rate was determined at 420 nm.

The chlorophyll amount was measured according to Vernon (1960).

2.4 Isolation of Chloroplasts and RNA Polymerase Assay

Chloroplasts were isolated by a modification of the procedures of Bottomley et al. (1971) and Dzokhadze and Balashvili (1976). The leaves (15 g) were harvested and homogenized in 45 ml of buffer A (0.05 M Tris-HCl, pH 7.8, containing 0.1 M sucrose, 0.01 M KCl, 0.001 M $MgCl_2$, 0.004 M mercaptoethanol, and 25 g/ml of chloramphenicol) in a "Komet" blendor (GDR) for 15 s. The homogenate was filtered through two layers of cheesecloth, two layers of flannel and one layer of nylon. The filtrate was centrifuged at 1000 g for 10 min. The pellet containing chloroplasts and nuclei was resuspended in 6 ml of buffer A, layered over discontinuous sucrose density gradients composed of 6 ml each of 0.5, 1.0, 1.5 and 2.0 M sucrose each in buffer B (0.05 M Tris-HCl pH 8.3, 0.01 M $MgCl_2$, 0.004 M mercaptoethanol) and centrifuged at 16,000 rpm in a Spinco SW-25 rotor for 1 h. The chloroplast fraction appeared as a discrete band in about 1 M sucrose, nuclei were sedimented. The chloroplasts were removed, suspended in buffer B and were centrifuged at 1500 g for 10 min. Chloroplast suspension was washed with buffer B twice and assayed for RNA polymerase activity.

Each 0.5-ml assay mixture contained 50 μmoles of Tris-HCl, pH 8.3, 8 μmoles of $MgCl_2$; 0.1 μmoles each of UTP, CTP and GTP; 0.0029 μmole [^{14}C]-ATP (321 μCi/μmole Czechoslovakia) and the chloroplast suspension containing 100 μg of chlorophyll. After incubation at 25°C for 15 min, the reaction was stopped by the addition of 2 ml of cold 10% trichloracetic acid containing 0.9% sodium pyrophosphate. After 30 min on ice, the solutions were filtered through RUFS filters (Czechoslovakia), which retained the acid-precipitable material. The filters were washed three times thoroughly with cold 5% acid with sodium pyrophosphate and once with ethanol. The discs were air-dried, placed in vials and counted in a Mark II USA scintillation spectrometer. The value (c.p.m) of a zero-time control was substracted from all values given in c.p.m.

3 Results and Discussion

3.1 Effect of Cytokinin on the Synthesis of Plastid and Cytoplasmic rRNA's in Excised Pumpkin Cotyledons

As previously described (Rybicka et al., 1977) the rapidity and character of response of excised pumpkin cotyledons to exogenous cytokinins are dependent on their physiological state before the hormonal treatment is started. Cotyledons excised from 4-day-old etiolated pumpkin seedlings and incubated on water for 24 h in darkness demonstrated relatively rapid response to BAP treatment. The difference in chlorophyll content between BAP-treated and water-treated cotyledons was observed in 6 h, while a very small BAP-stimulation of the growth was observed in 9-12 h (Mikulovich, 1978). We have chosen to use these experimental conditions for the investigation of the cytokinin influence on the plastid rRNA synthesis.

The carotenoid content was used as an indicator of cytokinin action on the formation of photosynthetic apparatus in excised cotyledons.

Table 1 shows that 24 h treatment of excised pumpkin cotyledons with BAP exerted more than three fold increase in carotenoid content in dark and more than nine fold increase in light. At the same time water-treated cotyledons in light contained markedly more carotenoids in comparison with BAP-treated cotyledons in dark. Hence, it indicates that the effect of light is more profound than the effect of cytokinin.

The highest carotenoid content was found in BAP-treated cotyledons in light. From these data one can conclude that accumulation of carotenoids in excised pumpkin cotyledons is very sensitive to BAP treatment and cytokinin and light cause a synergistic effect.

Table 1. Influence of BAP on the carotenoids contents in excised pumpkin cotyledons. Excised cotyledons from 4-day-old etiolated seedlings were placed on the water in darkness. After 24 h cotyledons were transferred on to the water or BAP (10 mg/l) and incubated in darkness or in light for next 24 h. Initial — before BAP was added

Conditions		Carotenoids	
		µg/cotyledon	Increase %
Initial		2.48	100
Darkness			
	H_2O	3.36	135
	BAP	7.87	317
Light			
	H_2O	10.03	412
	BAP	23.11	931

Table 2. Effect of BAP on the total RNA content in excised pumpkin cotyledons. Initial — before BAP (10 mg/l) was added

Conditions		Total RNA	
		µg/cotyledon	Increase %
Initial		481.8	100
Darkness			
	H_2O	558.5	115
	BAP	894.7	185
Light			
	H_2O	667.7	138
	BAP	1043.9	217

As shown in Table 2, the amount of total RNA in cotyledons increased with BAP treatment by 85% in the dark and by 117% in light over initial content. In contrast to carotenoid content, water-treated cotyledons in light contained a lesser amount of RNA than BAP-treated cotyledons in dark. It shows that BAP acted on the accumulation of total RNA more effectively than light. Nevertheless, again the highest total RNA content was in BAP-treated cotyledons in the light. In this case the effects of BAP and light were additive.

Fractionation of nucleic acids by polyacrylamide gel electrophoresis allowed us to distinguish between the accumulation of plastid and cytoplasmic rRNA's (Fig. 1). Electrophoregrams have demonstrated the changes in ratio of peak areas of different rRNA species. In all cases the absorbance ratio of 1.1×10^6 to 0.56×10^6 rRNA was 1.8, i.e., close to the theoretical value. No intermediary degradation products were detected. The cotyledons had contained significant amount of etioplast rRNA's before BAP treatment started (data not shown).

Cytokinin Action on RNA Synthesis in Chloroplasts

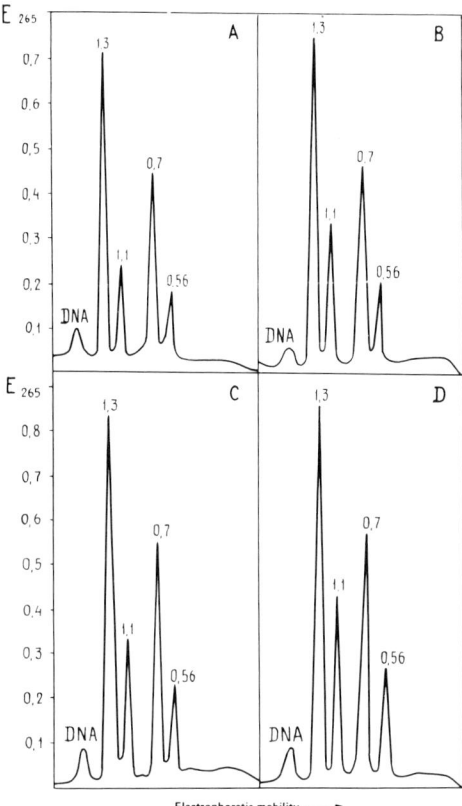

Fig. 1A-D. Gel-electrophoretic fractionation of nucleic acids extracted from excised pumpkin cotyledons. Cotyledons were incubated for 24 h in darkness on the water (**A**) or BAP (**B**) and in light on the water (**C**) or BAP (**D**). RNA components are referred to as 10^{-6} x molecular weight

On the basis of chemical measurement of the total RNA content (Table 2) and determination of the proportion of plastid rRNA's to total rRNA (after gel electrophoresis) the amounts of plastid and cytoplasmic rRNA's per cotyledon were calculated. From data presented in Table 3 the following conclusions can be drawn:

1. The accumulation of plastid rRNA's in etiolated or greening cotyledons was promoted in darkness and in light under the influence of BAP.
2. Comparing the water-treated cotyledons in light and in dark one can notice that light also stimulated the accumulation of chloroplast rRNA. However, under our experimental conditions, light effect was considerably less than that of BAP (compare water-treated cotyledons in light and BAP-treated cotyledons in dark).
3. The highest amount of plastid rRNAs was in BAP-treated cotyledons in light. The cytokinin and light do not substitute for one another, which indicates the different ways of their influence on the accumulation of plastid rRNA. On the other side their effect was not additive. Therefore it is possible to suggest that there are some common steps in a metabolic chain of reactions providing the activation of the plastid rRNA synthesis by BAP and by ligth.

To find out the interaction character of light and cytokinin in the regulation of the plastid rRNA synthesis, it is necessary to study the influence of the illumination of various intensities on this process, BAP concentration being changed.

Table 3. Accumulation of rRNA in water-treated and BAP-treated excised cotyledons. The amounts of plastid rRNA, cytoplasmic rRNA and low molecular weight RNA's were determined by measuring the relative proportions of these RNA's in electrophoregrams and calculating the amount of each in the total RNA determined by a quantitative method

Conditions		Plastid rRNA		Cytoplasmic rRNA	
		µg/cotyledon	Increase %	µg/cotyledon	Increase %
Initial		126.9	100	282.6	100
Darkness					
	H_2O	137.4	108	320.6	113
	BAP	285.6	225	466.0	164
Light					
	H_2O	185.9	146	375.0	132
	BAP	320.3	252	567.0	200

4. The presented data show that light and BAP increase the accumulation of plastid rRNA's as well as cytoplasmic rRNA's.

The effect of BAP on both species of RNA was also evident when the synthesis of different RNA's was determined in excised pumpkin cotyledons by the incorporation of $[^{32}P]_i$ into RNA in the light (Fig. 2).

It is necessary to note that in our experiments BAP has preferably stimulated the synthesis of plastid rRNA's rather than cytoplasmic rRNA's. However, our previous paper (Mikulovich et al., 1978) has demonstrated that the extent of cytokinin action either on plastid or on cytoplasmic rRNA synthesis is dependent on the physiological state of cotyledons and, perhaps, on the plastid development stage.

The increase in plastid rRNA content per cotyledon can be a result of a plastid number increase and RNA elevated content in each plastid. It was reported that cytokinin stimulated chloroplast replication in pumpkin cotyledons (Kulaeva et al., 1972). Therefore it is possible that BAP-stimulation of the synthesis of plastid rRNA's is partly due to an increase in the number of plastids in the cells.

However, using autoradiography of electron micrograph sections Drs. D. Neumann and W. Khokhlova (1981, this vol.) showed that BAP stimulated $[^3H]$-uridine incorporation into RNA in chloroplasts of excised pumpkin cotyledons in light.

It seems appropriate to relate the cytokinin-mediated increase of plastid and cytoplasmic rRNA synthesis to a cytokinin-mediated stimulation of the plastid and nuclear rDNA transcription.

3.2 Cytokinin Effect on RNA Synthesis in Isolated Barley Chloroplasts

A more direct approach to this problem is to study the influence of cytokinin on RNA synthesis in isolated chloroplasts. Such experiments were carried out with chloroplasts isolated from detached expended barley leaves.

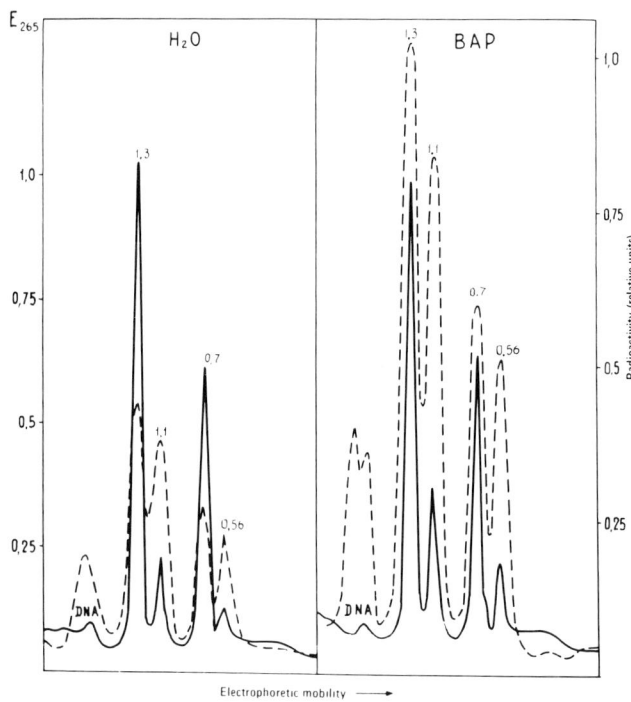

Fig. 2. Gel-electrophoretic fractionation of nucleic acids extracted from pumpkin cotyledon discs labeled for 6 h with [^{32}P]$_i$. Cotyledons were pre-incubated for 24 h in light on the water or *BAP*(10 mg/l). Discs (d=8 mm) were prepared for labeling. RNA components are referred to as 10^{-6} x molecular weight. *Solid line* absorbance; *broken line* radioactivity

Table 4. Characteristics of RNA polymerase reaction by isolated barley chloroplasts

Reaction mixture	Incorporation of [^{14}C]-AMP (c.p.m./100 μg DNA)
Complete[a]	18 438
−GTP, CTP, UTP	3 820
+Sodium pyrophosphate (10 μg/ml)	2 783
+Actinomycin D (10 μg/ml)	2 764
+Ribonuclease (100 μg/ml)	2 801

[a] (μmol); Tris-HCl (pH 8.3) − 50; MgCl$_2$ − 8; GTP, CTP, UTP − 0.1 of each; [^{14}C]-ATP − 0.0029; Total volume − 0.5 ml.

Table 4 demonstrates some characteristics of [^{14}C]-AMP incorporation by chloroplasts isolated from 9-day-old barley leaves. The incorporation was strictly dependent on the presence of the four nucleoside triphosphates and was significantly suppressed upon the addition of actinomycin D, ribonuclease or sodium pyrophosphate.

The RNA polymerizing process in isolated barley chloroplasts required for Mg^{2+} in the assay and was considerably reduced when Mg^{2+} was substituted for Mn^{2+} (Fig. 3).

The addition of (NH$_4$)$_2$SO$_4$ at the final concentration of 120 mM also inhibited [^{14}C]-ATP incorporation into the reaction product. As expected, α-amanitin did not

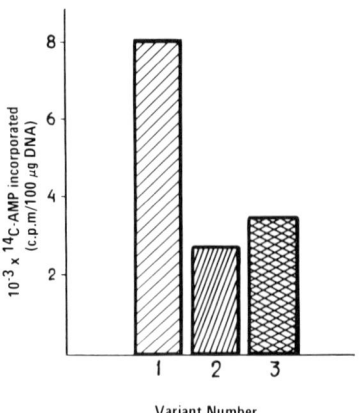

Fig. 3. Dependence on Mg^{2+}, Mn^{2+} and NH_4^+ of the RNA polymerase activity of chloroplast. *1* complete (16 mM Mg^{2+}); *2* free of Mg^{2+} + Mn^{2+} (2.8 mM); *3* complete (16 mM Mg^{2+}) + $(NH_4)_2SO_4$ (120 mM)

Table 5. Effect of α-amanitin on the RNA polymerase reaction by isolated barley chloroplasts

Concentration α-amanitin (µg/ml)	Incorporation of [^{14}C]-AMP (%)		
	Mg^{2+} (16 mM)	Mn^{2+} (2.8 mM)	Mg^{2+} (16 mM) + $(NH_4)_2SO_4$ (120 mM)
4	110	99	115
10	102	117	103
20	116	107	109

Incorporation by the system without inhibitor is called 100

affect RNA synthesis in chloroplasts (Table 5). These properties are similar to those reported for chloroplasts from other species (Bottomley et al., 1971; Horgen and Key, 1973; Schiemann et al., 1978; Wollgiehn and Munsche, 1972).

Hence, the incorporation of labelled precursor into acid-insoluble product by chloroplasts in vitro was due to chloroplast RNA polymerase activity and the reaction product is RNA.

To begin with, we have compared RNA synthesis in chloroplasts isolated from untreated and cytokinin-treated leaves. With this aim in view the excised barley leaves were pretreated with either water or BAP at 2 mg/l concentration at 23°C under illumination (30 Wt/m²) and after 24 h the chloroplasts were isolated.

As can be seen from Table 6, the RNA synthesis was increased in the chloroplast preparations from BAP-treated leaves in comparison with water-treated ones.

Thus, cytokinin treatment of excised barley leaves induces some changes in chloroplasts, which are retained after organelle isolation resulting in the enhancement of RNA synthesis in vitro.

Another aim was to examine whether cytokinin can effect RNA polymerase activity in chloroplasts after cell destruction. As previously noted (Selivankina et al., 1979), cytokinin added in the course of leaf homogenization increased the activity of

Table 6. Effect of pretreatment of barley leaves on the RNA polymerase reaction by isolated chloroplasts. Leaves were on the water or BAP (2 mg/l) during 24 h in light, then chloroplasts were isolated

Treatment	Incorporation of [^{14}C]-AMP (c.p.m./100 µg DNA)
H_2O	19760
BAP	30100 (151)[a]

[a] Result is expressed relative to the incorporation given by chloroplasts isolated from leaves pretreated with H_2O

chromatin-bound RNA polymerase. This effect was observed only with 9-12-day-old leaves, i.e., was related to the physiological state of leaves.

We have used the same conditions to investigate cytokinin effect on chloroplast RNA polymerase. BAP at 0.1 mg/l concentration was added to the homogenization buffer. Table 7 shows that BAP has induced the increase of chloroplast RNA polymerase activity. Such experiments were repeated many times and in every case BAP has stimulated RNA synthesis in chloroplast preparations. However, the period of chloroplast sensitivity to cytokinin varied in various experiments from 9 to 12 days of leaf age.

Similar experiments were carried out using another cytokinin, kinetin, which also retarded the yellowing of detached barley leaves and activated RNA synthesis.

Table 7. Effect of BAP on the RNA polymerase reaction by isolated barley chloroplasts. 6-BAP (0.1 mg/l) was added during homogenization

Treatment	Incorporation of [^{14}C]-AMP (c.p.m./50 µg DNA) Age of leaves (days)			
	9	10	11	12
−BAP	28600	13590	10335	5110
+BAP	27180 (95)[a]	24120 (175)[a]	15515 (150)[a]	8455 (165)[a]

[a] Results are expressed relative to the incorporation given by chloroplasts without BAP

Kinetin added in the course of leaf homogenization activated chloroplast RNA polymerase as well (Table 8). Therefore, cytokinins stimulate chloroplast RNA polymerase activity even when cell integrity is disturbed and no de novo synthesis of this enzyme can take place. Hence, cytokinin can influence the activity of preformed chloroplast RNA polymerase.

It was significant to determine whether cytokinin effect is specific for chloroplast RNA polymerase or whether it can be extended to other enzymatic systems of plastids.

Table 8. Effect of kinetin on RNA polymerase reaction by isolated barley chloroplasts. Kinetin (0.1 mg/l) was added during homogenization

Treatment	Incorporation of $[^{14}C]$-AMP (c.p.m./100 µg DNA)
−kinetin	5800
+kinetin	10400 (179)[a]

[a] Result is expressed relative to the incorporation given by chloroplast without kinetin

This series of experiments could assist in elucidating whether cytokinin effect upon chloroplast RNA polymerase is a part of its general beneficial influence on chloroplasts during their preparation.

Therefore, we have measured the rate of the Hill reaction and RNA polymerase activity in chloroplasts, when BAP was added to homogenization buffer. The buffer continely used for chloroplast preparation for the Hill reaction measurement was employed, which could be unfavorable for RNA polymerase activity. Chloroplasts were prepared in this experiment without purification on sucrose gradient. Table 9 shows that under these experimental conditions BAP did not exert any effect upon the rate of the Hill reaction and at the same time increased RNA polymerase activity in chloroplasts. Thus, cytokinin effect can hardly be attributed to the improvement of conditions of chloroplast preparation in the presence of phytohormone.

Table 9. Effect of BAP on Hill reaction and RNA polymerase reaction by isolated barley chloroplasts. 6-BAP was added during homogenization

Treatment per h	Hill reaction activity (µmole/mg chlorophyll/h)	Incorporation of $[^{14}C]$-AMP (c.p.m./50 µg DNA)
−BAP	123	7238
+BAP (2 mg/l)	127	14688 (203)[a]
−BAP	75	5541
+BAP (0.1 mg/l)	68	10714 (193)[a]

[a] Results were expressed relative to the incorporation given by chloroplasts without BAP

It should be mentioned that the chloroplasts used in our experiments were osmotically ruptured and did not possess intact outer membranes. Therefore, we have to exclude the possiblity of cytokinin exerting any influence on permeability of outer chloroplast membrane for $[^{14}C]$-ATP.

It appears reasonable to assume that cytokinin stimulates the RNA labeling in isolated chloroplasts through activation of RNA polymerase system of chloroplasts.

Summing up the results we presume that cytokinin controls the operation both of chloroplast and nuclear genomes in plant cells.

References

Bottomley W, Spencer D, Wheeler AM, Whitfeld PR (1971) Arch Biochem Biophys 143: 269
Burton KA (1956) Biochem J 62: 315
Dzokhadze DI, Balashvili MI (1976) Biokhimiya 41: 133
Horgen PA, Key JL (1973) Biochim Biophys Acta 294: 227
Karavaiko NN, Ohmann E, Kulaeva ON (1975) Fiziol Rast 22: 1031
Khokhlova VA (1977) Fiziol Rast 24: 1189
Khokhlova VA, Sveshnikova IN, Kulaeva ON (1971) Cytologia 13: 1074
Kulaeva ON (1973) Cytokinins – their structure and function (russ.). Isdat Nauk, Moscow
Kulaeva ON, Erkeev MI, Khokhlova VA, Sveshnikova IN (1972) Fiziol Rast 19: 1023
Kursanov AL, Kulaeva ON, Mikulovich TP (1969) Am J Bot 56: 767
Metzner H, Rau H, Senger H (1965) Planta 65: 186
Mikulovich TP (1978) Thesis, Moscow
Mikulovich TP, Khokhlova VA, Kulaeva ON, Sveshnikova IN (1971) Fiziol Rast 18: 98
Mikulovich TP, Wollgiehn R, Munsche D (1975) Fiziol Rast 22: 114
Mikulovich TP, Wollgiehn R, Khokhlova WA, Neumann D, Kulaeva ON (1978) Biochem Physiol Pflanz 172: 101
Munsche D, Wollgiehn R (1973) Biochim Biophys Acta 294: 106
Parthier B (1979) Biochem Physiol Pflanz 174: 173
Rybicka H, Engelbrecht L, Mikulovich TP, Kulaeva ON (1977) Fiziol Rast 24: 371
Schiemann J, Wollgiehn R, Parthier B (1978) Biochem Physiol Pflanz 1972: 507
Selivankina SY, Romanko EG, Kuroedov VA, Kulaeva ON (1979) Fiziol Rast 26: 41
Vernon LP (1960) Anal Chem 32: 1144
Wollgiehn R, Munsche D (1972) Biochem Physiol Pflanz 163: 137
Wollgiehn R, Parthier B (1964) Flora 154: 325

Effect of Cytokinin on Plastid Differentiation in Tobacco Cell Suspensions

A.-M. Lescure and P. Seyer[1]

1 Introduction

We are interested by the interaction of the chloroplast and nuclear genomes in the biogenesis of chloroplasts under cytokinin control. It is now well established, first that chloroplasts represent an extranuclear genetic system, second that a division of labor exists between the nuclear and the plastidial genomes, during the biogenesis of the organelles. Some chloroplast polypeptides are coded for by the chloroplast DNA (cpDNA) and synthesized inside the organelles, while others are made in the cytoplasmic compartment and must flow across the chloroplast envelope before being assembled into their functional configuration. In spite of the amount of work concerning the effect of cytokinin on chloroplast maturation, it is not clear whether this effect arises from a general stimulation of the metabolism by the hormone or from the control of specific events, either inside the organelles or in the cytoplasmic compartment.

In the case of the light-induced transition from etioplasts to mature chloroplasts, analyses of the membrane proteins have shown that almost the same polypeptide maps were obtained from etioplasts or from mature chloroplasts (Grebanier et al., 1979). Only some polypeptides changed during this transition: among these polypeptides, most were of cytoplasmic origin. However, several authors have reported that the synthesis of a thylakoid polypeptide of 32 Kd MW, chloroplastic in origin, is under a positive light control (Weinbaum et al., 1979; Grebanier et al., 1979). Furthermore, Bedbrook et al. (1978) have shown that the accumulation of this 32 Kd polypeptide would result from a photoregulation of the corresponding gene transcription.

Our purpose is to research if such specific changes of the expression of the chloroplast genome may be detected during the cytokinin-induced chloroplast differentiation, either at the translational or at the transcriptional level.

Suspension cultures of tobacco cells have been chosen for this work, taking advantage of the more homogenous response of suspended cells when compared with callus cultures. In the case of the AG_{14} cell line, selected for this study, the growth kinetics

---

Abbreviations: +CK cells: cells grown in medium supplemented with 0.5 μM kinetin (MK medium); −CK cells; cells grown in a cytokinin free medium (MO medium); cpDNA: chloroplast DNA; PAGE: polyacrylamide gel electrophoresis; SDS: sodium dodecylsulfate

[1] Laboratoire de Biochimie Fonctionnelle des Plantes, Faculté des Sciences de Luminy, 13228 Marseille, Cedex 9, France

were similar in the presence or in the absence of cytokinin, whereas the differentiation of mature chloroplasts under continuous light, occurred only in the cytokinin-supplemented medium (Seyer et al., 1975).

2 Evolution of Chloroplasts During a Cell Growth Cycle

Ultrastructural studies of plastids during a cell growth cycle either in the presence or in the absence of cytokinin have shown that in both media plastids underwent cyclic morphological transformations (Seyer et al., 1975). At the beginning of the log phase, thylakoids disintegrated and plastids stored starch. At the end of the cell division period, the organelles exhibited a dense stroma with some osmiophilic granules and only few lamellae. The differentiation of mature chloroplasts occurred in stationary phase in the cytokinin-supplemented medium. In the absence of cytokinin, only this last stage was modified: thylakoids did not differentiate and plastids remained similar to proplastids. Figure 1 illustrates the evolution of the chlorophyll contents during the cell growth cycle, in the presence of cytokinin (curve a) or in the absence (curve d): this evolution was well correlated with the described morphological changes. It is noteworthy that, when cytokinin was added at the beginning of the growth cycle to bleached cells previously grown without the hormone, chlorophyll synthesis occurred during stationary phase (curve b). However, to observe chlorophyll synthesis and thylakoid differentiation, it was necessary to add the hormone during the period of cell multiplication: when cytokinin was added after the end of the log phase, chlorophyll synthesis did not occur (curve c). These results indicated that some events which were necessary for subsequent lamellar assembly and chlorophyll accumulation were initiated during the log phase of cell growth.

3 Research of Chloroplastic Protein Markers of Cytokinin Activity

3.1 In Vitro Protein Synthesis by Crude Chloroplast Preparations[2], Isolated from Tobacco Cell Suspensions

Experiments were then undertaken in order to research if the synthesis of any of the polypeptides coded for by the cpDNA was specifically affected by the hormone (Lescure, 1978). The most direct way to establish which polypeptides are made on chloroplast ribosomes is to identify the products of protein synthesis by isolated organelles. Current methods allow discrete polypeptides to be synthesized by chloroplasts isolated from leaves of several higher plants (for a review, see Ellis, 1977). For these in vitro experiments, light is generally used as energy source to stimulate protein synthesis in intact chloroplasts. In the case of our experiments with dedifferentiated plastids isolated from cytokinin-starved cells, photophosphorylation could not occur and it was necessary to add ATP as energy source, to stimulate the in vitro pro-

---

[2] Crude plastids refer to sediments obtained by centrifugation of the filtrated cell homogenates, at low speed (2500 x g for 10 min, in our experiments)

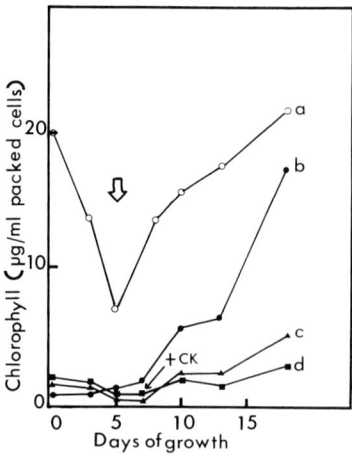

Fig. 1. Evolution of chlorophyll contents during a cell growth cycle. Curve a (○—○) MK→MK cells; curve d (■—■) MO→MO cells; curve b (●—●) MO→MK cells; curve c (▲—▲) MO cells transferred in MO medium at zero time; kinetin (0.5 μM) added at day 7th, e.g., two days after the end of log growth. The *arrow* at 5 days indicates the end of exponential growth

tein synthesis. When ATP was added to intact plastids suspended in the TS medium (50 mM Tricine-KOH, 0.33 mM sorbitol, pH 8.3) usually utilized for the light-driven experiments, no significant protein synthesis was observed. A good stimulation of the incorporation of labeled amino acids into chloroplast proteins was obtained, using the conditions described by Bottomley et al. (1974): according to this technique, intact organelles isolated in TS medium were resuspended in a low osmolarity medium without sorbitol; the resulting hypotonic shock induced ruptures of the plastid outer membranes, that allowed the penetration of the necessary co-factors. Table 1 shows the results obtained in these conditions, with crude plastids isolated in stationary phase from cells grown in the presence of cytokinin: a strong stimulation of the incorporation of radioactivity into the hot TCA-insoluble material was observed in the presence of ATP. Most of the incorporation was abolished by D-threo-chloramphenicol, a specific inhibitor of the translation on 70S ribosomes. On the contrary, the incorporation was little or not at all affected by cycloheximide, a specific inhibitor of protein synthesis on cytoplasmic ribosomes. These results argued for a protein synthesis on chloroplast ribosomes.

The efficiency of in vitro protein synthesis has been measured with plastids isolated at various stages of their evolution along the cell growth cycle (Lescure, 1978): the results have shown that the efficiency was very low when plastids were isolated during the exponential phase of cell growth. The protein synthesis efficiency then increased rapidly at the onset of the stationary phase, when chloroplasts differentiated: in the following experiments, plastids were always isolated at this step, that is at the beginning of the stationary phase.

3.2 Comparison of the Polypeptides Labeled in Vitro by Plastids Isolated from Cells Grown in Presence or in Absence of Cytokinin (Lescure, 1978)

The products of in vitro protein synthesis were then compared with plastids isolated from two cultures: the first one (+CK) has been repeatedly subcultured in the presence of the hormone; the second culture (−CK) was transferred in a cytokinin-free medium

Table 1. ATP-driven incorporation of [^{14}C]-amino acids into the TCA insoluble fraction

Incubation conditions[a]	cpm · mg plastid protein
1. Control without cofactors	20 x 10^3
2. + ATP 2 mM, GTP 0.2 mM creatine phosphate 10 mM creatine phosphokinase 200 µg · ml^{-1}	83 x 10^3
3. Same conditions as 2, + chloramphenicol 25 µg · ml^{-1}	24 x 10^3
4. Same condition as 2, + cycloheximide 25 µg · ml^{-1}	80 x 10^3

[a] 200 µl samples of crude chloroplast preparations (180 µg protein), suspended in the low osmolarity buffer, were incubated in the dark with 2.5 µCi [^{14}C]-aminoacids for 40 min. Each value corresponded to the average of the radioactivity into two samples

for one growth cycle just before the experiment: in this condition, chlorophyll synthesis in stationary phase did not occur. Plastids were isolated from each culture and the products of in vitro protein synthesis were analyzed on SDS-PAGE. Figure 2 shows the results of this comparisons: the left panel corresponds to the histogram of radioactive polypeptides synthesized by plastids, isolated from + CK cells. Seven peaks A to G, were separated, with well-defined electrophoretic mobilites; (peak H, shown in this pattern, migrated with free chlorophyll; this peak was run out of the gel in other experiments). The right panel corresponds to polypeptides synthesized by plastids isolated from —CK cells: the same series of radioactive peaks, A to G, were observed. However, it is clear that the lack of hormone did not affect to the same extent the synthesis of all these peptides: some polypeptides, like B and D, were as heavily labeled as in the case of the + CK sample. On the contrary, the labeling of other peptides, like F and G, was strongly reduced in the —CK sample. The low rate of in vitro labeling of these two polypeptides was repeatedly observed with plastids isolated from —CK cells. These peptides, which were little or not at all synthesized when chloroplast did not differentiate in the absence of hormone, would constitute useful markers of the effect of cytokinin on chloroplast maturation.

3.3 Protein Synthesis by Purified Preparations of Plastids

The foregoing experiments were performed with crude chloroplast preparations and did not allow the identification of the radioactive polypeptides with defined chloroplast protein subunits, stained by Coomassie blue on SDS-PAGE. In an attempt to identify the cytokinin markers, we decided to reexamine the in vitro protein

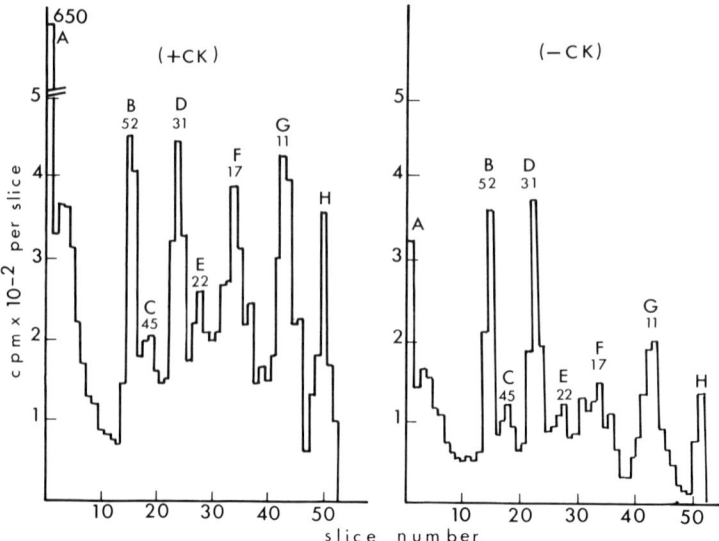

Fig. 2. Radioactive diagrams of polypeptides labeled in vitro by plastids isolated from + or −CK cells. Crude plastids were isolated at the onset of stationary phase. Chlorophyll contents were respectively 20 μg per g of fresh material for + CK cells and 11 μg per g for −CK cells. Radioactive incubations were performed in the dark, in the ATP-driven conditions, for 30 min. Proteins were solubilized in the presence of SDS 3% (w/v) and 2-mercatoethanol 5% (v/v). Polypeptides were separated by electrophoresis on SDS-PAGE (linear concentration gradient of acrylamide from 10% to 18%). Gels were loaded with 225 μg protein (15,700 cpm) for the + CK sample and 215 μg protein (7,000 cpm) for the −CK sample. After electrophoresis, gels were cut in 2 mm slices. Slices were solubilized in the presence of hydrogen peroxide; radioactivity was counted in Instagel scintillation fluid. (Lescure, 1978)

synthesis with purified organelles (Lescure, 1980). Highly purified chloroplasts may be prepared from tobacco cell suspensions, by centrifugation on sucrose gradient concentration (see Péaud-Lenoël this vol.). Unfortunately, such purified preparations were found inactive for protein synthesis. Morgenthaler and Mendiola-Morgenthaler (1976) have shown that centrifugation on silica concentration gradient (Ludox), allowed to purify functional chloroplasts from leaves. By adapting this method to our cell suspension cultures, we were able to isolate intact purified organelles, still capable of in vitro protein synthesis. Figure 3 shows the SDS-PAGE fractionation of polypeptides synthesized in the light-driven conditions, by such purified plastids. The densitometric pattern of the stained polypeptides was very similar to that reported by Axelos and Péaud-Lenoël (1980), after extensive purification of the organelles. The radioactive histogram showed seven peaks, migrating with very similar electrophoretic mobilities as peaks A to G, labeled in crude preparations.

We tried to identify the labeled peaks with defined peptides of the densitometric tracing. Peak B corresponded to the 52 Kd MW polypeptide of the densitometric pattern, which has been shown to co-migrate with the large subunit of the ribulose bisphosphate carboxylase. As expected for the large subunit, when polypeptides of stroma and membrane fractions were separately analyzed, peak B was found in the soluble

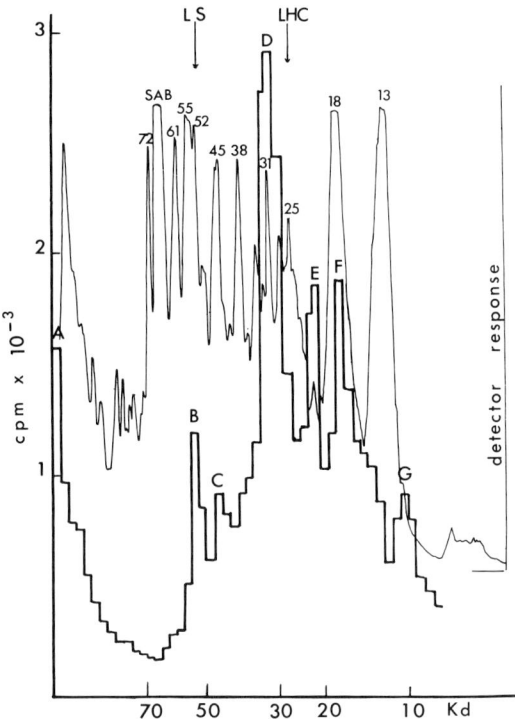

Fig. 3. SDS-PAGE analysis of polypeptides synthesized in the light-driven system by plastids purified on Ludox concentration gradient. Plastids were isolated at the onset of stationary phase. Radioactive incubations were performed in the presence of [^{35}S]-methionine for 35 min. Gel was loaded with 140 μg protein (50,000 cpm). —, 600 nm densitometric tracing of the Coomassie blue stained bands; –, (histogram): radioactivity. *Numbers* calculated MW of radioactive peaks; *SAB* bovine serum albumine added to Ludox gradient; *LS* large subunit of ribulose bisphosphate carboxylase; *LHC* subunit of the light-harvesting chlorophyll a/b complex

fraction. We have shown that the synthesis of this polypeptide was not significantly affected by the absence of cytokinin. All the other labeled polypeptides remained bound to the membrane fraction. Peak D, according to its apparent MW, would correspond very likely to the 32 Kd MW thylakoid polypeptide, which is the major product of in vitro protein synthesis by chloroplasts isolated from leaves. As we have already mentioned, the synthesis of this polypeptide in leaves is submitted to a positive light control. In our comparison of polypeptides synthesized by plastids isolated from cells grown in the presence or in the absence of cytokinin, the labeling of peak D did not appear to be early affected by the lack of hormone.

The other membrane-bound labeled peptides described with crude preparations, incubated in the ATP-driven conditions, have been repeatedly observed with defined electrophoretic mobilities when purified chloroplasts were incubated in the light-driven conditions. The identification of the two cytokinin markers with defined stained polypeptides still remained uncertain: peak F was located inside the wide 18 Kd MW region, which was very likely composed of several nonseparated polypeptides. Peak G, which migrated with an apparent MW of 11 Kd, was not coincident with any defined band.

The comparison of the densitometric patterns of plastid polypeptides prepared from cells grown with or without cytokinin (Fig. 4), did not allow to detect significant changes among the polypeptides which migrated in these low molecular weight positions. This indicated that peaks F and G might represent only minor components of the total plastid proteins. The identification of polypeptides F and G will require

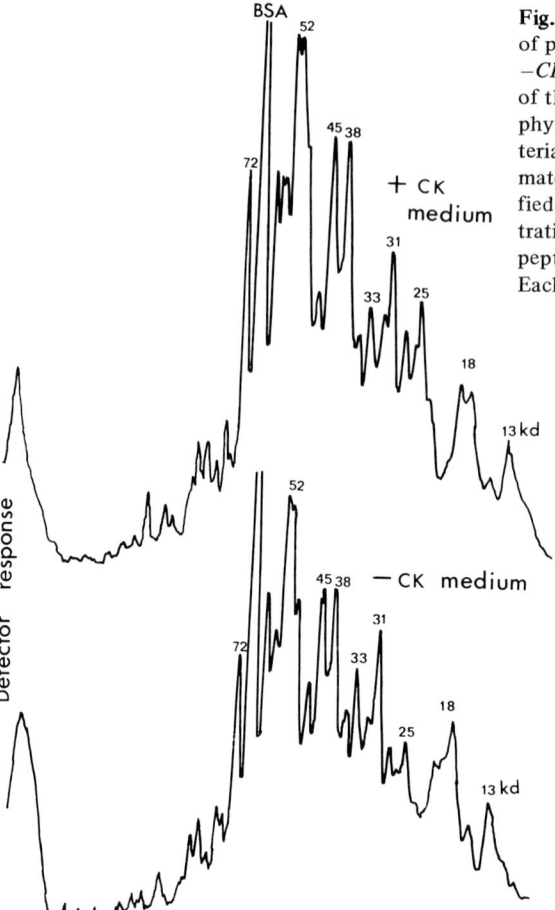

Fig. 4. Comparative densitometric patterns of plastid polypeptides prepared from + or −CK cells. Cells were harvested at the onset of the stationary phase of growth. Chlorophyll contents were 25 µg per g fresh material for + CK cells and 8 µg per g fresh material for −CK cells. Plastids were purified by centrifugation on a Ludox concentration gradient (cf. Lescure, 1980). Polypeptides were separated by SDS-PAGE. Each gel was loaded with 50 µg protein

more resolutive systems, like analysis of subchloroplast fractions by bidimentional electrophoresis and autoradiography.

One noteworthy observation has been made during these experiments: the radioactive low molecular weight polypeptides migrating in the 17 and 11 Kd MW positions were found to be actively synthesized only when plastids were isolated from + CK cells harvested in early stationary phase, e.g., at the onset of the development of the lamellar system. The labeling of these peptides was low when mature chloroplasts were isolated from cells in late stationary phase, whereas other peaks, like B and D, were still actively synthesized. This suggested that peaks F and G might have a transient developmental function during the thylakoid assembly, which would complicate their identification.

4 Mapping of the Recognition Sites for Restriction Enzymes on Tobacco cpDNA (Seyer, unpubl. results)

These results constitute preliminary requirements to experiments designed to identify tobacco cpDNA sequences, the transcription of which should be under cytokinin control. Such an approach implied (1) to be able to purify high molecular weight cpDNA from tobacco cell suspensions (2) to establish a restriction map of this DNA (3) to hybridize labeled RNA's, extracted from + or −CK cells, to defined restriction fragments. The two first steps are now available. The classical method described by Kolodner and Tewari (1975) to isolate cpDNA from leaves of several higher plants, was found unadapted to our cell cultures. In this technique, extrachloroplastic DNA which contaminates the crude chloroplast preparations is removed by a DNase treatment, before the lysis of the organelles. In the case of plastids from tobacco cells, the DNase digestion was inefficient. This situation was found to result from the formation of complexes between polysaccharides, chloroplasts, and chromatin: nuclear DNA which was enclosed inside such complexes was not accessible to the DNase. A new protocole has now been established, including isolation of chloroplasts in the presence of higher concentration of EDTA and further purification of the organelles by centrifugation on a Percoll concentration gradient. This procedure yields reproducible high MW cpDNA preparations from our tobacco cell suspensions (Seyer, unpubl. results).

Figure 6 shows the agarose gel electrophoresis of the restriction fragments of tobacco cpDNA from leaves or from cell suspensions, after digestion by various endonucleases. Whatever the origin of cpDNA, the restriction patterns were rigorously identical. This indicates that CpDNA prepared from tobacco cell cultures by the new procedure is well purified. The similarity of the restriction patterns of cpDNA prepared from leaves or from cell cultures also indicates that no noticeable changes of the nucleotide sequences have resulted from a long peroid (about 10 years) of tobacco cell subcultures in heterotrophic conditions.

The physical map of recognition sites has been established for 4 endonucleases (Seyer and Herrmann, in prep., tobacco cpDNA shares a structure common with cpDNA from *Zea mais* (Bedbrook and Bogorad, 1976) *Spinacia* (Crouse et al., 1978) and *Chlamydomonas* (Rochaix, 1978). This structure is characterized by two unique sequence regions of different sizes (15 and 60 Md) which separate inversely arranged copies of a duplicated segment (12.5 Md).

These results will allow the comparison of hybridizations between defined restriction cpDNA fragments and mRNA's extracted from + or −CK chloroplasts, with the hope of identifying specific messages.

5 Conclusions

Our experiments have shown that the protein synthesis efficiency of plastids was not deeply affected after a culture period of three cell generations in the absence of cytokinin, whereas lamellar system development and chlorophyll accumulation were strongly reduced in these conditions. Some polypeptides were about as actively labeled when plastids, extracted from cells grown in the presence or in the absence of cyto-

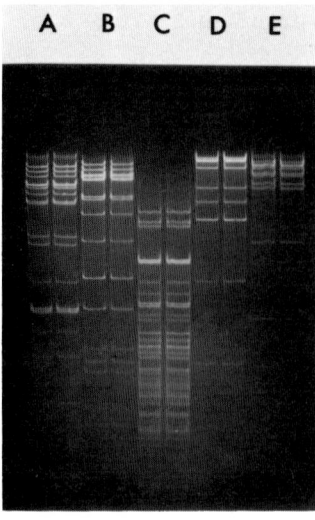

Fig. 5A-E. Agarose gel electrophoresis of tobacco cpDNA digested by various endonucleases: **A** Xho I, **B** Sal I + Bgl I, **C** Eco RI, **D** Bgl I, **E** Sal I. For each endonuclease, the left channel corresponds to cpDNA from leaves and the right channel to cpDNA from AG_{14} cell cultures. After endonuclease digestion (1 hr, 37°C, 10 to 20 units of enzyme per µg DNA), the fragments were separated by electrophoresis through 0.65% agarose gel, stained with ethidium bromide and photographed under U.V. light

kinin, were incubated with radioactive precursors: therefore, the synthesis of these polypeptides did not appear to be early affected by the lack of hormone. This was the case with polypeptide B, which has been shown to co-migrate with the large subunit of ribulose bisphosphate carboxylase and of peak D, which constituted the major labeled peak of the insoluble fraction. Peak D fits well with the 32 Kd MW thylakoid polypeptide of leaf chloroplasts, whose synthesis is under light control. The high labeling of peak D by plastids isolated from cytokinin-starved cells suggests that the two effectors, light and cytokinin, would have distinct impacts on the expression of the chloroplast genome.

On the contrary, other polypeptides like peaks F and G were little or not at all labeled by the cytokinin-starved plastids and therefore constitute markers of the hormone activity on chloroplasts. Whether this alteration in the polypeptides synthesized in vitro on plastid ribosomes reflects a similar alteration in vivo remains unknown. However, our results strongly suggest a specific control by cytokinin of only some of the polypeptides which are synthesized inside the organelles. These conclusions are in agreement with the results reported by Péaud-Lenoël and Axelos (this vol.), concerning the chloroplast polypeptides synthesized on cytosol ribosomes.

Specific modifications of the expression of the chloroplast genome by cytokinin may result from a control either of the transcription or of the translation process. The purposed hybridization experiments between mRNA's, extracted from plastids of + or −CK cells, and defined restriction fragments of cpDNA would help to precise if early modifications of the transcription occur after cytokinin addition.

Acknowledgments. This work benefited from financial support of the Centre National de la Recherche Scientifique (Contract ER 104). We are indebted to Prof. Dr. Péaud-Lenoel for useful disscusions and suggestions. We are also grateful to Mrs. M.C. Durand for her competent technical assistance.

References

Axelos M, Péaud-Lenoël C (1980) The apoprotein of the light-harvesting chlorophyll a/b complex of tobacco cells as molecular marker of cytokinin activity. Plant Sci Lett 19: 33-41

Bedbrook JR, Bogorad L (1976) Endonuclease recognition sites mapped on Zea mays chloroplast DNA. Proc Natl Acad Sci USA 73: 4309-4313

Bedbrook JR, Link G, Coen DM, Bogorad L, Rich A (1978) Maize plastid gene expressed during photoregulated development. Proc Natl Acad Sci USA 75: 3060-3064

Bottomley W, Spencer D, Whitfeld PR (1974) Protein synthesis in isolated spinach chloroplasts: comparison of light-driven and ATP-driven synthesis. Arch Biochem Biophys 164: 106-117

Crouse EJ, Schmitt JM, Bohnert HJ, Gordon K, Driesel AJ, Herrmann RG (1978) Intramolecular compositional heterogenity of Spinacia and Euglena chloroplast DNAs. In: Akoyunoglou G, Argyroudi-Akoyunoglou JH (eds) Chloroplast development. Elsevier, Amsterdam New York Oxford: p 565

Ellis RJ (1977) Protein Synthesis by isolated chloroplasts. Biochim Biophys Acta 463: 185-215

Grebanier AE, Steinbeck KE, Bogorad L (1979) Comparison of M.W. of proteins synthesized by isolated chloroplasts with those which appear during greening in Zea mays. Plant Physiol 63: 436-439

Kolodner R, Tewari KK (1975) The molecular size and conformation of the chloroplast DNA from higher plants. Biochim Biophys Acta 402: 372-390

Lescure AM (1978) Chloroplast differentiation in cultured tobacco cells: in vitro protein synthesis efficiency of plastids at various stages of their evolution. Cell Differ 7: 139-152

Lescure AM (1980) Polypeptides synthesized in vitro by plastids isolated from tobacco cell cultures: purification of the organelles on a silica density gradient. Plant Sci Lett 19: 181-191

Morgenthaler JJ, Mendiola-Morgenthaler L (1976) Synthesis of soluble, thylakoid and envelope membrane proteins by spinach chloroplasts purified from gradients. Arch Biochem Biophys 172: 51-58

Rochaix JD (1978) Restriction endonuclease map of the chloroplast DNA of *Chlamydomonas reinhardii*. J. Mol Biol 126: 597-617

Seyer P, Marty D, Lescure AM, Péaud-Lenoël C (1975) Effect of cytokinin on chloroplast cyclic differentiation in cultured tobacco cells. Cell Differ 4: 187-197

Weinbaum SA, Gresel J, Reisfeld A, Edelman M (1979) Characterization of the 32000 dalton chloroplast membrane protein. Plant Physiol 64: 828-832

Plastid Proteins of Cytoplasmic Origin as Molecular Markers of Cytokinin Activity

C. Péaud-Lenoël and M. Axelos[1]

1 Introduction

Apart from the impact of cytokinins on the mitotic activity of plant cells, the requirement of these hormones for chloroplast differentiation is one of the most striking subcellular effects (see Parthier, 1979). In spite of some useful contributions, the intimate nature of the molecular pathway(s) involving cytokinins in the development of plastids is still a matter of speculation. Tissue cultures (Stetler and Laetsch, 1965) and cell suspensions (Lescure, 1973) have been recognized as favorable biological systems to study this phenomenon, keeping in mind the specificity of growth and differentiation status of in vitro cultures compared with intact plants. Some years ago, we undertook to study tobacco cell suspensions in order to obtain stepwise answers to the questions raised by this aspect of cytokinin activities:

a) Is it possible to maintain tissue or cell suspension cultures with fully functional chloroplasts?
b) How could multiplication and differentiation of plastids be controlled in plant cell suspension cultures, apart from the light requirement?
c) Through which molecular activity do cytokinins promote chloroplast differentiation in cell cultures?

We shall summarize answers to the first two questions and, about the third, comment the present stage of our knowledge which is far from complete.

2 Chloroplast Evolution in Cell Suspension Cultures

Functional chloroplasts can be maintained in plant tissues and cell suspension cultures (see Lescure and Seyer, this vol.): it was shown that normal ultrastructures of plastids were positively controlled by the addition of cytokinins to many lines of tobacco cell cultures; photosynthetic membranes appeared in these plastids at the end of the growth phase and during stationary phase of the culture. Apart from CO_2 supply and from light energy which

---

Abbreviations: CK cells, O cells, respectively: AG_{14} tobacco cell suspensions grown with (CK) or without (O) kinetin 0,5 μM in the medium (see Lescure, 1973); CAP: D-threo-chloramphenicol; CH: cycloheximide; SDS-PAGE: polyacrylamide gel electrophoresis in the presence of sodium dodecylsulfate; LHCP: light-harvesting chlorophyll a/b-complex apoprotein.

[1] Laboratoire de Biochimie Fonctionnelle des Plantes (ER 104 CNRS) Département de Biologie Moléculaire et Cellulaire, Faculté des Sciences de Luminy, 13288 Marseille Cedex 9, France

are limiting factors for autotrophic growth, a cytokinin addition is also required (Tandeau de Marsac and Péaud-Lenoël, 1972) for most photosynthetic tobacco cell clones. It is worth mentioning that the cytokinin requirement was established for a number of cell clones which otherwise grew heterotrophically in a cytokinin-free medium (Péaud-Lenoël and Hanson, 1973). In order to determine the activity of cytokinins on the chloroplast evolution at the molecular level, our next step was to study the kinetics of plastid differentiation without interference of the light supply. It is underlined how important it is to distinguish clearly between light and cytokinin as parameters of plastid morphogenesis by appropriate experimental conditions. A large number of reports have shown that plastogenesis was phytochrome-controlled in superior plants: particularly, the equilibrium $P_r \leftrightarrows P_{fr}$ governed the change in the realization of development programs leading from etioplasts to chloroplasts (Mohr and Schopfer, 1977). This program shift is different from the transformation of proplastids to chloroplasts or, alternatively, from proplastids to etioplasts which are oberved, for instance, in meristems. In some tobacco cell lines like AG_{14} (Lescure, 1973) which we used in the present work, a program similar to (perhaps not identical with) the transformation of proplastids to chloroplasts is controlled by the exogenous cytokinin supply in addition to the non limiting illumination factor. With these cells we might also be able to study the cytokinin control of the program leading from proplastids to etioplasts in the dark. Seyer et al. (1975) detailed the morphological changes undergone by plastids during the culture. It is important to distinguish the two steps in the realization of the program, the first during the cell multiplication period, when dedifferentiation and multiplication of plastids occur, the second, at the end of the growth phase and during the stationary phase of the culture, leads to the development of photosynthetic membranes. Only this last step was affected by the cytokinin, as exemplified by the appearance of chlorophylls (chlorophyll a/b # 3). If cytokinin is omitted during the growth phase, chlorophyllian structures do not appear or appear very slowly (see Lescure and Seyer, this vol.).

3 Plastid Protein Evolution During the Differentiation Period

Following these findings, we undertook to characterize the polypeptide contents of the plastids during the differentiation period of the cell culture in the presence of cytokinins, compared with the leucoplasts of the cytokinin-depleted cells. Axelos and Péaud-Lenoël (1980) have described the procedure used to purify class I plastids from tobacco cell suspension cultures. This procedure was satisfactory only when the organelles contained little or no starch; for this reason, we are lacking information on the polypeptide contents of the starchy plastids which are present throughout most of the growth period of the cells. With this restriction, this method provided chloroplasts or leucoplasts in a good state of integrity or of cleanliness, as shown by their enzyme contents or TLC lipid analysis. For instance, the galactolipid contents and the lack of sterylglucoside or phosphatidylethanolamine were evidence for the low level of cytoplasmic membranes or mitochondrial contaminants in these plastid preparations (Mazliak, 1977; Eichenberger, 1977).

Total proteins extracted from the isolated plastids were submitted to electrophoresis on a polyacrylamide gel concentration gradient in the presence of sodium dodecyl sulfate, after Laemmli, as detailed by Axelos and Péaud-Lenoël (1980). In order to follow the differentiation process, plastids were analyzed after different cell culture periods in the presence of cytokinin. Figure 1 shows the photometric scannings of the Coomassie-blue stained electrophoresis gels loaded with plastid proteins of 8 days, 11 days, and 14 days cultures, at the time when thylakoids develop and chlorophylls are actively accumulated. Most of the polypeptide bands are present throughout the process; nevertheless, significant changes of the polypeptide map are observed: it is easy to notice that many bands decreased or disappeared in the region of M_r 30,000 to 33,000 and in the M_r 18,000 region while new bands emerge in the M_r 25,000 region. From these experiments, we concluded that during plastid maturation, specific proteins appeared or disappeared according to the considered culture period and that the set of proteins expressed in mature chloroplasts is only the last fraction of the proteins expressed along the realization of the program leading from proplastids to chloroplasts.

4 Comparison of Plastid Protein Synthesis in CK Cells or O Cells

The next question was whether the absence of cytokinin in the culture medium led to a lowered plastid protein synthesis as a whole or, alternatively, to specific alterations of the polypeptide electrophoretic map. In theory, the plastid protein synthesis rate in O cells cannot be reduced to nought, since this type of cells permanently maintains a plastid system which can resume photosynthetic activity in the presence of added cytokinin at any time and after a large number of generations. Therefore, the plastid protein synthesis rate of O cells can only be reduced down to a determined minimal value, compared to CK cells.

It was difficult to compare directly the amount of plastid proteins because the yield of isolated organelles may vary in an unknown manner from one extract to the other,

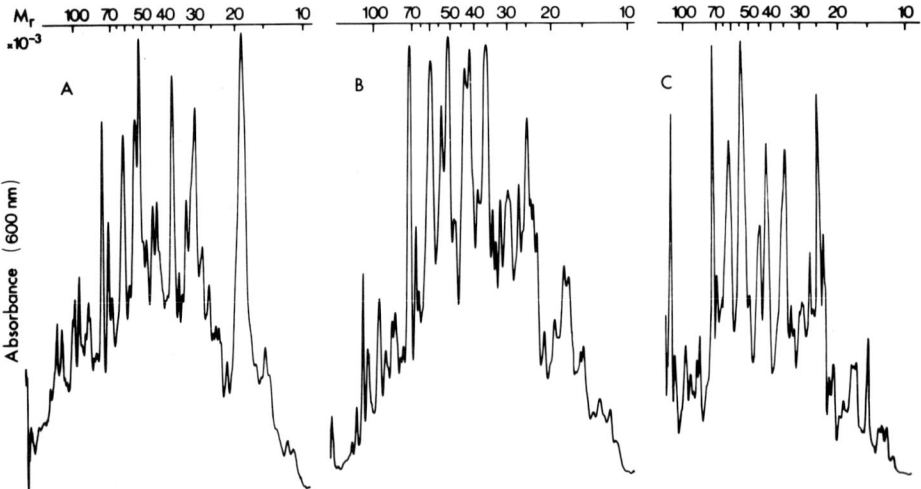

Fig. 1. A-C. Fractionation of plastid polypeptides prepared from CK cells at various stages of the culture: 8 days (**A**), 11 days (**B**), 14 days (**C**). Densitometric scannings of SDS-PAGE profiles. Linear acrylamide concentration gradients 10% to 18% (w/v). Gels stained by Coomassie blue. Molecular weights (MR) determined by marker proteins. (Axelos and Péaud-Lenoël, 1980)

but it was possible to estimate the specific incorporation rate of precursor labeled amino acids in each kind of culture. The use of appropriate antibiotics was also helpful in order to distinguish which of the labeled polypeptides were translated by 70S plastid ribosomes or by 80S cytoplasmic ribosomes. Figure 2 shows an example of the labeled polypeptide patterns obtained by SDS-PAGE after a 3h incubation of CK cells, with or without antibiotics, in the presence of a mixture of [^{14}C]-amino acids. These patterns were obtained from 15-day-old CK cell cultures in early stationary phase. There is a good fit between the radioactive polypeptide bands observed on the electrophoresis diagram of the cell sample incubated in the presence of cycloheximide and the polypeptide diagram observed after in vitro incorporation of labeled amino acids in isolated chloroplasts (see Lescure and Seyer, this vol.). Both in vivo and in vitro incorporation of labeled precursors have shown that, as a whole, the rate of protein synthesis on 70S chloroplast ribosomes was not strongly affected by the absence of cytokinin or the addition of the hormone to the culture medium.

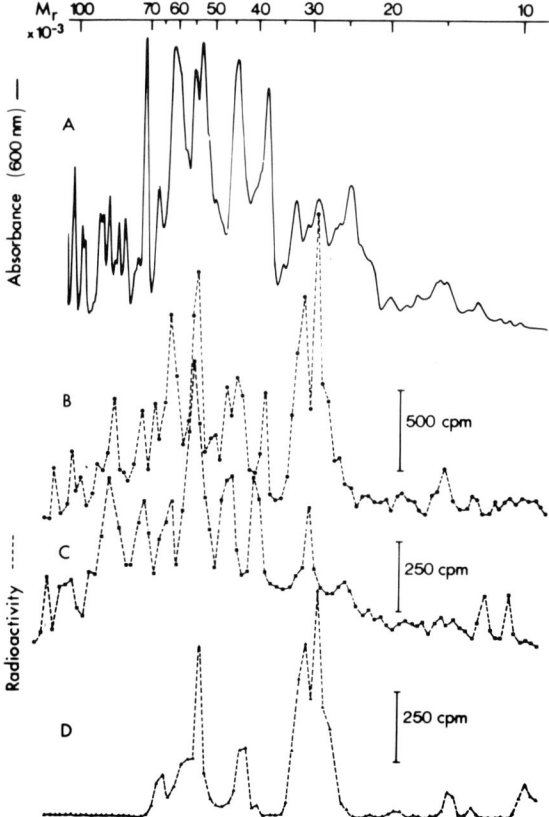

Fig. 2 A-D. SDS-PAGE analyses of in vivo incorporation of [^{14}C]-amino acids into plastid proteins of CK cells; techniques: see Fig. 1; radioactivity measurements: after densitometric scanning, the gels were sliced, each 1 mm slice was macerated in H_2O_2 and counted in a scintillation spectrometer. Curve A: densitometric scan; labeling patterns: curve B control cells, curve C cells incubated with CAP, curve D cells incubated with CH

In order to study the protein synthesis rate of plastids with special attention to cytoplasm-translated polypeptides, we analyzed the plastid proteins labeled by [^{14}C]-amino acids during 3h incubations of CK and O cells in the presence of chloramphenicol (Fig. 3). When the specific radioactivity of total plastid protein was measured in these samples, no significant difference was detected between CK or O cell plastid proteins. When particular radioactive peaks were considered, the SDS-PAGE diagrams were found similar although local differences still require further investigation. Chloramphenicol induced a general lowering of the incorporation rates in CK cell as well as in O cell samples.

It was noticed that some polypeptides were very rapidly labeled whereas others exhibited very low biosynthesis rates, suggesting that the turn over of different proteins was not the same in a given plastid and/or that different polypeptides have different biosynthetic periods during the evolution of the organelles in the cell culture. We concluded from this study that the evolution of plastid proteins in tobacco cell suspensions is characterized by a gradual change in the qualitative protein contents when the plastids differentiate from the proplastid to the chloroplast stage at the end of exponential growth. Most polypeptide subunits of the plastid proteins are present throughout all stages of differentiation, both in cells cultivated in the presence of cytokinin or in cytokinin-depleted cells. However, significant differences were observed in some polypeptide bands. The rate of plastid protein synthesis ist not, as a whole, tightly related to the presence of the hormone.

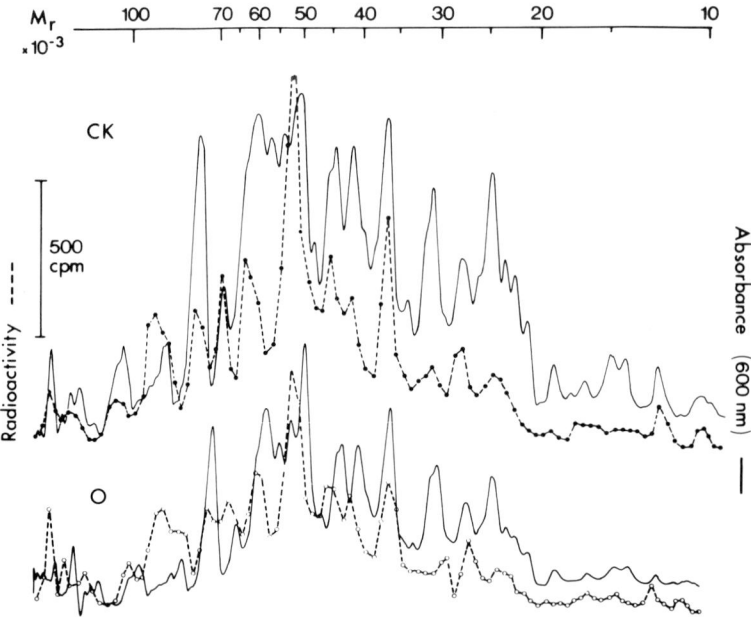

Fig. 3. In vivo [^{14}C]-amino acid incorporation into plastid proteins of *(CK)* cells, grown in the presence of kinetin 0.5 μM or cells transferred in cytokinin-free medium *(O)* and grown over three generations in this medium. SDS-PAGE analysis techniques: see Figs. 1 and 2

5 Compared Polypeptide Composition of Plastids Prepared from CK or O Cells

If the SDS-PAGE electrophoresis scannings of the Coomassie blue stained gels are compared, significant differences appear between the patterns of plastids extracted from CK or O cells. It was noticed that polypeptide bands located at Mr 25,000 and Mr 23,000 on the CK electrophoretic diagram were absent or very faint on the diagram of the leucoplasts extracted from O cells (Axelos and Péaud-Lenoël, 1980), as shown by Fig. 4. From the reports of many authors (see Thornber et al., 1979; Burke et al., 1979), it is known that the light-harvesting chlorophyll a/b-protein complex contains polypeptides in this range of molecular weights. We undertook to verify if the LHCP of tobacco cells contained the same polypeptides which appeared on the SDS-PAGE diagram of the chloroplasts extracted from CK cells. The method used to isolate the light-harvesting chlorophyll a/b-protein complex from tobacco cell cultures has been described (Axelos and Péaud-Lenoël, 1980). According to the visible spectrum of this complex and its SDS-PAGE fractionation, this material corresponded to the light-harvesting chlorophyll a/b-protein complex isolated from tobacco leaves, as described by Burke et al. (1979). Figure 5 shows the identity of absorption spectra of the complexes extracted from leaves or cell suspensions. On Fig. 6, a slab SDS-PAGE analysis of CK and O cell plastids is presented to compare the patterns of LHCP extracted from tobacco leaves (Burke et al.,

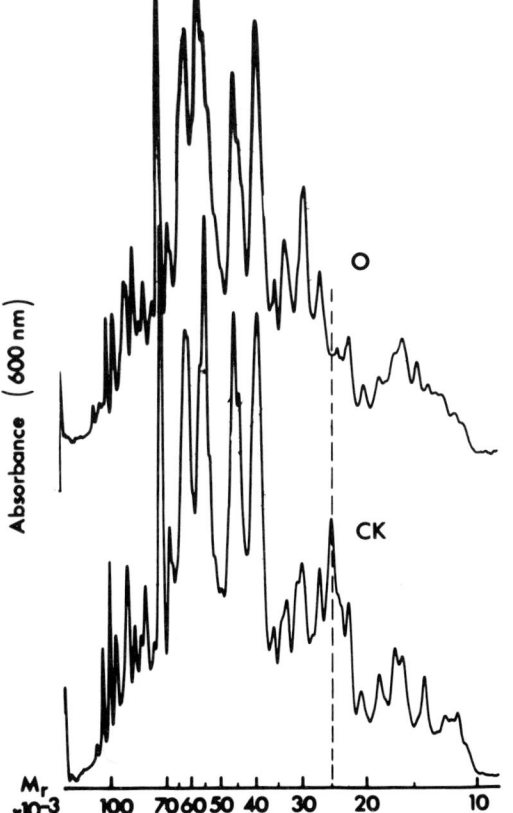

Fig. 4. Compared polypeptide composition of *CK* and *O* plastids. SDS-PAGE analysis as in Fig. 1. *Dashed line* position of 25,000 Mr polypeptide absent from *O* plastids. *CK* cells contained 21 µg chlorophyll, *O* cells 1 µg chlorophyll (per g fresh material); *O* cells maintained over 15 transfers in cytokinin-free medium

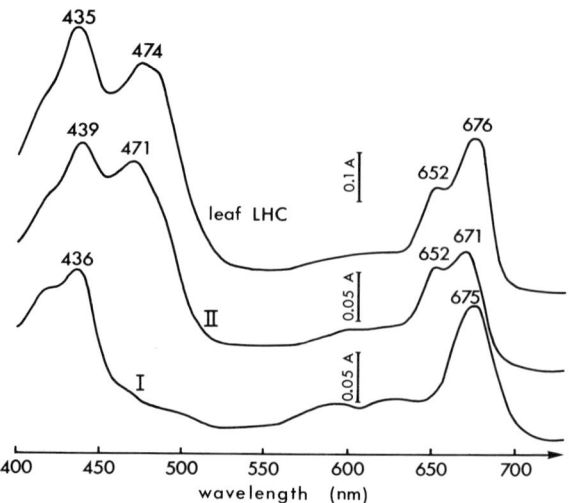

Fig. 5. Absorption spectrum of tobacco leaf *LHC* compared to spectra of chlorophyll-protein complexes I and II, isolated from tobacco cell suspensions. (Axelos and Péaud-Lenoël, 1980); absorption maxima as indicated in nm

1978) (lane 2), total CK chloroplast polypeptides (lane 3), LHCP from tobacco cell suspensions (lane 4) and the total O cell leucoplast polypeptides (lane 5). This experiment strongly suggests that the M_r 25,000 polypeptide seen on lanes 2 and 4 is the same polypeptide present only on the electrophoresis of CK cell chloroplasts (lane 3). The nature of the band $Mr = 23,000$ also present on lanes 3 and 4 is still controversial. The light-harvesting chlorophyll a/b complex has often been reported to contain two polypeptides of neighboring molecular weights (Burke et al., 1979; Thornber et al., 1979). The Mr 23,000 polypeptide might be derived from the Mr 25,000 subunit of LHCP by splitting in the living cells or during the extraction process. Local differences between the electrophoreses of lanes 3 and 5 appear outside of the Mr 23,000-25,000 region. These differences are alternatively positive or negative if the electrophoresis of lane 3 is taken as the standard. This indicates that the effect of cytokinins, whether direct or indirect, on the polypeptide composition of plastids is expressed both as a stimulation (for instance in the case of the Mr 25,000 polypeptide) or as a repression effect.

We are now driven to the hypothesis that cytokinins may control the biosynthesis of LHCP at the level of its cytosolic precursor, a polypeptide of Mr 29,500 (Apel and Kloppstech, 1978) or alternatively, that the effect of the hormone takes place during the processing of this precursor and/or its transport to the chloroplast. Our first attempt to answer these questions was to measure the labeling rate of LHCP and to compare this rate with the labeling of the corresponding area on the SDS-PAGE diagram of plastids extracted from cytokinin depleted cells (Fig. 3). Unfortunately, the labeling rate of LHCP is very slow and many extraneous polypeptides interfere in this region of the radioactivity mappings, in such a way that no conclusive evidence was obtained by this method. We have to wait for the results of other experiments to provide an answer.

Whatever the case, LHCP seems a useful molecular marker to trace the impact of cytokinins on the evolution of plastid polypeptides which are translated on the 80S ribosomes. Moreover, in conformity with the suggestion by Walter et al. (1977), the cytokinin-controlled presence of LHCP might stimulate the chlorophyll accumulation in the plastids by complexing the free chlorophylls as soon as they are biosynthesized, thereby

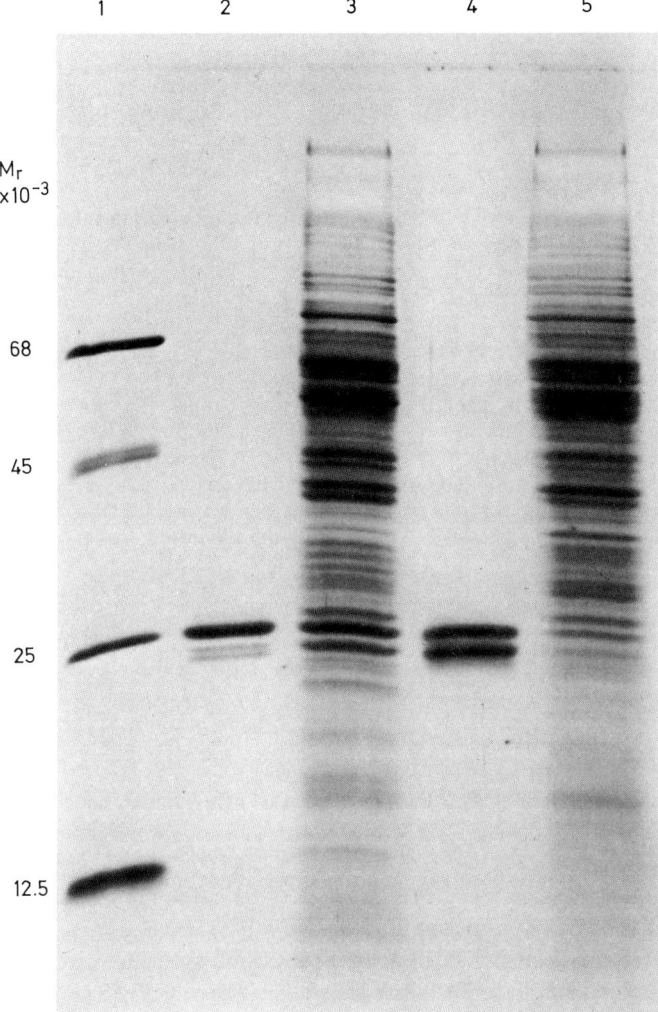

Fig. 6. Slab gel electrophoresis of plastid polypeptides (modified after Axelos and Péaud-Lenoël, 1980); lane *1* molecular weight markers; *other lanes* see text. O cells transferred twice in a cytokinin-free medium

eliminating the final products from the biosynthetic route of chlorophyll and, accordingly, increasing the efficiency of this route. If it was substantiated, this mechanism would provide a nonexclusive explanation to the well-known effect of cytokinins on chlorophyll biosynthesis and antisenescence reaction.

Moreover, it will be of interest to study the interactions between the phytochrome-stimulated biosynthesis of the LHCP precursor mRNA (Apel and Kloppstech, 1978; Apel, 1979) and the stimulation of LHCP accumulation in the presence of cytokinins.

Acknowledgments. The present work has been sponsored by the CNRS (ER 104) and benefited of many scientific contributions, especially from A.D. Hanson, A-M. Lescure, P. Seyer and N. Tandeau de Marsac. We gratefully acknowledge the technical help of M-C. Blot-Durand.

References

Apel K (1979) Phytochrome-induced appearance of mRNA activity for the apoprotein of the light-harvesting chlorophyll *a/b* protein of barley (*Hordeum vulgare*). Eur J Biochem 97:183-188

Apel K, Kloppstech K (1978) The plastid membranes of barley (*Hordeum vulgare*). Light-induced appearance of mRNA coding for the apoprotein of the light-harvesting chlorophyll *a/b* protein. Eur J Biochem 85:581-588

Axelos M, Péaud-Lenoël C (1980) The apoprotein of the light-harvesting chlorophyll *a/b* complex of tobacco cells as a molecular marker of cytokinin activity. Plant Sci Lett 19:33-41

Burke JJ, Ditto CL, Arntzen CJ (1978) Involvement of the light-harvesting complex in cation regulation of excitation energy distribution in chloroplasts. Arch Biochem Biophys 187:252-263

Burke JJ, Steinback KE, Arntzen CJ (1979) Analysis of the light-harvesting pigment-protein complex of wild type and a chlorophyll-*b*-less mutant of barley. Plant Physiol 63:237-243

Eichenberger W (1977) Steryl glycosides and acylated steryl glycosides. In: Tevini M, Lichtenthaler HK (eds) Lipids and lipid polymers in higher plants. Springer, Berlin Heidelberg New York, pp 169-182

Lescure A-M (1973) Selection of markers of resistance to base-analogues in somatic cell cultures of *Nicotiana tabacum*. Plant Sci Lett 1:375-383

Mazliak P (1977) Glyco- and phospholipids of biomembranes in higher plants. In: Tevini M, Lichtenthaler HK (eds) Lipids and lipid polymers in higher plants. Springer, Berlin Heidelberg New York, pp 48-74

Mohr H, Schopfer P (1977) The effect of light on RNA and protein synthesis in plants. In: Bogorad L, Weil JH (eds) Nucleic acids and protein synthesis in plants. Plenum Publ Comp, New York, pp 239-260

Parthier B (1979) The role of phytohormones (cytokinins) in chloroplast development. Biochem Physiol Pflanz 174:173-214

Péaud-Lenoël C, Hanson AD (1973) Une exigence en kinétine pour le développement de chloroplastes fonctionnels dans les cals de tissus de *Nicotiana tabacum* croissant par photosynthèse. CR Acad Sci Paris D 277:1853-1855

Seyer P, Marty D, Lescure A-M, Péaud-Lenoël C (1974) Effect of cytokinin on chloroplast cyclic differentiation in cultured tobacco cells. Cell Differ 4:187-197

Stetler DA, Laetsch WM (1965) Kinetin-induced chloroplast maturation in cultures of tobacco tissue. Science 149:1387-1389

Tandeau de Marsac N, Péaud-Lenoël C (1972) Cultures photosynthétiques de lignées clonales de cellules de tabac. CR Acad Sci Paris D 274:1800-1802

Thornber JP, Markwell JP, Reinman S (1979) Plant chlorophyll-protein complexes: recent advances. Photochem Photobiol 29:1205-1216

Walter G, Averina NG, Meister A (1977) Protochlorophyllid-Resynthese unter dem Einfluß von Kinetin. Spektrophotometrische Untersuchungen an Weizenpflanzen (*Triticum aestivum*) in vivo. Biochem Physiol Pflanz 171:409-417

Section 6
Animal Systems Responding to Cytokinin or Cytokinin Analogs

6-Benzylaminopurine as a Growth Factor for *Drosophila melanogaster* Cells Grown in Vitro

J.L. Becker[1] and J. Roussaux[2]

1 Introduction

Cytokinins are powerful activators of the plant growth and morphogenesis (Skoog and Armstrong, 1970; Hall, 1973). Most of them are adenine derivatives with a substitution on the 6th position of the purinic nucleus (Skoog and Armstrong, 1970). Numerous experiments have reported the biological effects of cytokinins, especially on cellular division, and described some biochemical modifications induced by such molecules. However, their precise action remains unknown. An essential property of the cytokinins will be considered: some of them are potent biosynthetic activators and, for instance, allow the growth *in vitro* of auxin or thiamin-dependent tissues on medium free of these components (Digby and Skoog, 1966; Stoutemyer and Britt, 1969; Jordan and Skoog, 1971; Syono and Furuya, 1972). 6-substituted adenines, as bases or nucleosides, also interfere in the division of animal cells (Buckley et al., 1962; Grillo and Polsky, 1966; Gallo et al., 1969; Rebhun and Monroy, 1970; Dorée and Guerrier, 1975), often by inhibiting it. Experiments on animal cells were mostly performed *in vitro* on malignant cell lines or on the whole animals using high cytokinin concentrations. In order to look for activator effects on animals similar to those described on plants, we have tested the action of small cytokinin concentrations on nonmalignant cells with important nutritional requirements. This paper demonstrates that 6-benzylaminopurine (6-BAP) is a growth factor for *Drosophila melanogaster* cells grown *in vitro*, and describes some effects of the cytokinin.

2 Materials and Methods

2.1 Biologicals

2.1.1 In Vitro Cell Line

The *Drosophila melanogaster* line used is Ko. This line derives from the K line obtained by Echalier and Ohanessian (1970). In contrast with the original line, the studied line is able to grow in a fetal calf serum free medium (Ko % line) but remains dependent on yeast extract and lactalbumine hydrolysate (D22 medium).

---

[1] Laboratoire de Zoologie – ERA 615, Université Pierre et Marie Curie, 7, Quai Saint-Bernard, 75005 Paris, France
[2] Laboratoire de Biologie Végétale IV, Université Pierre et Marie Curie, 12, rue Cuvier, 75005 Paris, France

2.1.2 Particularities of Purine Metabolism in Drosophila melanogaster Cell Lines

These lines as well as the fly "oregon" do not possess spontaneously measurable hypoxanthine-guanine-phosphoribosyltransferase (HGPRT) and 5'-nucleotidase activities (Becker, 1974a, b). However, these activities are detected when purine bases or an inhibitor of the *de novo* purine biosynthesis (azaserine) are added to the culture medium. These activities are also measurable in cell extracts (from untreated cultures) free of nucleotides after dialysis or treatment by activated charcoal (Becker, 1978).

2.1.3 Culture Medium and Growth of Cells

Cells are grown in the D 22 medium (Echalier and Ohanessian, 1970) with a slight modification: lactalbumine hydrolysate is replaced by a definite amino-acid mixture (D 51 P medium). As the D 22, D 51 P medium contains yeast extract (1.4 g/l).

In order to appreciate the growth and the survival of *Drosophila* cells in a chemically defined medium, about 6.10^6 cells are transferred in plastic flasks; after their attachment, they are washed twice with the synthetic medium then incubated in it. The cell proliferation is appreciated by counting the cells.

2.2 Incorporation of Labeled Nucleosides and Amino Acids

The medium of cell cultures is removed during exponential growth and replaced by a fresh medium containing $1\mu Ci/ml$ of labeled precursor. The cells are incubated for 1 h then washed. The radioactivity incorporated into DNA, RNA or proteins is determined according to Scheffler and Buttin (1973).

2.3 Determination of Hypoxanthine-Guanine-Phosphoribosyl Transferase, Nucleoside-Kinase and 5'-Nucleotidase Activities

All assays are performed on sonically disintegrated cells from exponentially growing cultures. Enzymatic activities are assayed according to Albertini and De Mars (1970) for HGPRT and to Robert de Saint Vincent and Buttin (1973) for nucleoside-kinases. 5'-nucleotidase activity is estimated by the conversion of [^{3}H]5'-guanosine monophosphate (GMP) by cell extracts in the absence of phosphoribosyl pyrophosphate (PRPP). The reaction products are separated by ascending chromatography on Polygram 300 UV_{254} sheet with 5% $Na_2 H PO_4$ water solution. Spots detected by UV light are cut out and counted in a scintillation mixture. The amount of proteins in the extracts is determined according to Lowry et al. (1951).

2.4 [^{14}C]-6-Benzylaminopurine ([^{14}C]-6-BAP) Incorporation and Research of Purine Metabolites

2.4.1 [^{14}C]-6-BAP Incorporation in Drosophila Cells

Cells are incubated for 0, 0.5, 2 and 6 h in the culture medium supplied with $1\mu Ci/ml$ of [^{14}C]-6-BAP and 6-BAP (10^{-5} M). Cultures previously grown in the presence of azaserine (5.10^{-5} M), hypoxanthine (10^{-4} M) and uridine (10^{-3} M) are submitted to the same treatment; uridine is necessary to avoid hypoxanthine toxicity (Bagnara et al. 1974). The radioactivity in the TCA-soluble or TCA-precipitable fractions is determined according to Scheffler and Buttin (1973). To localize the [^{14}C]-6-BAP the extracts from treated cells are incubated with DNase (90 min.; pH 7.0) or hydrolyzed by KOH (0.3 M; 50°C, 50 min.) prior to trichloracetic acid (TCA) precipitation.

2.4.2 Research of 6-BAP Derivatives in Cell Extracts

After treatment or not by dialysis or activated charcoal in order to remove nucleotides (Becker, 1978), cell extracts obtained according to Robert de Saint Vincent and Buttin (1973) are incubated with [^{14}C]-6-BAP and with ribose or phosphate donors. (PRPP; ATP; Ribose-1-P; Ribose-5-P). The extracts are fractionated as described above (2.3).

2.5 Chemicals

[^{3}H]-thymidine (27 Ci/mmol); [^{3}H]-uridine (3.1 Ci/mmol); [^{3}H]-GMP (5 Ci/mmol); [^{3}H]-methionine (71 Ci/mmol); [^{3}H]-leucine (40 Ci/mmol); [8-^{14}C]-6-benzylaminopurine (13.4 mCi/mmol) were purchased from Amersham. Non radioactive 6-BAP from Fluka. Carnitine, PRPP, R-1-P, R-5-P, ATP, thymidine triphosphate (TTP), azaserine and all purines from Calbiochem. RNase-free DNase from Sigma. Kinetine, isopentenyladenine, furoyladenine, thenyl and thenoyladenine were a gift from the Laboratoire des Substances Naturelles (Gif/Yvette).

3 Results

3.1 Action of Substituted Adenines on Drosophila Cell Growth

3.1.1 Construction of a Chemically Defined Medium by Addition of Carnitine and 6-BAP

Carnitine is able to replace yeast extract in the diet of *Tenebrio molitor* (Fraenkel et al., 1948). When added to the culture medium of *Drosophila* cells at a concentration of 1 mg/l, the yeast extract can be omitted, but fetal calf serum remains an obligatory supplement to cell proliferation (Table 1). The addition of 6-BAP (10^{-7} M) to a culture medium containing carnitine permits the growth of *Drosophila* cells in the absence of both yeast extract and fetal calf serum (Table 1). 6-BAP is active on cell proliferation in a range of concentrations varying from 10^{-8} M to 10^{-6} M, without any effect at 10^{-5} M and toxic at higher concentrations (Fig. 1). A maximal growth is obtained when adenine (10^{-4} M) and uridine (10^{-3} M) are added three times weekly in the medium.

In the presence of 6-BAP, the cell morphology is slightly modified: the cells are more flattened on the substrate. This may reflect the absence of sterols in the medium and eventually a modification of the cellular membrane. Moreover, the cells are more fragile, and a part of the population is disrupted during the scraping transfer process. To avoid the effect of the proteases released by the disrupted cells on the intact population, the cultures are washed twice with fresh medium just after the cell attachment.

Table 1. Growth of *Drosophila* cells in a medium containing carnitine, 6-BAP, and natural extracts

Basic medium		Fetal serum	Carnitine	6-BAP	Survival of the cultures
D_{22}	Complete	+	0	0	Unlimited (> 10 years)
	Complete	0	0	0	Unlimited (> 5 years)
	Yeast extract free	+	0	0	15 days
	Yeast extract free	0	0	0	4 days
D51P	Complete	+	0	0	Unlimited (> 3 years)
	Complete	0	0	0	17 days
		0	0	0	11 days
	Yeast extract free	+	+	0	Unlimited (> 6 months)[a]
	Yeast extract free	0	+	0	11 days
		0	+	+	Unlimited (> 6 months)[a]

[a] Cultures were deliberately stopped.

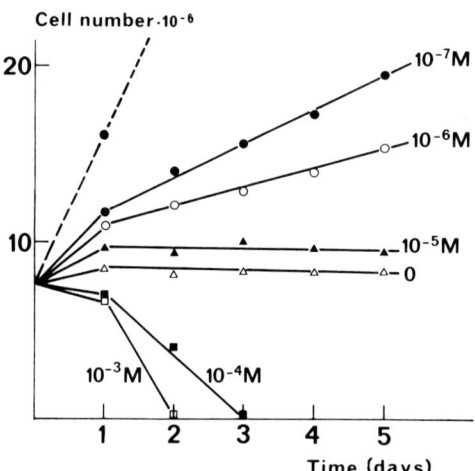

Fig. 1. Action of various concentrations of 6-BAP on the growth of *Drosophila* cells in the synthetic medium (- - - • - - -: growth in the complete medium D 22 + fetal calf serum)

3.1.2 Action of Other Substituted-Adenines

In the synthetic medium (D 51 P + carnitine), 6-BAP has been replaced by other adenines substituted on the 6-position, whose biological activity is known (Skoog et al., 1967; Roussaux et al., 1977). None of them has an activity similar to that of 6-BAP and some are clearly unfavorable to the cell growth (Table 2).

3.2 Action of 6-BAP on Some Metabolic Activities in Drosophila Cells

3.2.1 Action of 6-BAP on the Incorporation of Exogenous Nucleosides and Amino Acids

In order to characterize the action of 6-BAP on the growth of the *Drosophila* cells, an analysis of the nucleoside and amino acid incorporation has been performed on the precipitable and soluble TCA fractions.

6-Benzylaminopurine as a Growth Factor for *Drosophila melanogaster* 323

Table 2. Growth of *Drosophila* cells in the synthetic medium supplied with some substituted adenines

Compounds (10^{-7}M)	Cytokinin effect on the plant tissues[a]	Stimulation of *Drosophila* cells growth	Survival of cultures (number of transfers)
Isopentenyladenine	+ + + +	+	3
Furfuryladenine (kinetin)	+ + + +	0	0
Furoyladenine	+	±	1
Thenyladenine	+ + +	0	0
Thenoyladenine	+ +	0	0
Benzyladenine	+ + + +	+ +	Unlimited

[a] Tobacco-pith test according to Skoog et al. (1967) and Roussaux et al. (1977)

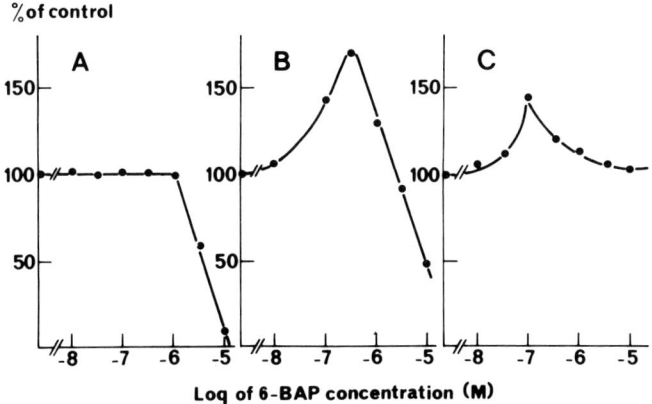

Fig. 2 A-C. Action of concentrations of 6-BAP on the incorporation of labeled precursors into the TCA precipitable fraction from *Drosophila* cells. A: [^{3}H]-thymidine; B: [^{3}H]-uridine; C: [^{3}H]-leucine

The effect of the regulator is detectable within 20 min after its addition to the culture medium. The [^{3}H]-thymidine incorporation in the TCA-precipitable fraction is inhibited by 6-BAP concentrations higher than 10^{-6} M, lower concentrations are without effect (Fig. 2 A). 6-BAP concentrations in a range of 10^{-8}, 10^{-6} M enhance the incorporation of [^{3}H]-uridine in the TCA precipitable fraction (Fig. 2 B). A slight but reproducible enhancement of the [^{3}H]-leucine (or [^{3}H]-methionine) incorporation is also observed at the concentration of 10^{-7} M (Fig. 2 C). In the TCA soluble fraction, the [^{3}H]-leucine and [^{3}H]-methionine incorporations are not affected by 6-BAP, whereas high cytokinin concentrations inhibit the incorporation of [^{3}H]-uridine and [^{3}H]-thymidine (Table 3).

3.2.2 Action of 6-BAP on HGPRT, 5'-Nucleotidase and Nucleoside-Kinase Activities

Cell cultures grown for 24 h in a medium supplemented by various concentrations of 6-BAP display spontaneous measurable HGPRT and 5'-nucleotidase activities (Table 4). Moreover, the extracts from such treated cells possess reduced nucleoside-kinase activ-

Table 3. Radioactivity (cpm) in the TCA-soluble fraction isolated from *Drosophila* cells incubated in a medium containing 6-BAP and labeled precursors

6-BAP Concentration	[^{3}H]-Thymidine	[^{3}H]-Uridine	[^{3}H]-Leucine	[^{3}H]-Methionine
10^{-5} M	550	416	816	1465
10^{-6} M	1397	1306	845	1788
10^{-7} M	1602	1215	807	1406
Control	1650	1286	805	1380

Table 4. Hypoxanthine-guanine phosphoribosyl transferase and 5'-nucleotidase activities in the 6-BAP treated cells

	6-BAP [M]			
	0	10^{-7}	10^{-6}	10^{-5}
HGPRT activity[a]	< 0.1	1.8	2.4	2.7
5'-nucleotidase activity[a]	< 0.1	2.5	3.1	3.3

[a] nmol/h/mg prot. of inosinic acid + inosine (HGPRT) or guanosine (5' nucleotidase)

ities compared to the control cultures. Nucleoside-kinase activities of the treated cell extracts are found similar to the control when 5'-nucleotidase inhibitor (TTP 0.1 M) is added to the reaction mixture (Murray and Friedrichs, 1969).

The addition of 6-BAP in the cell-free extracts induces the inhibition of the thymidine-kinase and uridine-kinase activities. In the two cases, the inhibition is non competitive toward both the nucleoside and ATP (Fig. 3). This inhibitory effect has also been found in extracts of cells grown with or without 6-BAP.

3.3 Behavior of [^{14}C]-6-BAP

3.3.1 Incorporation of [^{14}C]-6-BAP in Drosophila Cells

When cells are incubated in the presence of [^{14}C]-6-BAP, radioactivity is only detected in the TCA soluble fraction. In contrast, a slight incorporation is observed if cells are previously treated with azaserine, hypoxanthine, and uridine. To localize the [^{14}C]-6-BAP, cell extracts are treated by DNase, then precipitated by TCA. Radioactivity remains associated to the precipitate. Alkaline hydrolysis of these extracts eliminates radioactivity from the TCA precipitable fraction showing the linkage of the radioactivity to the RNA fraction (Table 5, Fig. 4).

3.3.2 6-BAP Metabolism in Cell Extracts

No metabolite is detected by chromatography in cell extracts incubated with [^{14}C]-BAP and ribose or phosphate donors. However, the extracts free of nucleotides incubated with [^{14}C]-6-BAP, PRPP and 5'-nucleotidase inhibitor (TTP) display a radioactive spot

6-Benzylaminopurine as a Growth Factor for *Drosophila melanogaster* 325

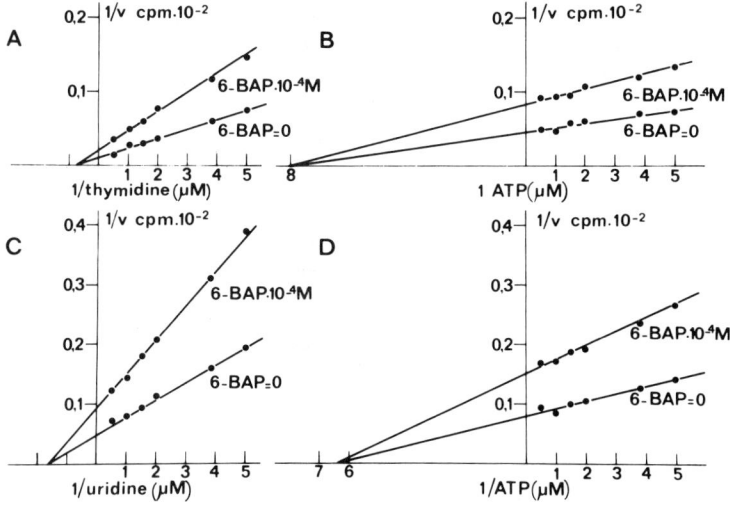

Fig. 3 A-D. Action of 6-BAP on thymidine kinase (**A-B**) and uridine kinase (**C-D**) activities in *Drosophila* cell extracts. Interactions between 6-BAP-nucleoside (A-C) and 6-BAP-ATP (B-D)

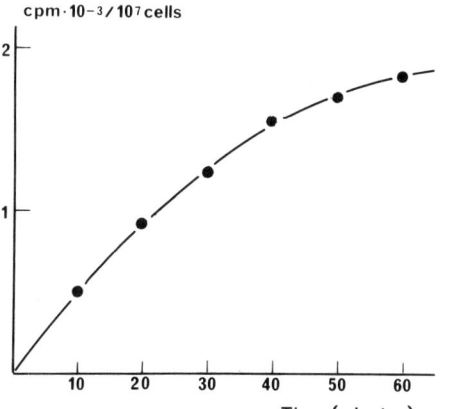

Fig. 4. Incorporation of [^{14}C]-6-BAP into the TCA precipitable fraction from *Drosophila* cells grown in a medium supplemented with azaserine, hypoxanthine and uridine

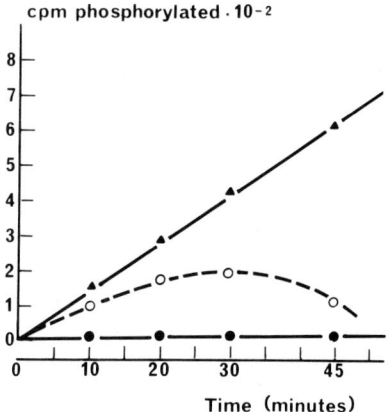

Fig. 5. Phosphorylation of [^{14}C]-6-BAP by *Drosophila* cell extracts. (Nucleotide-free extracts with —▲— or without — o — exogenous TTP; extracts with endogenous nucleotides — • —)

Table 5. Incorporation of [^{14}C]-6-BAP into TCA precipitable fraction

Treatment prior to TCA precipitation	c.p.m.
None	1380
DNase (90 min)	1340
Alkaline hydrolysis (60 min)	36

supposed to be the nucleotide of 6-BAP. A HGPRT activity is also detected. No radioactivity is associated with AMP (Fig. 5).

4 Discussion

These results indicate that it is possible to obtain the infinite proliferation of *Drosophila* cell lines on a synthetic medium containing carnitine and 6-BAP. This observation offers suitable interest because the complexity of usual culture media is unfavorable for the development of the somatic genetics of insect cells. One supposes that 6-BAP can substitute constituents from fetal calf serum essential for cell proliferation. As alternative hypothesis, 6-BAP could also modify the permeability of *Drosophila* cells as previously described in other materials (Marré et al., 1974; Penot, 1974). If an enhanced permeability occurred, a better utilization of some nutrients from the medium could be considered. The penetration of some ions could be promoted and induce biosynthetic activations which release the cells from nutritional requirements.

Studies on the action of several substituted adenines do not permit to detect any correlation between their effect upon insect cell multiplication and their cytokinin activity on plant material. That observation suggests that there is no specificity or unity of action of these substances in the two kingdoms. Exogenous cytokinins may be supposed to act directly, or through metabolized forms; some of them may be storage metabolites (Laloue et al., 1975, 1977). Nevertheless, as the possibilities of metabolization are unequal according to the nature of the cytokinin and the biological material used, it is not surprising to find some differences between the results obtained with plant tissues and *Drosophila* cells.

In the presence of small quantities of 6-BAP, an enhanced uridine, methionine, and leucine incorporation into the TCA-precipitable fraction is observed, while the incorporation of thymidine remains equal; the incorporations in the TCA-soluble fraction are not modified by the cytokinin. On the other hand, high 6-BAP concentrations induce the inhibition of the [^{3}H]-thymidine and [^{3}H]-uridine incorporation, as well in the TCA-precipitable as in the TCA-soluble fractions. These results suggest that low 6-BAP contrations do not modify the cell permeability toward the used markers, but stimulate the synthesis of proteins and ribonucleic acids as in other materials (Roussaux et al., 1976). On the contrary, higher concentrations should produce the global reduction of nucleosides permeability. Such an inhibitory effect of 6-BAP has been observed in the case of ion penetration during the senescence of leaves (Penot, 1974).

However, the 6-BAP induces HGPRT and 5'-nucleotidase activities in the cells, which are inhibited by endogenous nucleotide. This effect is not specific, because some bases or nucleosides give the same result more efficiently (Becker, 1978); it suggests that 6-BAP disturbes the endogenous nucleotide pool. This effect is probably independent of the 6-BAP action on the cell growth because the addition of bases or nucleosides to the D 51 P + carnitine medium does not stimulate the development of cultures (unpubl. result).

The inhibition of nucleoside-kinase activities (in vivo) may be produced by the combined effects of the 5'-nucleotidase (which hydrolyses the reaction product during the assay) and of the 6-BAP which acts as a negative effector on these activities.

6-BAP seems to be slightly metabolized by *Drosophila* cell extracts, because no degradation products could be ever found. A similar situation was described in *Acer* cells (Dorée and Guern, 1973). The [^{14}C]-6-BAP is not incorporated in the TCA precipitable fraction in normal conditions of culture; then the HGPRT activity in the cells is too low. By contrast, a treatment by azaserine induces a significant incorporation: 6-BAP is phosphorylated, as in other materials (Pethe-Sadorge et al., 1972; Chen and Eckert, 1977) but here by HGPRT.

This phosphorylation is also observed in the extracts free of endogenous nucleotides and the presence of 6-BAP-nucleotide has been supposed. The DNase action on the TCA precipitable fraction of 6-BAP treated cell extracts suggests the incorporation of [^{14}C]-6-BAP in the RNA's, as described in other observations (Armstrong et al., 1976; Teyssendier de la Serve and Jouanneau, 1979).

In conclusion, these results indicate that 6-BAP can sustain *Drosophila* cell multiplication in a synthetic medium, but its mode of action in this phenomenon is unknown.

Acknowledgments. Miss E. Moreau is gratefully acknowledged for her technical assistance.

References

Albertini RJ, De Mars R (1970) Diploid azaguanine resistant mutants of cultured human fibroblasts. Science 169:482-485

Armstrong DJ, Murai N, Taller BJ, Skoog F (1976) Incorporation of cytokinin N^6-benzyladenine into tobacco callus transfer ribonucleic acid and ribosomal ribonucleic acid preparations. Plant Physiol 57:15-22

Bagnara AS, Letter AA, Henderson JF (1974) Multiple mechanisms of regulation of purine biosynthesis *de novo* in intact tumor cells. Biochim Biophys Acta 374:259-270

Becker JL (1974a) Purine metabolism pathways in Drosophila cells grown in vitro: phosphoribosyltransferase activities. Biochimie 56:779-781

Becker JL (1974b) Metabolisme des purines dans des cellules de *Drosophila melanogaster* en culture in vitro: interconversion des purines. Biochimie 56:1249-1253

Becker JL (1978) Regulation of purine biosynthesis in cultured *Drosophila melanogaster* cells. I: Conditional activity of HGPRT and 5'-nucleotidase. Biochimie 60:619-625

Buckley WB, Witkus ER, Berger CA (1962) Kinetin, as a mitotic stimulant in *Triturus viridescens*. Nature (London) 194:1200-1201

Chen CM, Eckert RL (1977) Phosphorylation of cytokinin by adenosine kinase from wheat germ. Plant Physiol 59:443-447

Digby J, Skoog F (1966) Cytokinin activation of thiamine biosynthesis in tobacco callus cultures. Plant Physiol 41:647-652

Dorée M, Guern J (1973) Short-time metabolism of some exogenous cytokinins in *Acer pseudoplatanus* cells. Biochim Biophys Acta 304:611-622

Dorée M, Guerrier P (1975) Action spécifique de la N-6-isopenténylad́enine sur la division cellulaire des oeufs de l'oursin *Sphaerechinus granulis*. CR Acad Sci Ser D 281:1421-1424

Echalier G, Ohanessian A (1970) In vitro culture of *Drosophila melanogaster* embryonic cells. In Vitro 6:162-172

Fraenkel G, Blewett M, Coles M (1948) B$_t$, a new vitamin of the B-group and its relation to the folic acid group and other anti-anoemia factors. Nature (London) 161:981

Gallo RC, Whang-Peng J, Perry S (1969) Isopentenyladenosine stimulates and inhibits mitosis of human lymphocytes treated with phytohemagglutinin. Science 165:400-402

Grillo RS, Polsky R (1966) Karyokinetic interference in *Triturus* following kinetin treatment. Exp Cell Res 44:375-381

Hall RH (1973) Cytokinins as a probe of developmental processes. Annu Rev Plant Physiol 24:415-444

Jordan WR, Skoog F (1971) Effects of cytokinins on growth and auxin in coleoptiles of derooted *Avena* seedlings. Plant Physiol 48:97-99

Laloue M, Gawer M, Terrine C (1975) Modalités de l'utilisation des cytokinines exogènes par les cellules de tabac cultivées en milieu liquide agité. Physiol Veg 13:781-796

Laloue M, Terrine C, Guern J (1977) Cytokinins: metabolism and biological activity of N^6 (Δ^2-isopentenyl) adenosine and N^6 (Δ^2-isopentenyl) adenine in tobacco cells and callus. Plant Physiol 59:478-483

Lowry OH, Rosebrough NJ, Farr AL, Randall RJ (1951) Protein measurement with the folin phenol reagent. J Biol Chem 193:265-270

Marré E, Colombo R, Lado P, Rasi-Caldogno F (1974) Correlation between proton extrusion and stimulation of cell enlargment. Effects of fusicoccin and of cytokinins on leaf fragments and isolated cotyledons. Plant Sci Lett 2:139-150

Murray AW, Friedrichs B (1969) Inhibition of 5'-nucleotidase from Ehrlich ascite-tumor cells by nucleoside triphosphates. Biochem J 111:83-89

Penot M (1974) Evolution de la perméabilité au molybdate chez des disques de feuille de tabac au cours du vieillissement. Influence de la benzylaminopurine. Plant Sci Lett 3:399-405

Pethe-Sadorge P, Signor Y, Guern J (1972) Sur la synthèse des nucléosides 5'-monophosphates de cytokinines par les cellules d'*Acer pseudoplatanus*. CR Acad Sci Ser D 275:2493-2496

Rebhun LI, Monroy E (1970) Stimulation and inhibition of cell division in mammalian cells by 6-substituted adenines. J Cell Biol 47:168a

Robert de Saint Vincent B, Buttin G (1973) Studies on 1-β-D arabinofuranosyl-cytidine-resistant mutants of chinese hamster fibroblasts. Eur J Biochem 37:481-488

Roussaux J, Hoffelt M, Farineau N (1976) Evolution des RNA ribosomaux au cours du verdissement de cotylédons de Concombre en présence de 6-benzylaminopurine. Can J Bot 54:2328-2336

Roussaux J, Tran Thanh Van, Adeline MT, Adou P, Dat Xuong N (1977) Activité cytokinine de dérivés acylés, alkylés, acyl-alkylés de l'adénine et d'isostères azotés de la γ, γ-diméthylallyladénine. Phytochemistry 16:1865-1869

Scheffler IE, Buttin G (1973) Conditionally lethal mutations in chinese hamster cells. J Cell Physiol 81:199-216

Skoog F, Armstrong DJ (1970) Cytokinins. Annu Rev Plant Physiol 21:359-384

Skoog F, Hamzi HQ, Szweykowska AM, Leonard NJ, Carraway KL, Fujii T, Helgeson JP, Loeppky RN (1967) Cytokinins: Structure/activity relationships. Phytochemistry 6:1169-1192

Stoutemyer VT, Britt OK (1969) Growth and habituation in tissue cultures of english ivy, *Hedera helix*. Am J Bot 56:222-226

Syono K, Furuya T (1972) Effects of cytokinins on the auxin requirement and auxin content of tobacco calluses. Plant Cell Physiol 13:843-856

Teyssendier de la Serve B, Jouanneau JP (1979) Preferential incorporation of an exogenous cytokinin, N^6-benzyladenine, into 18 S and 25 S ribosomal RNA of tobacco cells in suspension culture. Biochimie 61:913-922

Use of N⁶-Benzyladenosine as an Analog of Internal Methylation (N⁶-Methyladenosine) in Studies of Post-Transcriptional Regulation of RNA in Human Cells Infected by Herpes Simplex Virus

A. Garcia, A. Epstein, and B. Jacquemont[1]

1 Introduction

In animal DNA viruses and in eukaryotic cells, a substantial body of evidence shows that RNA is first transcribed as a larger precursor molecule which gives rise to messenger RNA by a maturation process. This processing includes capping and methylation of the 5' end, internal methylation, polyadenylation of the 3' end, cleavage with exclusion of symmetrical sequences and introns (Revel and Groner, 1978; Abelson, 1979). DNA viruses are useful tools to study this type of processing because, in infected cells, viral messenger RNA and viral polypeptides can be readily isolated and analyzed.

For the past few years, we have been studying the maturation of viral RNA, more precisely the cleavage and methylation of viral RNA in cultures of human cells infected by Herpes simplex virus type 1 (Jacquemont and Roizman, 1975b; Jacquemont et al., 1980). Two specific types of methylation were observed in viral RNA, one corresponding to the methylation of the 5'-end in the cap structure and the other to the internal methylation of adenosine at N^6. When methylation of the cap was examined in vivo, it was found that it was essential for RNA translation (Jacquemont and Huppert, 1977). It was previously reported that the proportion of internal methylation decreased during infection (Bartkoski and Roizman, 1976). Since no role has been attributed so far to internal methylation, we investigated this problem with N^6-methyladenosine and its structural analogs such as N^6-ethyladenosine and N^6-benzyladenosine. With N^6-methyladenosine itself or with N^6-ethyladenosine, no effect was observed on virus multiplication, viral DNA replication, viral RNA synthesis, viral RNA methylation and viral polypeptide synthesis. When labeled N^6-methyladenosine was added to infected cells, it was found that this analog was rapidly demethylated into adenosine. N^6-benzyladenosine, a cytokinin, was used for further experiments, since it was found that it was not degraded and was incorporated into RNA.

In this paper, we describe its action on infected cells and its effect on post-transcriptional events involved in viral macromolecular synthesis.

2 Fate of Tritiated N^6-Benzyladenosine

In all experiments N^6-benzyladenosine (Bzl⁶ Ado) synthesized in Dr Imbach's laboratory (Montpellier, France) was used at 20 µg/ml, corresponding to 60 µM. Human

---

[1] Unité de Virologie Fondamentale et Appliquée, I.N.S.E.R.M. – U.51, Groupe de Recherche C.N.R.S. 33, 1 place Professeur Joseph Renaut, 69371 Lyon, Cedex 2, France

epidermoid carcinoma no. 2 (HEp2) cells were infected with 100 pfu/cell of Herpes Simplex type 1 (HSV-1), F strain, and labeled with 20 µg/ml of tritiated Bzl^6 Ado (750 mCi/mmol, 0.35 mCi/ml). The radioactive analog was obtained after an isotopic exchange with tritium gas and the tritium was fixed in C^8 of purine ring. Six hours after infection, the cells were incubated for 2 h with the tritiated analog. The distribution of the analog in the soluble pool as well as its incorporation into RNA were determined by the amount of N^6 benzylaminopurine (BAP) found in acid-soluble and acid-insoluble components after formic acid hydrolysis followed by cellulose paper chromatography. In the soluble pool, 100% of the label was found as BAP showing that no degradation took place at this level. A small amount of radioactivity was incorporated into RNA (estimated approximatively as 0.5% of the normal incorporation of [^{14}C]-labeled adenosine into RNA under the same conditions). However only 20% of that incorporation was found as BAP, and the rest was found mostly as adenosine, suggesting a partial degradation of the cytokinin after its incorporation in RNA molecules.

3 Effect of N^6-Benzyladenosine on Cell Growth and on HSV Multiplication

HEp2 cells exponentially growing in monolayers were incubated from 0 to 4 days with different concentrations (from 1 to 20 µg/ml) of Bzl^6 Ado; cell growth was determined by protein estimation. The cytokinin inhibited cell growth at a concentration of 5 µg/ml. However, if after 10 h contact with 20 µg/ml, the cytokinin was removed, the surviving cells started to grow again and multiplied as the controll cells.

To test the effect of cytokinin on HSV-1 multiplication, confluent HEp2 cells were incubated with the cytokinin 1 h before infection with the virus at a multiplicity of 10 pfu/cell. The amount of virus was measured 14 h after infection and compared to the virus produced by control cells containing no cytokinin. We found 93% reduction in viral multiplication. When the cytokinin was removed after 14 h, virus multiplication started again but the yield was inhibited by 74% after 24 h and by 62% after 72 h.

4 Macromolecular Syntheses

HEp2 cells were infected and labeled as described in the legend of Fig. 1. Because Bzl^6 Ado reduced cell permeability for nucleosides and amino acids, macromolecular syntheses were estimated from the ratio of acid precipitable radioactivity to total radioactivity uptaken by cells. From these data, only total protein synthesis seemed to be affected. In the treated cells, protein synthesis decreased at a later time (Fig. 1 A, B, C).

In order to analyze viral RNA synthesis, total, nuclear, and cytoplasmic RNA, extracted and purified at different times after infection (Jacquemont et al., 1980) were hybridized to viral DNA fixed on filters (Jacquemont and Roizman, 1975b). In the presence of cytokinin, the synthesis of viral RNA increased slightly until 8 h and thereafter the difference with controls steadily increased reaching a maximum around 14 h (Fig. 1 D). When nuclear and cytoplasmic RNA syntheses were separately observed, it appeared that up to 8 h, all viral RNA was normally translocated to the cytoplasm. On the contrary, the large amount of later observed excess viral RNA remained in the nucleus (Fig. 1 E, F).

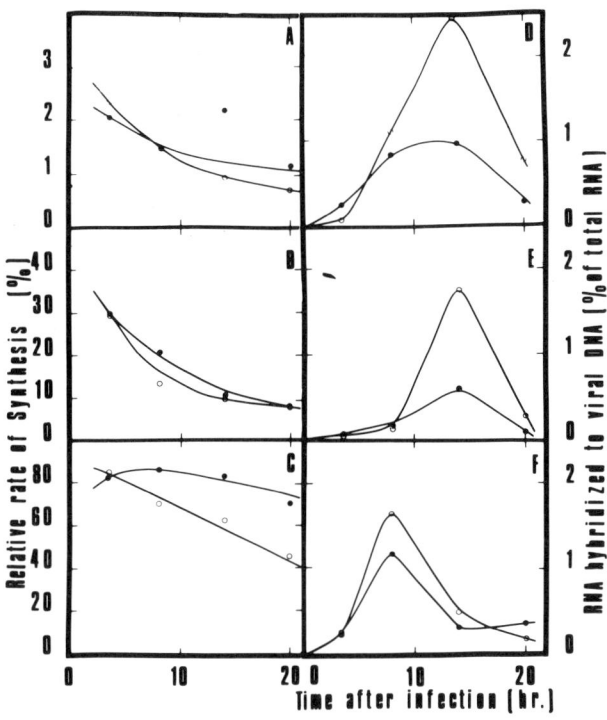

Fig. 1 A-F. Synthesis of total DNA, RNA, proteins of viral RNA. Cells were infected with 50 pfu/cell of HSV-1. At different times after infection, cells were labeled for 2 h in the absence or in the presence of 20 µg/ml of Bzl⁶ Ado, added 30 min before labeling. Relative rate of synthesis corresponding to the ratio of acid-insoluble radioactivity to intracellular total radioactivity was estimated. A Cells were labeled with [³H]-thymidine (10 µCi/ml); B, D, E, F Cells were labeled with [³H]-uridine (10 µCi/ml); C Cells were labeled with the mixture of [¹⁴C]-valine, [¹⁴C]-isoleucine, [¹⁴C]-leucine (0.5 µCi/ml). For D, E and F, nuclei and cytoplasms were separated and the proportion of specific viral RNA determined by hybridization to viral DNA fixed on filter. D total viral RNA. E nuclear viral RNA. F cytoplasmic viral RNA. —•——•— control infected cells; —o——o— Bzl⁶ Ado infected cells

5 Nuclear Cleavage of Viral RNA

Because the most significant differences appeared in the nuclei, we compared the size of viral RNA synthesized in the presence or absence of Bzl⁶ Ado.

In order to compare early and late infection, one sample of cells was infected in the presence of cycloheximide which allowed any immediately early RNA synthesized to accumulate in the absence of protein synthesis (Kozak and Roizman, 1974). A second sample of cells was harvested at a later time. Cells were labeled and treated as indicated in the legend of Fig. 2. The analysis of the size of viral RNA under denaturing conditions (Jacquemont et al., 1980) yielded two sets of results (Fig. 2).

1. The proportion of viral RNA was increased by the cytokinin at early time and decreased at later time. This last result seems at variance with that described on Fig. 1 E: this was probably due to the period of labeling: 2 h in a first experiment and 0.5 h in the last. These differences suggested that the increase of viral RNA in nuclei (Fig. 1 E) at late times corresponded to a storage of unmatured transcripts rather than to a real stimulation of RNA synthesis.
2. In the presence of cycloheximide, the cytokinin introduced but slight differences in size of viral RNA. However, at later times, the size of RNA molecules increased in the presence of cytokinin (Fig. 2 B). The proportion of synthesized viral RNA decreased in the presence of cytokinin and the differences were more pronounced when the size of the viral RNA molecules were larger (Fig. 2 B). Consequently our data sug-

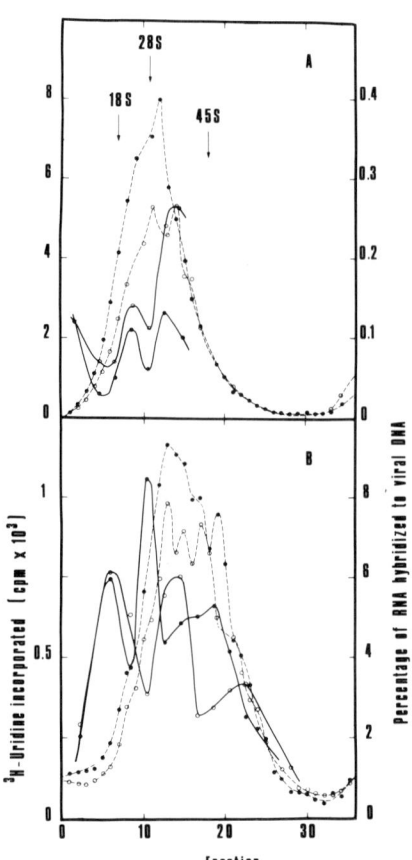

Fig. 2 A, B. Distribution of nuclear viral RNA in a formamide sucrose gradient. Cells were infected with 100 pfu/cell of HSV-1 and labeled with [^{3}H]-uridine (100 μCi/ml) for 30 min. Nuclei RNA were isolated, denaturated and centrifuged in a 5%-20% sucrose gradient containing 70% formamide. Acid-insoluble radioactivity was measured and RNA fractions were pooled, repurified and hybridized to viral DNA. *Arrows* indicate rRNA sedimentation determined by optical density at 260 nm. **A** Cells were incubated with cycloheximide (100 μg/ml) 1 h before infection. One and a half hours after infection, cells were labeled in the absence or in the presence of 20 μg/ml of Bzl^6 Ado added 30 min before labeling. **B** Cells were labeled at 9.5 h in the presence or in the absence of Bzl^6 Ado, as in A. -●--●- total RNA from control infected cells determined by acid-insoluble radioactivity; -●—●- viral RNA from control infected cells determined by hybridization to viral DNA; -○--○- total RNA from Bzl^6 Ado infected cells determined as in the control cells; -○—○- viral RNA from Bzl^6 Ado infected cells determined as in the control cells

gested that the nuclear cleavage of viral RNA was disturbed mainly at later time. Since Bzl^6 Ado was an analog of the internal methylated nucleotide, the disturbance of nuclear cleavage could be related to a modification of mRNA structure.

To investigate this possibility, two types of experiments were carried out, the first to determine the secondary structure of viral mRNA and a second to analyze the distribution of methylated nucleotides into mRNA.

6 Secondary Structure of Viral RNA

Viral cytoplasmic RNA was purified from cells infected as described in the previous experiment. Since symetrical transcripts of RNA were found in the nuclei (Kozak and Roizman, 1974; Jacquemont and Roizman, 1975a), the investigation of secondary structure of RNA was carried out on cytoplasmic mRNA. The proportion of single-stranded sequences in mRNA was estimated by kinetic analyses using S1 nuclease digestion under conditions which left double stranded sequences unsplit (Kozak, 1980). When the viral cytoplasmic RNA was digested as shown in Table 1:

Table 1. Analysis of secondary structure of viral mRNA by nuclease S1 digestion. Cells were infected as in Fig. 2. Cytoplasms were separated and viral RNA purified by hybridization to viral DNA on filter. The viral RNA were eluted, self-reannealed and digested with nuclease S1 at 21°C in 0.3 M NaCl, 30 mM sodium acetate pH 4.5, 2 mM Zn SO$_4$ and 5% glycerol

Time of labeling	Percentage of hydrolysis by S1 nuclease	
	After 60 min of hydrolysis	After 90 min of hydrolysis
1.5 to 2 h in the presence of cycloheximide[a]	92.8	93.8
2 h in the presence of cycloheximide[a] and Bzl^6 Ado[b]	95.6	96.2
9.5 to 10 h	81.5	80.0
9.5 to 10 h in the presence of Bzl^6 Ado[b]	84.3	88.6

[a] 100 µg/ml of cycloheximide were added 1 h before infection and maintained
[b] 20 µg/ml of Bzl^6 Ado were added 30 min before labeling

1. The proportion of secondary structure was much more important at a later time than at the immediately early time corresponding to a difference of 13% secondary structures in favor of the late RNA.
2. The presence of the cytokinin in the cells decreased slightly the percentage of mRNA resistant to S1 nuclease action in the immediately early RNA but did so more significantly for late RNA (9%).
3. From these and other data, we concluded that the kinetics of digestion of viral RNA synthesized at a later time in the presence of cytokinin was slower than that of control RNA suggesting some modification of secondary structure.

7 Methylation of Viral RNA

In order to visualize methylation of RNA molecules, cells were labeled with [^{3}H]-methyl-methionine and ^{32}P (orthophosphate). The distribution of methyl groups into viral RNA at early and late times is shown in Table 2:

1. It appeared that the distribution of methyl groups changed during the infection. Early internal methylation represented 54% and cap 46% of the total methyl groups incorporated. However, late in infection, when viral RNA was increased 10-fold, the internal methylation decreased to 22% and the ratio methyl/phosphate decreased from 13 to 0.45. These results suggested a temporal control in viral RNA methylation related to a decrease of internal methylation as described by Bartkoski and Roizman (1976) and to an increase of the proportion of cap methylation.
2. Late in infection in the presence of cytokinin, the amount of [^{32}P] radioactivity incorporated into RNA showed that the amount of viral RNA was roughly the same as that for control infected cells. At the same time, cap methylation was not inhib-

Table 2. Distribution of methyl groups in viral RNA. Cells were infected with 100 pfu/cell of HSV-1. At 1 and 7 hours, cells were labeled with [^{3}H]-methyl-methionine (1 mCi/ml) and ^{32}P (orthophosphate) (1 mCi/ml). RNA was extracted and viral RNA selected by hybridization to viral DNA on filter. Viral RNA was digested by pancratic and T2 RNases. Hydrolysis products separated on DEAE Sephadex column were eluted by 0.01 M to 0.4 M NaCl linear gradient in 10 mM Tris-HCl, pH 7.4, 7 M Urea. Distribution of methyl groups between internal methylation, cap were estimated from [^{3}H]-radioactivity found at −2, and −5, −6 charges. A known mixture of nucleotides with different phosphate charges was added as controls

Time of labeling	Designation of results	Internal methylation	Cap
1 to 4 h	Methyl radioactivity (cpm)	3136	2772
	Phosphate radioactivity (cpm)	246	
	Ratio methyl/phosphate	13	11
	Distribution of methyl (%)	54	46
7 to 10 h in the absence of Bzl^6 Ado	Methyl radioactivity (cpm)	780	2795
	Phosphate radioactivity (cpm)	1690	
	Ratio methyl/phosphate	0.45	1.7
	Distribution of methyl (%)	22	78
7 to 10 h in the presence of Bzl^6 Ado	Methyl radioactivity (cpm)	1586	3703
	Phosphate radioactivity (cpm)	1851	
	Ratio methyl/phosphate	0.90	2.0
	Distribution of methyl (%)	30	70

[a] 20 µg/ml of Bzl^6 Ado were added 30 min before labeling
[b] Phosphate radioactivity corresponding to nucleoside-3'-monophosphate at the level of internal methylation

ited but internal methylation was activated. This was obvious from the ratio of [^{3}H]-methyl to [^{32}P] in viral RNA.

All the modifications previously described at a level of RNA synthesis and RNA processing led us to analyze viral mRNA translation.

In HSV-1 replication, three types of viral polypeptides specified as α, β and γ were successively synthesized (Honess and Roizman, 1974): α polypeptides corresponded to immediately early mRNA, β polypeptides were the polypeptides synthesized in the absence of viral DNA synthesis and γ polypeptides followed DNA synthesis.

Two types of experiments were carried out.

8 Functional Stability of mRNA

We estimated the ability of mRNA to synthesize viral polypeptides. This was obtained by an experiment with actinomycin D as reported in Fig. 3. At an early time (Fig. 3 A) the functional stability of mRNA synthesized in the presence of actinomycin D increased about twofold in the presence of cytokinin. At a later time, the functional stability of mRNA molecules was twofold higher than at an early time, whereas in the presence of cytokinin, it decreased fourfold. The effect at early time could be due either to a longer

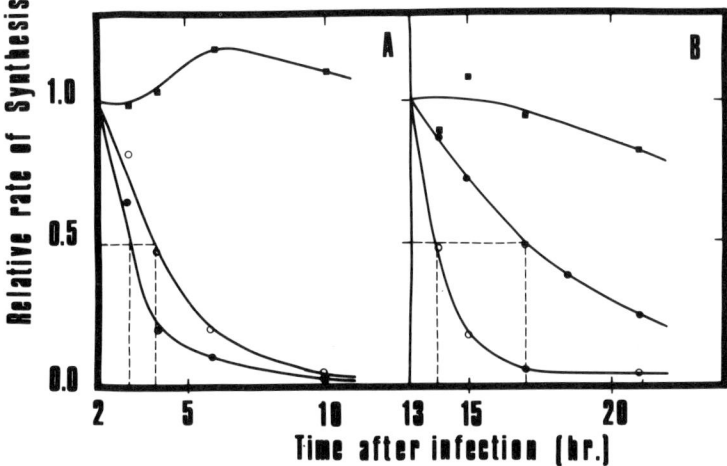

Fig. 3 A, B. Relative rates of synthesis of viral polypeptides exposed to actinomycin D. Cells were infected with 20 pfu/cell of HSV-1. At 1 h (**A**) or 12 h (**B**), three samples of cells were respectively treated as follows: a first sample with 20 μg/ml of Bzl[6] Ado and 10 μg/ml of actinomycin D 1 h later, a second sample with 10 μg/ml of actinomycin D only and an untreated third sample. At different times, cells were labeled with the mixture [^{14}C]-valine, [^{14}C]-isoleucine, and [^{14}C]-leucine (2 μCi/ml) for 1 h. Relative rate of synthesis corresponding to the ratio of acid-insoluble radioactivity to intracellular total radioactivity was estimated. The stability was determined by comparing half-lives (50% decrease). —■—■— control; —○—○— actinomycin D; —●—●— actinomycin D in the presence of Bzl[6] Ado

half-life of viral mRNA or to a difference in the translational capacity. Our data did not allow a final conclusion. At a later time, autoradiograms of polypeptides separated by electrophoresis (data not shown) showed that these proteins corresponded mostly to γ polypeptides. Consequently, this observed synthesis in the presence of actinomycin D is not due to survival of β mRNA.

From these three experiments, we concluded that the parallel observation of the increase of secondary structure of mRNA and the functional ability (to synthesize polypeptides) was related to a decrease of internal methylation. This was in agreement with the observation of Perry and Scherrer (1975) who showed that the lack of internal methylation of different mRNA gave rise to more stable RNA. Our results, observed late in infection, provided direct evidence that the increase of internal methylation by cytokinin did lead to RNA with less secondary structure and lower functional stability.

9 Analysis of Viral Polypeptide Synthesis

Viral polypeptide synthesis was analysed in the presence of the cytokinin added 1 h before infection as described in the legend of Fig. 4. The relative rate of polypeptide synthesis decreased progressively in the presence of cytokinin (data not shown). Autoradiogram data showed that this decrease correspond to an inhibition of some γ polypeptides

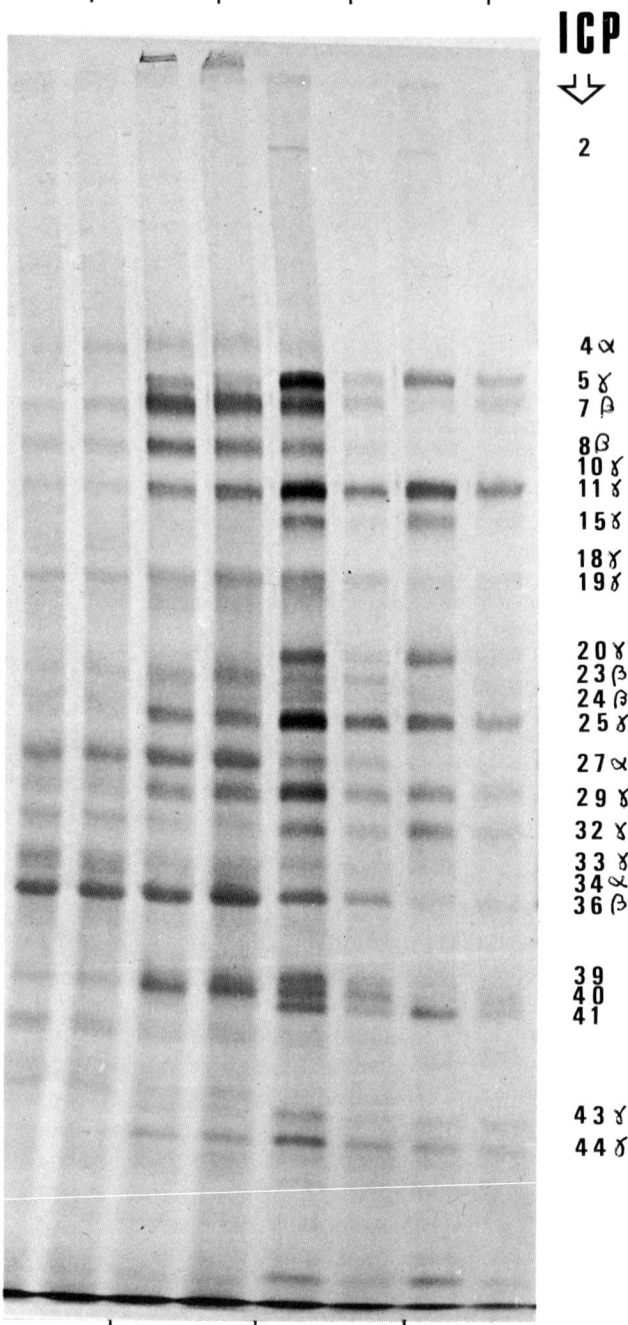

Fig. 4. Autoradiograms of infected cell polypeptides electrophoretically separated in polyacrylamide gels. Cells were infected with 20 pfu/cell of HSV-1 in the absence or in the presence of 20 μg/ml of Bzl[6] Ado added 1 h before infection and maintained. At different times, cells were labeled, as indicated at the bottom of the figure with the mixture [^{14}C]-valine, [^{14}C]-isoleucine, [^{14}C]-leucine (2 μCi/ml) for 1 h, then lysed. To the right of the figure, infected cell polypeptides (*ICP*) were indicated using Honess et al. (1974) designation: (−) Control infected cells; (+) Bzl[6] Ado infected cells

(Fig. 4): the third population of viral polypeptides to be synthesized in infected cells (Honess and Roizman, 1974).

It seems unlikely that the inhibition of viral DNA synthesis after a long contact with the analog (data not shown) was responsible for this inhibition, since short contact with the analog very late in infection has essentially the same effect. On the other hand, α and β polypeptide synthesis seemed to be normal both in amount and in kinetics (Fig. 4).

Roizman and Morse (1978) suggested that the decrease of internal methylation in mRNA during the infection is a viral function corresponding to a β polypeptide. As we observed that these polypeptides were normally synthesized and that the effect on γ polypeptides was dose-dependent (data not shown), it is possible that the cytokinin may competitively block the enzyme which demethylated mRNA.

Acknowledgments. A. Garcia was a fellow of the Ligue Nationale Française contre le Cancer. We are grateful to Y. Michal for excellent technical assistance and A. Mary for careful typing. These studies were supported by contracts INSERM no. 78 50 97-1, ATP no. 52 77 84 and ASR no. 1003 for Franco-American Cooperative Research Program

References

Abelson J (1979) RNA processing and the intervening sequence problem. Annu Rev Biochem 48: 1035-1069

Bartkoski M, Roizman B (1976) RNA synthesis in cells infected with Herpes simplex virus. XIII. Differences in the methylation patterns of viral RNA during the reproductive cycle. J Virol 20:583-588

Honess RW, Roizman B (1974) Regulation of Herpes virus macromolecular synthesis. I. Cascade regulation of the synthesis of three groups of viral proteins. J Virol 14:8-19

Jacquemont B, Huppert J (1977) Inhibition of viral RNA methylation in Herpes Simplex type 1 infected cells by 5' S-isobutyl adenosine. J Virol 22:160-167

Jacquemont B, Roizman B (1975a) RNA synthesis in cells infected with Herpes Simplex virus. X. Properties of viral symetric transcripts and double stranded RNA prepared from them. J Virol 15:707-713

Jacquemont B, Roizman B (1975b) Ribonucleic acid synthesis in cells infected with Herpes Simplex virus. XI. Characterization of viral high molecular weight nuclear RNA. J Gen Virol 29:155-165

Jacquemont B, Garcia A, Huppert J (1980) Processing of viral high molecular weight RNA in cells infected by Herpes Simplex type 1. J Virol 35:282-289

Kozak M (1980) Influence of mRNA secondary structure on binding and migration of 40S ribosomal subunits. Cell 19:79-90

Kozak M, Roizman B (1974) Regulation of Herpes virus macromolecular synthesis: nuclear retention of none translated viral RNA sequences. Proc Natl Acad Sci USA 71:4322-4326

Perry RP, Scherrer K (1975) The methylated constituants of globin mRNA. FEBS Lett 66:217-228

Revel M, Groner Y (1978) Post transcriptional and translation of controls of gene expression in eukaryotes. Annu Rev Biochem 47:1079-1126

Roizman B, Morse LS (1978) Human Herpes virus 1 as a model of regulation of Herpes virus macromolecular metabolism: a review. Oncogenesis and herpes viruses III. Int A f Res Cancer (Lyon) 1:269-297

Calcium-mediated Transduction of the Hormonal Message in 1-Methyladenine-induced Meiosis Reinitiation of Starfish Oocytes

M. Dorée[1] and T. Kishimoto[2]

In most animals, oocytes develop to the prophase stage of meiosis and then stop. In starfish meiosis is reinitiated by a hormone produced by the follicle cells (Hirai et al., 1973), which has been identified as 1-methyladenine (1-MeAde) (Kanatani et al., 1969). Fifteen to 20 min after 1-MeAde addition to a fully grown prophase-blocked oocyte, nuclear membrane breaks down, then chromosomes condense and the meiotic spindle is formed. When cytoplasm taken from starfish oocytes treated for more than 7 min with 1-MeAde is injected into prophase-blocked oocytes, meiosis is reinitiated (Kishimoto and Kanatani, 1976). The maturation-promoting factor (MPF) present in cytoplasm is not depleted through multiple serial transfers, whereas it loses its activity when diluted in its extracted state: this suggests that it exhibits "autocatalytic properties". MPF lacks specificity since for example amphibian MPF can induce meiosis when injected into starfish oocytes (Kishimoto, unpubl. data). On the other hand, cytoplasmic extracts of synchronized Hela cells can induce meiosis when injected into Amphibian oocytes. Extracts from G_1 and S phase cells have no activity whereas MPF activity increases rapidly during late G_2 to reach a peak in mitosis (Sunkara et al., 1979). These findings support the view that in the animal kingdom a common cytoplasmic factor might control both the release of oocytes from prophase block and the progression of mitotic cells from G_2 to M. This paper is concerned with the analysis of some significant events occurring in starfish oocytes between interaction of 1-MeAde with the cell membrane and appearance of MPF.

1 Localization and Specificity of 1-MeAde Receptors

Microinjection of 1-MeAde into the inside of an oocyte fails to induce the breakdown of nuclear membrane, whereas its external application brings about meiosis reinitiation (Kanatani and Hiramoto, 1970). However this result would also be expected if 1-MeAde conversion to an active metabolite requires passage across the cell membrane. If 1-MeAde receptors are intracellular, inhibition of 1-MeAde uptake must result in decreasing its biological effect if the hormone is used at a limiting concentration. In contrast dose-response to 1-MeAde is not modified when hormone uptake is severely inhibited by 1,9-diMeAde addition, which cannot by itself substitute for 1-MeAde in inducing meiosis reinitiation (Doree and Guerrier, 1975). This confirms that 1-MeAde receptors are local-

---

[1] Station biologique de Roscoff, 29211 Roscoff, France
[2] National Institute for Basic Biology, Okazaki 444, Japan

ized on the cell surface. Moreover, several properties of plasma membrane are modified within seconds in intact oocytes, and isolated plasma membranes release Ca^{2+} following hormone addition (Doree et al., 1976a, 1976b; Doree, 1980a; Doree et al., 1978; Guerrier et al., 1979).

The implication is that the compound itself, and not a metabolite originating from l-MeAde, is the molecule that interacts with the hormone receptor to trigger the meiosis program. Only N_1-substituted adenines among adenine derivatives are active compounds. Although l-MeAde is the most active alkyladenine, synthetic l-benzyladenine is still more active than the natural hormone. Not only replacement of the methyl group at the N_1 position by a much larger group, but addition of a second substituent as large as a benzyl group at the N_6 position, as in $1,N_6$-dibenzyladenine or N_6-benzyl-l-methyladenine does not suppress biological activity. By contrast, a second substituent at N_7 or N_9 suppresses biological activity, even when it is as small as a methyl group. This suggests that the hormone binds with the receptor in the N_9 or N_7-N_9 region. On the other hand, all the active compounds are sufficiently strong bases to exist appreciably in cationic form at pH 8 (pH of sea water), with distribution of positive charge mainly in the pyrimidine ring. Thus protonation might be a prerequisite for binding to the membrane receptors. Alternatively proton displacement might be involved during hormone-receptor interaction. (Doree et al., 1976a).

2 l-MeAde Induced Ca^{2+} Release from the Plasma Membrane

2.1 Occurrence and Timing of Ca^{2+} Release in Intact Oocytes

Prophase-blocked oocytes injected with the Ca^{2+}-sensitive photoprotein aequorin emit light less than 2 s following external application of l-MeAde (Fig. 1). In contrast no light is emitted when l-MeAde is added to a suspension containing both 10^6 oocytes and aequorin in chelex 100-treated Ca^{2+}-free sea water. This means that Ca^{2+} is liberated only inside the oocyte. Specificities of both Ca^{2+} and biological response are identical (Moreau et al., 1978a). Changes in light intensity tend to give an exaggerated impression of the changes in Ca^{2+} responsible for them. It seems likely, however, that steady-state equilibrium between Ca^{2+} release and buffering in cytoplasm is already reached within the first minute after hormone application.

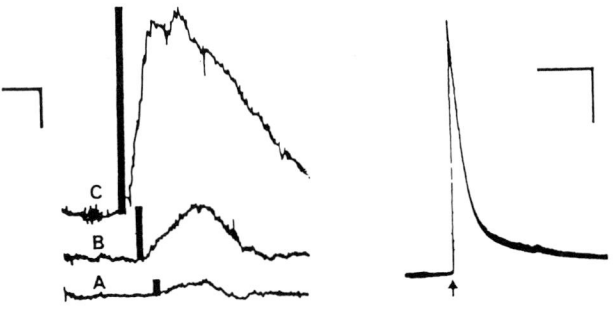

Fig. 1. l-MeAde-induced Ca^{2+} release as demonstrated by aequorin luminescence. *Left* light emission recorded from 30 aequorin microinjected oocytes following addition (*vertical bars*) of 10^{-8} M (*A*), 2×10^{-7} M (*B*) or 2×10^{-6} M l-MeAde (*C*). *Right* light response recorded when 2×10^{-7} M l-MeAde is added to a plasma membrane-rich fraction obtained from about 2×10^5 oocytes. *Vertical* and *horizontal segments on the left* and *on the right* correspond to 0.12 nA anode current and to 12 s respectively

2.2 Ca^{2+} Release from Isolated Plasma Membranes

Figure 1 shows an original record of the light response when 2×10^{-7} M l-MeAde is added to a plasma membrane-rich fraction obtained from isolated cortices (Doree et al., 1978). Less than 0.1 s after hormone addition, the anode current increases by 0.5 nA (corresponding to about 6×10^8 Ca^{2+} ions released from 1 oocyte cortex). Endoplasmic membranes only release small amounts of Ca^{2+} which can be accounted for by plasma membrane contamination.

2.3 Localization of Ca^{2+} Release in Intact Oocytes

When oocytes are injected with both aequorin and l-MeAde, light may be emitted, but only after a 1-2 min delay, readily accounted for by the time the hormone needs to pass across the cell membrane (Moreau et al., 1978b). Thus all l-MeAde receptors are localized on the cell membrane and exclusively accessible from outside. One can ask, however, whether in intact oocytes Ca^{2+}-induced Ca^{2+} release might not be involved, as in sarcoplasmic reticulum (Fabiato and Fabiato, 1978). It has been shown, however, in animal cells of the same size, that inwards leaks of Ca^{2+} through cell membrane due to impalement with microelectrodes were confined to the immediate vicinity of the plasma membrane (less than 1 μ) even when external Ca^{2+} concentration was as high as 2 mM (Rose and Loewenstein, 1975). The implication is that transient change of intracellular Ca^{2+} following l-MeAde addition to intact oocytes (diameter: about 150 μ) is probably restricted to the cortical region (thickness: about 3 μ). Even involvement of Ca^{2+}-induced Ca^{2+} release within the oocyte cortex itself seems unlikely since oocytes which have already discharged most of the Ca^{2+} stored in cortical granules, during activation induced either by sperm or ionophore A 23187, demonstrate the same sensitivity to l-MeAde as unactivated oocytes (Schuetz, 1975). Possible Ca^{2+}-induced Ca^{2+} release at the level of the plasma membrane itself cannot, however, be ruled out.

2.4 Origin of the Released Ca^{2+}

The total amount of Ca^{2+} which can be released in vitro following l-MeAde addition is about 2×10^{-14} ion · g per oocyte (Doree et al., 1978; Fig. 5). Total cell surface of sea urchin oocytes (150 μ diameter) is about 10^5 μ^2 (Schroeder, 1978). Assuming the same value, and that 1 Ca^{2+} serves as a bridge between each phospholipid and the following separated by 12 Å (Vandenheuvel, 1963), the total amount of Ca^{2+} bound on the internal side of the cell membrane would be about 1.2×10^{-13} ion · g in starfish oocyte. However in physiological conditions phosphatidylcholine and phosphatidylethanolamine are nearly electrically neutral and cannot bind Ca^{2+} appreciably. On the other hand only chelation-coordination bonds between Ca^{2+} and phospholipids might be expected to break in response to exogeneous stimuli. A two-dimensional coordination complex involving Ca^{2+} and carboxyl groups could not exist because of the relatively large area occupied by the phospholipid molecules (Deamer and Cornwell, 1966). Finally only phosphatidylserine, phosphatidic acid and polyphosphoinositides have the ability to complex with Ca^{2+} through coordination-chelation binding involving phosphate groups (Papahadjopoulos, 1968). The total amount of Ca^{2+} bound in such a way might correspond to the pool of Ca^{2+} which can be released following hormonal stimulation.

2.5 Role of Ca^{2+} Release

Injection of EGTA into starfish oocytes suppresses meiosis reinitiation if performed before or during the transient increase of free Ca^{2+} which follows hormone application, whereas meiosis occurs when EGTA is injected later (Moreau et al., 1978a). In as much as EGTA does not inhibit Ca^{2+} release, this suggests that sufficient transfer of Ca^{2+} from the plasma membrane to some Ca^{2+}-binding target (CaBT), probably localized in oocyte cortex, is an essential event for meiosis to occur, and that this transfer is completed within 30 s following hormone addition. On the other hand it has been shown, by eliminating this compound from the medium at various times after its application, that presence of 1-MeAde was required during 4 min 30 s (hormone-dependent period: HDP), in the same conditions, for meiosis to occur (Doree et al., 1976a). Recent reinvestigation of the effects of EGTA microinjection confirm that no effect is observed when EGTA is injected after the end of HDP. However, nuclear membrane breakdown is delayed by several minutes if EGTA is injected after the transient peak of increased free Ca^{2+} but before the end of HDP (Kishimoto and Doree, unpubl. result). This suggests that Ca^{2+}-CaBT would be a short-lived complex, whose decay depends on the rate at which free Ca^{2+} decreases following hormonal stimulation, and whose level has to remain higher than a critical value at least during HDP for meiosis to be triggered.

An alternative hypothesis would be, however, that EGTA simply suppresses Ca^{2+}-induced Ca^{2+} release on the inner side of the plasma membrane. In agreement with this, it has been shown that in certain conditions, meiosis can be reinitiated simply by raising extracellular Ca^{2+} concentration (Moreau et al., 1978a). The effect is not due to a permanent increased Ca^{2+} influx since the steady-state intracellular concentration of free calcium does not change, but rather to Ca^{2+}-induced Ca^{2+} release, since only a *transient* increase of intracellular Ca^{2+} occurs. Moreover, effect of increased Ca^{2+} concentration is potentiated by eliminating Na^+ from the medium, as also observed for Ca^{2+}-induced Ca^{2+} release in sarcoplasmic reticulum (Endo and Kitazawa, 1980).

Theophyllin competitively inhibits 1-MeAde-induced meiosis reinitiation (Doree et al., 1976a) even when given only 1 min before the end of HDP, whereas it does not when given 1 min later (Fig. 2). Identical results are obtained with caffein, procain, $MnCl_2$ and N-ethylmaleimide, which like theophylline have been proved to inhibit competitively 1-MeAde induced Ca^{2+} release in intact oocytes (Moreau et al., 1978a) as well as from isolated plasma membranes (Doree et al., 1978). These results suggest that biochemical changes responsible for the switch to the irreversibly stimulated state cannot occur in the presence of excess membrane-bound Ca^{2+} (or alternatively that plasma membrane competes with CaBT for Ca^{2+}). In agreement it has been reported that caffein enhances Ca^{2+} binding to isolated plasma membranes of beef heart muscle (Pang, 1980).

2.6 Coupling of Ca^{2+} Release with Proton Binding

Ca^{2+} binding on isolated sarcolemma strongly increases with pH (Pang, 1980). To check the possibility that protons might provide the positive charges required to exchange with

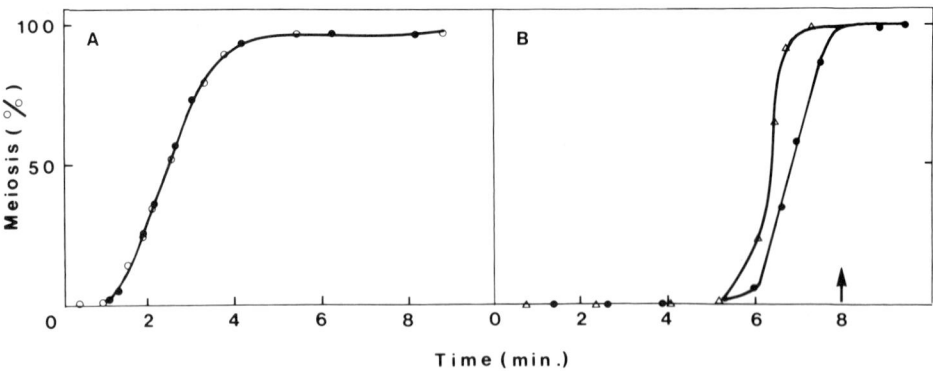

Fig. 2 A, B. Effect of 4 mM theophyllin (**A**) and 10 mM NH$_4$Cl (**B**) on meiosis reinitiation induced by 2×10^{-7} M l-MeAde. l-MeAde is added at zero time to prophase-blocked oocytes. % meiosis estimated 1 h later. **A** aliquotes of cell suspension are collected at the various indicated times and either the hormone concentration is decreased to 4×10^{-9} M by dilution in hormone-free sea water (●) or it remains unchanged but theophylline is added to 4 mM final concentration (○). **B** experiment performed with a different batch of oocytes (*HDP* about 7 min instead of 2 min 30s in **A**) (●) same experiments as in **A** (for the *same symbol*). (△) aliquots receive 10 mM NH$_4$Cl at the varous indicated times and both NH$_4$Cl and 1-MeAde are diluted 50 times at 8 min (*arrow*)

Ca^{2+}, a sink of protons was established near the plasma membrane by adding 10 mM NH$_4$Cl in sea water (pH 8.2). Cell membranes are only permeable to the uncharged form of weak bases (Jacobs, 1940). Since the intracellular pH of the starfish oocyte is 6.3 and pKa of NH$_4$Cl 9.25, more than 99% of the uncharged base that enters the oocyte will pick up a proton, resulting in a local increase of pH. It was found that NH$_4$Cl competitively inhibits the hormone-induced meiosis reinitiation (Doree et al., unpubl. data). Inhibitory effect of NH$_4$Cl was strongly dependent on pH, and clearly dependent only upon the amount of the uncharged form of the weak base. When NH$_4$Cl was applied after the end of HDP, no inhibition occurred, whereas meiosis was blocked even when it was applied only for 2 min just before the end of HDP (Fig. 2). Intracellular pH decreases by less than 0.01 unit during the first 5 min which follow application of 10 mM NH$_4$Cl to sea urchin eggs (Steinhardt et al., 1978). Therefore it is very likely that significant decrease of pH is limited to the immediate vicinity of the plasma membrane when NH$_4$Cl is applied for only 2 min to starfish oocytes.

In as much as local increase of pH suppresses meiosis reinitiation only if it occurs during HDP, it seems likely that membrane-bound Ca^{2+} is exchanged against H$^+$ following hormone application. After the burst of Ca^{2+} release, a steady-state low level of membrane-bound Ca^{2+} must be kept for several minutes to prevent plasma membrane from reversing to its initial state. If pH increases, Ca^{2+} is allowed to bind back to the plasma membrane and transition to the irreversibly stimulated state will stop before completion.

3 cAMP-Dependent Modulation of Hormonal Message Transduction

Table 1 shows that injection of the A subunit of choleratoxin, which stimulates adenyl cyclase activity (Gill, 1977), increases oocyte sensitivity to l-MeAde. This suggests that

Table 1. Enhancement of l-MeAde-induced maturation by choleratoxin (CHT) A subunit in *Asterina* oocytes

Exp. No.	Meiosis reinitiation by l-MeAde treatment		Concentration of l-MeAde (M)
	Oocytes injected with CHT A subunit[a]	Oocytes injected with vehicle (control)	
I	$\frac{15}{19}$[b]	$\frac{0}{13}$	2×10^{-7}
II	$\frac{9}{15}$	$\frac{1}{16}$	10^{-7}
III	$\frac{7}{10}$	$\frac{0}{10}$	3.3×10^{-8}

[a] Oocytes were injected with CHT A subunit (18 μg/ml, final concentration in oocytes) and then treated with 1-MeAde 2 to 3 h after injection
[b] Number of oocytes reinitiating meiosis/Number of oocytes receiving injection

binding of cAMP to, probably, a cAMP-binding protein, localized on the cell membrane and accessible to cAMP from the inside, controls either the affinity of l-MeAde for its receptor or the transduction of the hormonal message. A recent report that cAMP strongly inhibits Ca^{2+} binding to isolated plasma membranes clearly favors this last hypothesis (Pang, 1980). No changes in cAMP levels could be demonstrated following l-MeAde addition. On the other hand Ca^{2+} release occurs within 100 ms when l-MeAde is added to isolated plasma membranes (Fig. 1): it seems therefore unlikely that increased binding of cAMP to the plasma membrane might be involved between hormone addition and Ca^{2+} release. However changes in membrane-bound cAMP might be a tool for the oocyte to modulate its responsiveness to the follicular hormone throughout the breeding season, in addition to rendering cAMP resistant to hydrolysis by cyclic nucleotides phosphodiesterases (O'Dea et al., 1971).

Maller and Krebs (1977) proposed that prophase block in amphibian oocytes might be due to the cAMP-dependent phosphorylation of a specific protein. On the other hand, phosphorylation of the cAMP-binding subunit of protein kinase facilitates the cAMP-dependent dissociation of protein kinase into its catalytic and regulatory components, which suggests that phosphorylation or dephosphorylation of cAMP-binding protein might regulate their activity (Erlichman et al., 1974). Since the regulatory subunit of a membrane-bound cAMP-dependent protein kinase proved to be one of the best substrates for cAMP-dependent protein kinases, in synaptic membranes (Uno et al., 1977), the effects of intracellular injection of purified heterologous catalytic subunit (C) of cAMP-dependent protein kinase were investigated in starfish oocytes (Doree et al., unpubl. data). It was found that meiosis failed to be reinitiated even when injection occurred as late as 1 min before the end of HDP. In contrast injection had no effect 2 min later. Inhibition was competitive with respect to l-MeAde concentration and increased with time elapsed between C injection and l-MeAde addition (injection performed first). Since these findings supported the view that C might act on the cell membrane to phosphorylate a protein involved in the mechanism of action of l-MeAde, oocytes cortices were isolated and incubated with ATP-γ-[^{32}P] in the presence of C. Most of the radioactivity was incorporated into a single protein of 39,000-40,000 daltons comigrating with C (Fig. 3). It

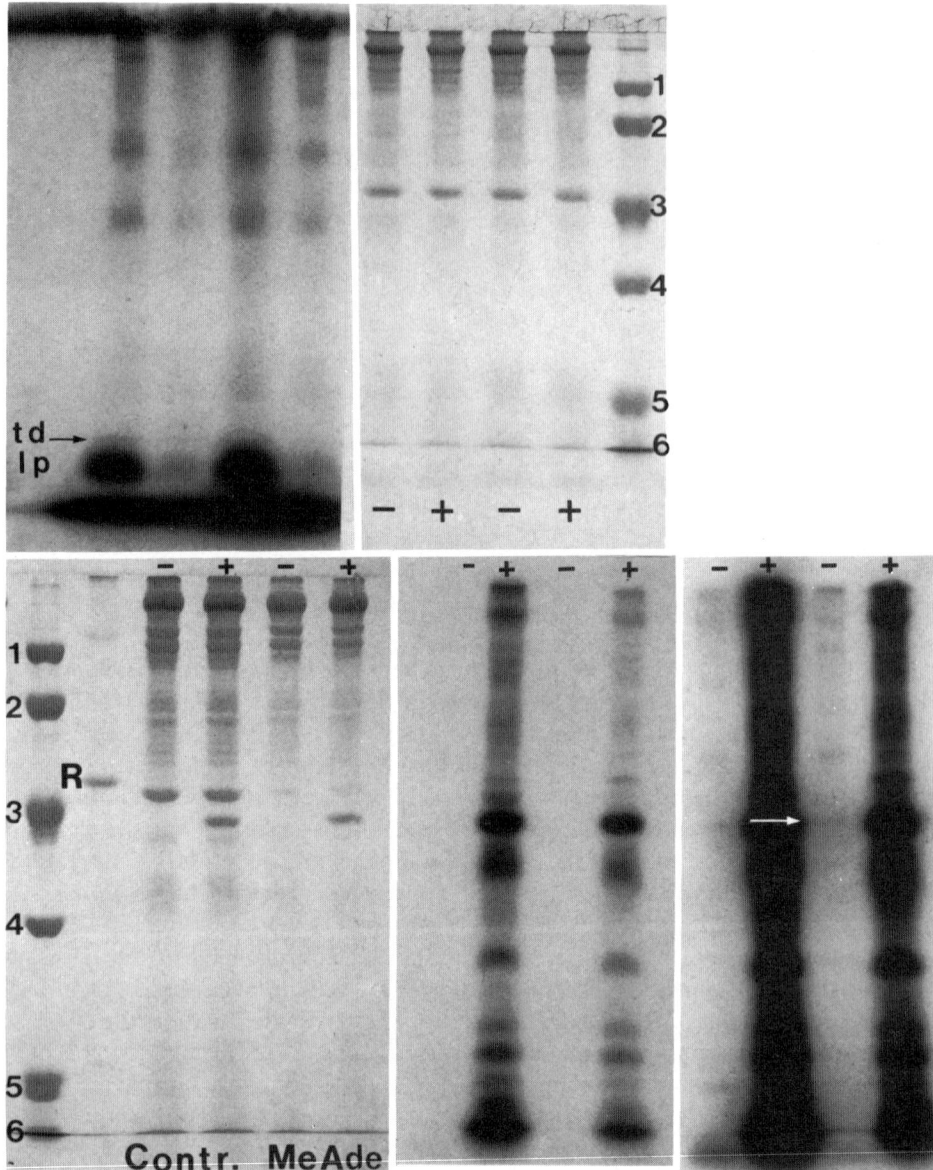

Fig. 3. Effect of Ca^{2+} (*top*) and catalytic subunit (*bottom*) on endogenous phosphorylation of isolated cortices incubated for 3 min with ATP-γ-[^{32}P] (10% SDS polyacrylamide gels). *Top* Coomassie blue staining of proteins is shown on the *right*, and the corresponding autoradiograms on the *left*. Incubation was performed in duplicate in the presence (+) or absence (−) of 0.1 mM Ca^{2+}. *Td* tracking dye (bromophenol blue); *lp* lipoproteins. Molecular weight markers are phosphorylase b: 94,000 daltons (*1*), bovine serum albumin: 67,000 (*2*), ovalbumin: 43,000 (*3*), carbonic anhydrase: 30,000 (*4*), soybean trypsin inhibitor: 20,100 (*5*) and α-lactalbumin: 14,400 (*6*). *Bottom* Coomassie blue staining of proteins on the *left*; the corresponding autoradiograms exposed for 1 day or for 1 month are shown respectively in the *middle* or on the *right*. Isolated cortices were prepared either from control oocytes (*contr.*) or from oocytes treated for 15 min with 10^{-6} M l-MeAde. Cortices were incubated with ATP-γ-^{32}P either in the presence (+) or absence (−) of pure catalytic subunit of beef heart cAMP-dependent protein kinase. *R* regulatory subunit of rabbit skeletal muscle cAMP-dependent protein kinase. *1-6* same molecular weight markers as above

has been shown that C does not catalyze its own phosphorylation (Peters et al., 1977; Le Peuch et al., 1979; Bechtel et al., 1977). On the other hand, the same 40 Kd protein was still labeled even after boiling the cortices, eliminating possible phosphorylation of C due to endogenous protein kinases. Finally radioactivity was also found to be incorporated into a 39,000-40,000 daltons protein in the absence of exogenous catalytic subunit, although to a much lower extent. These results indicate that a minor protein component of the oocyte cortex is potentially a very good substrate for C. Although evidence is lacking, there is a temptation to suggest that this cortical protein might be the cAMP-binding protein involved in the modulation of 1-MeAde action: phosphorylation of this protein would be expected to decrease its affinity for cAMP and therefore to decrease the sensibility of the plasma membrane toward the hormone. Plasma membranes of *Dictyostelium discoideum* have been shown to contain a single species of cAMP-binding protein, whose molecular weight is just 40,000 (Wallace and Frazier, 1979).

4 1-MeAde-Induced Stimulation of Protein Kinase Activity: its Relationships to Ca^{2+} Release from the Plasma Membrane

The sequence of events between 1-MeAde binding to the membrane receptors and nuclear membrane breakdown does not include synthesis of any specific protein (Guerrier and Doree, 1975). On the other hand, the rate of [^{32}P] incorporation into proteins increases by 10-20 fold following 1-MeAde addition, due to the stimulation of protein kinase activity (Guerrier et al., 1977; Doree et al., unpubl. data). Doses-responses are identical for enhancement of protein phosphorylation and meiosis reinitiation. Moreover, any treatment which decreases the former below a critical value precludes the latter from occurring. These findings support the view that protein phosphorylation plays a key role in the post-traductional control of meiosis reinitiation. Although protein kinases are maximally stimulated very early, they remain under plasma membrane control until the end of HDP: either elimination of 1-MeAde or addition of theophyllin before the end of HDP results some minutes later in the reversal of protein kinase activity to the resting level of control oocytes. On the contrary no change is observed when the hormone is eliminated or the drug added after the end of HDP. This suggests that stimulation of protein kinase activity might be directly related to displacements of Ca^{2+} from the plasma membrane, which are the main feature of HDP.

It has been shown that the Ca^{2+}-ionophore A 23187, which decreases Ca^{2+} gradients across biological membranes, can stimulate protein phosphorylation (Haslam and Lynham, 1977). On the other hand, ionophore reinitiates meiosis of prophase-blocked oocytes in several forms (Masui and Clarke, 1979). In starfish, however, ionophore fails to induce both stimulation of protein phosphorylation and meiosis reinitiation, although it releases 20 times more Ca^{2+} than 1-MeAde inside oocytes (Moreau et al., 1978a). 1-MeAde still stimulates protein phosphorylation and meiosis reinitiation following ionophore application (Fig. 4). These results show that at least in starfish oocytes, increase of free Ca^{2+} is not by itself the signal for stimulation of protein phosphorylation, and even suggest that it might be detrimental if we consider that 1-MeAde-induced stimulation of protein phosphorylation is reduced after ionophore treatment. Direct demonstration of the inhibitory effect of Ca^{2+} is provided by in vitro experiments: short preincubation of

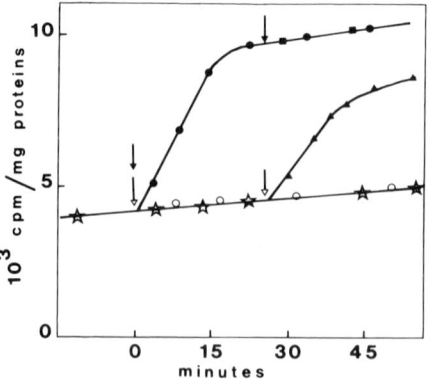

Fig. 4. Effect of ionophore A 23187 and 1-MeAde on protein phosphorylation. Oocytes of *Asterias rubens* were incubated for 2 h in sea water containing 2 μCi/ml labeled phosphate, washed and resuspended in unlabeled sea water and divided at zero time into three batches. One received 1-MeAde to the final concentration 2×10^{-7} M (●), the second received ionophore A 23187 at the final concentration 2 μM (○), the third served as control (☆). Aliquotes were removed at the specified times and incorporation of [^{32}P] into proteins were measured. After 25 min, 1-MeAde-treated oocytes received (■) or not (●) 2 μM ionophore, whereas ionophore-treated oocytes received (▲) or not (○) 2×10^{-7} M 1-MeAde, and effects on protein phosphorylation was estimated. *Arrows* point to the succession addition of 1-MeAde (↓) or ionophore (↓)

starfish homogenates (but not 105,000 g supernatants) in the presence of Ca^{2+}, with or without inhibitors of Ca^{2+}-dependent proteases, strongly inhibits protein kinase activity, even if the protein kinase assay itself is performed in the presence of excess EGTA (Doree et al., unpubl. data). Stimulation of [^{32}P] incorporation into cortical proteins levels off just at the end of HDP. Using isolated cortices it was found that phosphorylation of proteins and lipoproteins was strongly inhibited in the presence of micromolar concentrations of Ca^{2+} (Fig. 3). At least in the cortical region, including the plasma membrane of the oocyte, stimulation of protein phosphorylation might be due to the decrease of membrane-bound Ca^{2+} rather than to the concomitant increase in cytosolic free Ca^{2+}. This would suggest that in starfish oocytes a membrane-bound protein kinase is inhibited by some interaction with the phospholipid bilayer, and that Ca^{2+} release decreases this interaction, as it does in other biological membranes (Takai et al., 1979).

According to one of the two possible interpretations of EGTA microinjection experiments (see Sect. 2.5) a short-lived unidentified Ca^{2+}-CaBT complex would play a key role in meiosis reinitiation. Ca^{2+}-calmodulin would be a good condidate since it has been shown, in mammals, to exhibit multiple calcium-dependent regulatory activities, including activation of protein kinase (Schulman and Greengard, 1978). Although the total amount of calmodulin does not change in starfish oocytes following 1-MeAde addition (Doree, 1980b), a transient change in the amount complexed with Ca^{2+} might provide a signal for stimulation of protein phosphorylation. However, the lack of effect of EGTA when injected after the end of HDP would not be expected if Ca^{2+}-calmodulin-dependent protein kinase were involved in the phosphorylation of endoplasmic proteins, since the Ca^{2+}-calmodulin complex dissociates readily following EGTA addition.

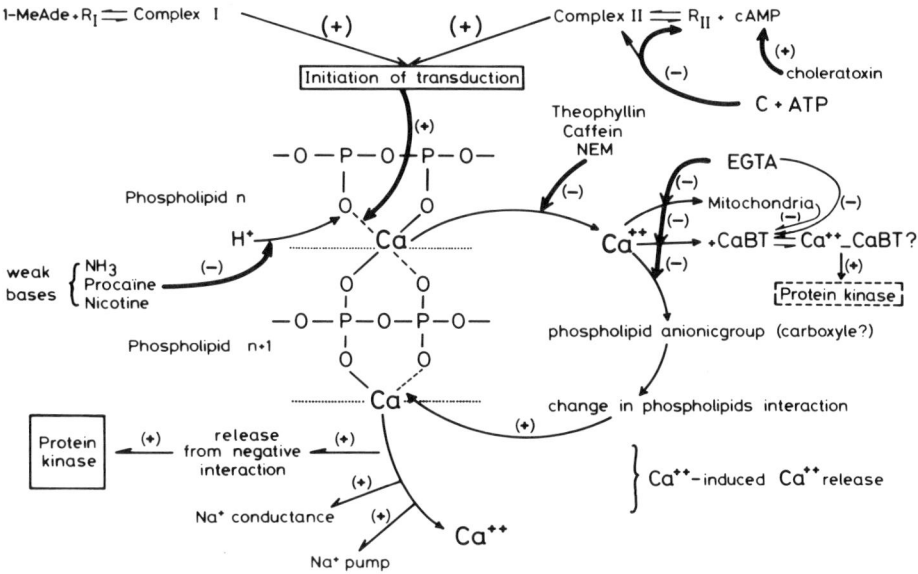

Fig. 5. Transduction of l-MeAde message: a possible model

References

Bechtel PJ, Beavo JA, Krebs EG (1977) Purification and characterization of catalytic subunit of skeletal muscle adenosine 3':5'-monophosphate-dependent protein kinase. J Biol Chem 252: 2691-2697

Deamer DW, Cornwell DG (1966) Calcium action on fatty acid and phospholipid monolayers and its relation to the cell membrane. Biochim Biophys Acta 116:555-562

Dorée M (1980a) Calcium ions release induced by l-methyladenine increases membrane permeability to Na^+. Exp Cell Res 131

Dorée M (1980b) Calmodulin content does not change following hormone-induced meiosis reinitiation in starfish oocytes. Experientia

Dorée M, Guerrier P (1975) Site of action of l-methyladenine in inducing oocyte maturation in starfishes: kinetical evidences for receptors localized on the cell membrane. Exp Cell Res 91:296-300

Dorée M, Guerrier P, Leonard NJ (1976a) Specificity of the l-methyladenine receptors in starfish oocytes. Proc Natl Acad Sci USA 73:1669-1673

Dorée M, Guerrier P, Moreau M (1976b) Etude de quelques proprietés fondamentales du recepteur de la l-methyladenine. In: CNRS (ed) Actualites sur les hormones d'invertebres. Coll Int CNRS 251:199

Dorée M, Guerrier P, Moreau M (1978) In vitro release of Ca^{2+} from the plasma membrane in starfish oocytes. Exp Cell Res 115:251-260

Endo M, Kitazawa T (1980) Calcium-induced calcium release in cardiac muscle. The Seiriken Conference on cardiac cellular response and divalent cation, May 1980, Okazaki, JAPAN (Abstr)

Erlichman J, Rosenfeld R, Rosen OM (1974) Phosphorylation of a cyclic adenosine 3':5-monophosphate-dependent protein kinase from bovine cardiac muscle. J Biol Chem 249:5000-5003

Fabiato A, Fabiato F (1978) Calcium-induced release of calcium from the sarcoplasmic reticulum of skinned cells. Ann NY Acad Sci 307:491-522

Gill DM (1977) Mechanism of action of cholera toxin. In: Greengard P, Robinson GA (ed) Advances in cyclic nucleotide research, vol 8. Raven Press New York, p 85

Guerrier P, Dorée M (1975) The requirement of l-MeAde during nuclear maturation. Dev Biol 47: 341-348

Guerrier P, Moreau M, Dorée M (1977) Stimulation of protein phosphorylation induced by l-methyladenine. Mol Cell Endocrinol 7:137-150

Guerrier P, Moreau M, Dorée M (1979) Relationships between hormone-induced Ca^{2+} release and ^{86}rubidium uptake stimulation in starfish oocytes. Rev Can Biol 38:145-156

Haslam RJ, Lynham JA (1977) Relationships between phosphorylation of blood platelet proteins and secretion of platelet granule constituents. Biochem Biophys Res Commun 77:714-722

Hirai S, Chida K, Kanatani H (1973) Role of follicle cells in maturation of starfish oocytes. Dev Growth Differ 15:21-31

Jacobs MH (1940) Cold Spring Harbor Symp Quant Biol 8:30-39

Kanatani H, Hiramoto H (1970) Site of action of l-methyladenine in inducing oocyte maturation in starfish. Exp Cell Res 61:280-284

Kanatani H, Shirai H, Nakanishi K, Kurokawa T (1969) Isolation and identification of meiosis-inducing substance in starfish *Asterias amurensis*. Nature (London) 221:273-277

Kishimoto T, Kanatani H (1976) Cytoplasmic factor responsible for germinal vesicle breakdown and meiotic maturation in starfish oocyte. Nature (London) 260:321-322

Le Peuch C, Haiech J, Demaille JG (1979) Concerted regulation of cardiac sarcoplasmic reticulum calcium transport by cAMP-dependent and Ca^{2+}-calmodulin dependent phosphorylations. Biochemistry 18:5150-5157

Maller JL, Krebs EG (1977) Progesterone-stimulated meiotic cell division in *Xenopus* oocytes. J Biol Chem 252:1712-1718

Masui Y, Clarke HJ (1979) Oocyte maturation. Internal Rev Cytol 57:185-281

Moreau M, Guerrier P, Dorée M, Ashley CC (1978a) l-methyladenine induced release of intracellular calcium triggers meiosis in starfish oocytes. Nature (London) 272:251-253

Moreau M, Guerrier P, Dorée M (1978b) New evidences for the absence of intracellular receptors for l-methyladenine recognition. Exp Cell Res 115:245-249

O'dea RF, Haddox MK, Goldberg ND (1971) Interaction with phosphodiesterase of free and kinase-complexed cyclic adenosine 3', 5'-monophosphate. J Biol Chem 246:6183-6190

Pang DC (1980) Effect of ionotropic agents on the calcium binding to isolated cardiac sarcolemma. Biochim Biophys Acta 598:528-542

Papahadjopoulos D (1968) Surface properties of acidic phospholipids: interaction of monolayers and hydrated liquid crystals with uni- and bivalent metal ions. Biochim Biophys Acta 163:240-254

Peters KA, Demaille JG, Fischer EH (1977) cAMP-dependent protein kinase from bovine heart. Characterization of the catalytic subunit. Biochemistry 16:5691-5697

Rose B, Loewenstein WR (1975) Calcium ion distribution in cytoplasm visualized by aequorin: diffusion in cytosol restricted by energized sequestering. Science 190:1204-1206

Schroeder TE (1978) Microvilli on sea urchin eggs: a second burst of elongation. Dev Biol 64:342-346

Schuetz AW (1975) Cytoplasmic activation of starfish oocytes by sperm and divalent ionophore A 23187. J Cell Biol 66:86-94

Schulman H, Greengard P (1978) Ca^{2+}-dependent protein phosphorylation system in membranes from various tissues, and its activation by "calcium-dependent regulator". Proc Natl Acad Sci USA 75:5432-5436

Steinhardt RA, Shen SS, Zucker RS (1978) Direct evidence for ionic messengers in the two phases of metabolic derepression at fertilization in the sea urchin egg. In: Dirksen ER, Prescott D, Fox CF (eds) Cell reproduction, ICN-UCLA Symp Mol Cell Biol, vol 12. Academic Press, London New York, p 1

Sunkara PS, Wright DA, Rao PN (1979) Mitotic factors from mammalian cells induce germinal vesicle breakdown and chromosome condensation in amphibian oocytes. Proc Natl Acad Sci USA 76:2799-2802

Takai Y, Kishimoto A, Iwasa Y, Kawahara Y, Mori T, Nishizuka Y (1979) Calcium-dependent activation of a multifunctional protein kinase by membrane phospholipids. J Biol Chem 254:3692-3695

Uno I, Ueda T, Greengard P (1977) Adenosine 3':5'-monophosphate-regulated phosphoprotein system of neuronal membranes. J Biol Chem 252:5164-5174

Vandenheuvel H (1963) Model for lipid arrangement in membranes. Ann NY Acad Sci 122:57-76

Wallace LJ, Frazier WA (1979) Photoaffinity labeling of cAMP and AMP-binding proteins of differentiating Dictyostelium discoideum cells. Proc Natl Acad Sci USA 76:4250-4254

Subject Index

Abcisic acid and cytokinin interaction 267-274
Actinomycin D inhibitor of RNA synthesis
    212-217, 252-260
Adenine-oligonucleotide precursor of cytokinin
    27-33
Affinity chromatography of receptor-proteins
    172-178, 218-227
Agrobacterium tumefaciens 3-16, 44-55
Amanitin (α-) inhibitor of RNA-polymerase I
    218-227, 287-297
Aminoacyl tRNA synthetases 275-286
Aminolevulinate (δ-) synthetase 275-286
Apical dominance release by cytokinin 140-150
Autoradiography 162-171, 193-211, 267-274
Auxin and cytokinin interaction 252-260
Azaserine inhibitor of protein synthesis 319-328
Azido-(8-)benzyladenine (^{14}C-labeled)
    162-171, 179-190
Azido-(2-)substituted cytokinins 153-161
Azido-(8-)substituted cytokinins 153-171,
    179-190

Barley plants 218-227, 287-297
Benzyladenine (N^6-) 80-96, 129-139, 153-161,
    218-227, 261-266, 275-297, 319-328
Benzyladenine incorporation into rRNA
    228-238
Benzyladenosine (N^6-) 80-96, 129-139, 329-337
Benzyladenosine 5'-di, triphosphate 80-96
Benzyladenosine 3'-phosphate 228-238
Benzyladenosine 5'-phosphate 80-96, 228-238
Benzyladenosine incorporation into RNA
    329-337
Binding of calcium 338-348
– protons 338-348
Biosynthetic pathways of cytokinins 34-43
Bud induction by cytokinins 105-114, 162-171

Calcium ion release 338-348
Calmodulin 338-348
Carnitine 319-328
Caulonema of Mosses 105-114, 162-171

Cell cultures (human) 329-337
– – of *Drosophila* sp. 319-328
– expansion by fusicoccin 261-266
Cell-free systems for cytokinin biosynthesis
    34-43
Cell suspension cultures (Rose) 193-211
– – – (Tobacco) 80-96, 228-238, 298-316
Ceratodon purpureus 212-217
Chloramphenicol 275-286
Chloronema of Mosses 105-114
Chloroplast DNA 298-307
– enzymes 252-260
– gene expression 257-307
– proteins (polypeptides) 298-316
Chloroplasts (isolated) 287-316
Chlorophyll biosynthesis 241-251, 267-286,
    298-307
Chromatin-bound RNA polymerase 218-227
^{14}C-labeled adenine as cytokinin precursor
    27-33, 34-43
– amino-acids as protein precursors 308-316
– kinetin 105-114
– mevalonate as cytokinin precursor 44-55
Cordycepin as transcription inhibitor 212-227
Corynebacterium fascians 17-26
Cucumber cotyledons 212-217, 261-286
Cyclic-AMP 338-348
Cyclic-AMP receptor-protein association
    338-348
Cycloheximide 275-286
Cytokinin activity on animal cells 319-328
– – on bud development 105-114, 140-150,
    162-171
– – on cell division 193-211, 228-251,
    319-328
– – on cotyledon development 212-217,
    261-286
– – of diphenylurea derivatives 97-104
– – on DNA replication 252-260
– – on enzyme formation 252-286
– – on ion movements 267-274
– – on metabolite transport 252-260
– – on microbody differentiation 252-266

Subject Index

- – on mitochondrial functions 179-190, 261-266
- – on plastid differentiation 252-316
- – on protein synthesis 193-227, 252-260, 275-286, 298-316
- – on sex determination 129-139
- – on transcription 212-227, 267-297, 329-337
- – on translation 172-178, 212-227, 275-286
- analogues 115-128, 153-171, 179-190
- antagonists 80-96, 140-150
- antisenescence activity 252-260
- autonomous tissue cultures 34-43, 97-104, 241-251
- binding sites 153-161, 179-190, 267-274
- biosynthesis by *Agrobacterium* sp. 3-16, 44-55
- – in chloroplasts 241-251
- – by *Corynebacterium* sp. 17-26
- – de novo 3-66, 105-114
- biosynthetic pathways 44-55
- chemical synthesis 153-171
- concentration as discriminating factor 261-266
- dependent tissue cultures 97-104
- derivatization 3-16
- extraction procedures 3-26, 56-79

Cytokinin-7-glucosides 80-96
Cytokinin HPLC-MS analysis 3-16, 56-79
- incorporation into RNA 228-238, 329-337
- inhibitor of viral development 329-337
- level in Crown-gall 3-16
- – in leaves 97-104, 252-260
- – in plant organs 34-43, 69-79
- mass spectrometry 3-16, 56-79, 129-139
- metabolic activation 80-96
- metabolism in cell cultures 80-96
- – in *Drosophila* cells 319-328
- – in *Lupinus* sp. plants 69-79
- – in *Mercurialis* sp. 129-139
- – in Mosses 105-114
- – in *Phaseolus* sp. 97-104
- – in tissue cultures 56-66, 241-251
- metabolite extraction 80-96, 105-114
- nucleotides 319-328
- from oligonucleotides 27-33
- production by bacteria 17-26, 69-79

Cytokinin-receptor-protein association 105-114, 129-139, 153-190, 218-227
Cytokinin requirements for growth 80-104, 115-128, 252-266, 319-328
- target cells 162-171
- TLC analysis 69-79, 105-114
- translocation in *Lupinus* sp. plants 69-79
- from tRNA 3-34, 44-66, 212-217
- uptake 80-96, 162-171

Cytoplasm RNA analysis 329-337

Deuterium-labeled cytokinin 56-79
Dihydrozeatin (riboside) 69-79
Diphenylureas as cytokinin analogues 97-104, 115-128
DNA replication and cytokinins 252-260
Drosophila melanogaster 319-328

Enzyme formation and cytokinins 252-260

Funaria hygrometrica 105-114, 162-171
Fusicoccin and cytokinin interaction 261-266

Gene analysis of plasmid Ti 3-16
Genetic regulation 97-104, 129-139, 241-251
Glucosaminides 129-139
Glyceraldehyde 3-phosphate dehydrogenase 252-260

Herpes simplex virus 329-337
Hill reaction 287-297
Hormone message transduction 338-348
Hydroxybenzyladenine (N^6-), ortho, meta, para 179-190
Hydroxypyruvate reductase 261-266
Hypoxanthine-guanine phosphoribosyltransferase 319-328

Immunodetection of cytokinin-binding protein 179-190
Incorporation of ^3H-benzyladenine in tRNA 228-238
Induction of cytokinin autonomy 241-251
Isopentenyladenine (N^6-Δ^2-) 3-26, 34-55, 80-114, 129-139, 153-161, 172-178, 212-227
Isopentenyladenosine (N^6-Δ^2-) 34-55, 80-104, 172-178
Isopentenyladenosine kinase 34-43
Isopentenyladenosine-5'mono, di- or triphosphate 27-33, 80-96
Δ^2-Isopentenyl diphosphate: 5'-AMP isopentenyltransferase 17-43
Isopentenyl diphosphate: tRNA isopentenyltransferase 27-33

Kinetin (N^6-furfuryladenine) 105-114, 252-260, 287-297

Lactobacillus acidophilus 27-33
Leaf homogenate as cytokinin receptor source 218-227
Leucoplast proteins (polypeptides) 298-316
Light and cytokinin interaction 252-286
- harvesting chlorophyll a/b complex apoprotein 308-316
Lipid breakdown 261-266

Subject Index

Lupinic acid 69-79
Lupinus albus, luteus, angustifolius 69-79

Meiosis reinitiation 338-348
Mercurialis annua 129-139
Mercurialis tissue cultures (Crown-gall) 129-139
Mercurialis tissue cultures (normal) 129-139
Metabolite transport and cytokinin 252-260
Methyladenine (1-) 338-348
Methyladenine receptor-protein association 338-348
Methyladenosine (N^6-) 329-337
Methylation of RNA 329-337
Methylpurine (6-) as transcription inhibitor 212-217
Microtubules 193-211
Mitochondrial functions 261-266
Molecular markers of cytokinin activity 308-316

N-acetylglucosaminidase 129-139
Nicotiana sp. plants 34-43
^{15}N-labeled cytokinin 56-66
Nodules and roots 69-79
N-phenyl-N'(4-pyridyl) ureas 115-128
Nuclear gene expression 275-286, 308-316
Nuclei isolation 218-227
Nucleosidase 34-43
Nucleoside kinase 319-328
Nucleotidase (5'-) 34-43, 319-328
Nucleus RNA analysis 329-337

Oligoribonucleotide fractionation 228-238
Organogenesis in plants 140-150, 241-251

Permeability of cells to cytokinins 80-96, 261-266
Peroxisomes (microbodies) differentiation 261-266
Phaseolus sp. as cytokinin receptor source 27-33, 179-190
Phaseolus sp. tissue cultures 97-104
Phosphoribosyltransferase 34-43
Phosphorylase of cytokinin nucleosides 34-43
Photoaffinity labeling 153-171
Pisum sp. bud formation 140-150
Plasma membranes 338-348
Plasmid presence and virulence of bacteria 3-26
Plasmids of *Agrobacterium* sp. 3-16
– of *Corynebacterium* sp. 17-26
Plastid differentiation 241-266, 275-286, 308-316
– function in cytokinin production 241-251
– RNA polymerase 267-274
Plastids (etiolated) 287-297

Polyacrylamide gel electrophoresis 193-211, 298-316, 329-337
Polysome-cytokinin interactions 212-217, 275-286
Polysome isolation 193-211
Post-transcriptional RNA maturation 329-337
Protein kinase 338-348
– phosphorylation 338-348
– synthesis by isolated plastids 298-307
– – (viral) 329-337
– – in vitro 193-227
– – in vivo 193-211, 308-316
Protonema of Mosses 105-114, 162-171, 212-217
Pumpkin cotyledons 218-227, 287-297

Receptor of cytokinins in cytosol 179-190
– – on ribosomes 179-190
Receptor-protein purification 179-190
Regeneration of plants from tissues 241-251
Regulation of RNA transcription 329-337
Restriction enzyme mapping of ctDNA 298-307
– – – of TiDNA 3-16
Ribosome-mRNA binding 212-217
Ribulose-bisphosphate carboxylase 252-260, 275-286
RNA biosynthesis 212-217, 267-274, 287-297
RNA detection by autoradiography 267-274
RNA polymerase I, II or III 218-227, 287-297
RNA transcription in chloroplasts 287-297
Rose suspension cultures 193-211
rRNA fractionation on sucrose gradients 228-238, 287-297
rRNA oligonucleotide pattern 228-238
Rye seedlings 252-260

Starfish oocytes 338-348
Streptomycin action on plant tissues 241-251
Structure activity relationship 115-128, 140-150, 162-171, 179-190, 252-260
Synchronization of mitoses 193-211

Tobacco cell suspension cultures 80-96, 228-238, 298-316
– Crown-gall tissue cultures 3-16, 44-55
– pith bioassay 115-128, 140-150
– tissue cultures 44-55, 115-128, 241-251
Tritium-labeled adenine as cytokinin precursor 44-55
– adenosine as cytokinin precursor 69-79
– benzyladenine (N^6-) 80-96, 228-238
tRNA of *Agrobacterium tumefaciens* 3-16
tRNA of *Corynebacterium fascians* 17-26
tRNA cytokinin contents 27-33
tRNA of Tobacco Crown-gall 3-16

tRNA of *Vinca* tissue cultures 56-66
tRNA turn-over rate 27-33
Tubulin assembly 193-211
— biosynthesis 193-211

Vinblastin 193-211
Vinca Crown-gall or normal tissue cultures 56-66

Wheat germ as cytokinin receptor source 172-190
— — as protein synthesis system 193-211

Zeatin (cis or trans) 3-79, 129-139, 218-227
— glucosides 56-66
— (2-methylthio) 3-26
— riboside (cis or trans) 3-79, 129-139
— riboside 5'-phosphate 56-66

Encyclopedia of Plant Physiology

New Series

Editors: A. Pirson, M. H. Zimmermann

Volume 7
Physiology of Movements
Editors: W. Haupt, M. E. Feinleib

With contributions by numerous experts

1979. 185 figures, 19 tables. XVII, 731 pages
ISBN 3-540-08776-1

"... Most of the authors have done an outstanding job of providing broad perspective and critical analysis. Therefore many of the papers should prove valuable as material for the classroom... The wealth of information, up-to-the-minute structuring of ideas, and variety of opinions expressed in the volume are bound to elicit in almost everyone a sense of excitement about plant movements and quite a few ideas for experimentation.
Physiology of Movements is equally useful for browsing or for cover-to-cover study. It is indeed encyclopedic rather than conglomeratic. Like its still useful predecessor in the old series, it will long remain an outstanding guide to the stage of historical development it covers." *Science*

Volume 8
Secondary Plant Products
Editors: E. A. Bell, B. V. Charlwood

With contributions by numerous experts

1980. 176 figures, 44 tables and numerous schemes and formulas. XVI, 674 pages
ISBN 3-540-09461-X

The first comprehensive exposition of this important and timely field, illuminates recent research results on a variety of secondary plant by-products. Using numerous illustrations and tables, Drs. Bell and Charlwood discuss the biochemical and physiological phenomena involved in the synthesis and accumulation of compounds such as alkaloids, isoprenoids, plant phenolics, non-protein amino acids, amines, cyanogenic glycosides, glucosinolates, and betalains.
This volume is sure to become a standard reference to all botanists, biochemists, pharmacologists, and pharmaceutical chemists.

Volume 9
Hormonal Regulation of Development I
Molecular Aspects of Plant Hormones
Editor: J. MacMillan

With contributions by numerous experts

1980. 126 figures, XVIII, 681 pages
ISBN 3-540-10161-6

This initial volume of the threepart *Hormonal Regulation of Development* concentrates on the molecular and subcellular aspects of the main classes of native plant hormones. It contains a survey of the different chemical groups of plant hormones, their structures, homologues, occurrence, purification and identification. In a break with tradition, the main groups of hormones are treated together within each topic, and their characteristics compared and contrasted. The role and importance of bio-assays is evaluated. Progress in our knowledge of biosynthesis and metabolism of plant hormones, as well as present data and theories on the molecular mechanisms of hormone action and effects on tissues are described. The necessary chemical and biochemical background for the succeeding two volumes is provided.
Topics are reviewed authoritatively and in a comprehensive yet readable manner. The text includes numerous illustrations and tables, as well as an extensive bibliography and subject index.

Springer-Verlag
Berlin
Heidelberg
New York

Plant Growth Substances 1979

Proceedings of the 10th International Conference on Plant Growth Substances, Madison, Wisconsin, July 22–26, 1979

Editor: F. Skoog

1980. 209 figures, 62 tables.
XVI, 527 pages
(Proceedings in Life Sciences)
ISBN 3-540-10182-9

This latest volume in the *Proceedings in Life Sciences* Series adheres to the high standards of scientific excellence, covering the full spectrum of research in this rapidly growing field. The proceedings of two previous conferences on plant growth substances – also published by Springer-Verlag – met with great critical acclaim. Significant work by leading international authorities has been included. The 50 papers present vital information on plant growth substances and hormonal regulation. This volume, with its wealth of data, will prove useful not only to all plant physiologists, but botanists in general.

Transport in Plants

U. Lüttge, N. Higinbotham

1979. 1 portrait, 180 figures, 33 tables.
X, 468 pages
ISBN 3-540-90383-6

"This elegant volume is a very precise synthesis of a large and complicated aspect of the physiology of plants. Its great strength is breadth of coverage of transportation of minerals and organic substances (plus, inevitably, water) along with the depth of integration of the subject of other fundamtenal physiological processes in plants. There is coverage of transport that ranges from small ions to large molecules (from minute distances to hundreds of meters), involves both the smallest parts of cell organelles and the largest organ systems, and, of course,details out-to-in versus in-to-out directionality. Another asset is the elaboration of models that are then tested or criticized by presentation of experimental evidence. The authors have gone to considerable length to represent alternate points of view, and they are to be commended for it. Liberally illustrated, it has approximately 1,400 references and an exceptionally complete and detailed subject index. Advanced undergraduates, graduate students, plant physiologists, and even researchers in the field will greet the book with enthusiasm." *Choice*

Springer-Verlag Berlin Heidelberg New York

WITHDRAWN